THE SPECTRUM IN CHEMISTRY

THE SPECTRUM IN CHEMISTRY

J. E. CROOKS
Chemistry Department, King's College
University of London

1978

ACADEMIC PRESS
London New York San Francisco
A Subsidiary of Harcourt Brace Jovanovich, Publishers

ACADEMIC PRESS INC. (LONDON) LTD.
24/28 Oval Road,
London NW1

United States Edition published by
ACADEMIC PRESS INC.
111 Fifth Avenue
New York, New York 10003

Library of Congress Catalog Card Number: 77-81375
ISBN Casebound Edition 0-12-195550-8
ISBN Paperback Edition 0-12-195552-4

Set by Eta Services (Typesetters) Ltd., Beccles, Suffolk
Printed in Great Britain by Galliard Ltd., Great Yarmouth

Preface

Many chemistry journals carry full-page advertisements for such items as "a computer-based energy-dispersive X-ray fluorescence analyser", "an atomic absorption spectrophotometer with a 6-lamp turret", or "a 90 MHz FT NMR featuring fully automatic T_1 measurements and automated selective decoupling sequences". It is clear that chemists obtain a great deal of useful information by studying the interaction between matter and electromagnetic radiation, but the beginning student finds it difficult to relate the various regions of the electromagnetic spectrum. This book is intended to be a guide-book or map of the spectrum for the benefit of chemistry students. Each chapter describes the methods used and the information available from a particular region of the spectrum. The chapters are in general arranged in order of regions of increasing frequency, with the exception of the final two chapters which discuss energy levels only seen in the presence of magnetic fields.

It is hoped that this book will make spectroscopy appear as a less awesome topic to the average student. Spectroscopy is shown to be a group of some of the most powerful methods for obtaining information about the world we live in, rather than treated as a field of operation of the mathematical methods of quantum mechanics. The book lays stress on methods and applications, although sufficient fundamental theory is given to enable the reader to understand how the data may be interpreted. Mathematical proofs are omitted, since the typical user of spectroscopic data is content, for example, to take the selection rules for granted. It is intended that the reader will gain an appreciation of what it is like to use spectroscopic techniques in each region of the spectrum. The apparatus is described in sufficient detail to give an understanding of what is necessary to produce the data. The reader is then shown how the raw data, usually in the form of a wavy line on a section of chart paper, is interpreted.

Further insight will be obtained by working through the problems, which are mostly drawn from recent literature. Experimental observations are given, and the student is asked to obtain the same information as was deduced by the original authors. The problems are thus drawn from the real world of the practising spectroscopist, and are not mere text-book exercises.

Several important topics discussed in this book are rarely encountered in other texts at this level. X-ray fluorescence is an analytical tool of major industrial utility, although rarely encountered in university chemistry departments, and is discussed in Chapter 11, which deals with the X-ray region of the spectrum. Optical activity is a phenomenon familiar to every first-year chemistry student, but very few students, and perhaps not a much higher proportion of university chemistry lecturers, understand the relationship between optical activity and molecular asymmetry, even at the most descriptive and least mathematical level discussed in Chapter 9.

It is assumed that the reader has previously attended a lecture course or studied a text giving an introductory account of theories of atomic and molecular structure. A Check-List is provided of the topics with which the reader should have some acquaintance before starting this book. There seemed little point in adding to the length of this book for instance, with yet another discussion of the Heisenberg Uncertainty Principle.

The book is in SI units throughout. As much spectroscopic literature is in non-SI units, a discussion of the conversions between SI and other units is given in the Appendix. The units of magnetism, as required for example in the study of nuclear magnetic resonance present an especially tricky problem, and are discussed at length.

The author would like to acknowledge many helpful comments and criticisms from his colleagues, notably Professor S. F. Mason, Dr. M. P. Melrose and Dr. D. Couch.

1977 *J.E.C.*

Check-List for Prior Knowledge

The wave nature of light
Determination of the charge to mass ratio of the electron
Black-body radiation
Planck's quantum theory
The photoelectric effect, and Einstein's interpretation
The spectrum of atomic hydrogen
The Bohr-Sommerfeld atom
s, p, d, f atomic orbitals
Spin quantisation
The Pauli Exclusion Principle
Wave-particle duality
The Heisenberg Uncertainty Principle
The de Broglie relationship
The Schroedinger Equation
Wave-mechanical interpretation of atomic orbitals
Molecular orbitals; σ and π, σ^* and π^*

Contents

1 The Principles of Spectroscopy

1.A The Electromagnetic Spectrum

A beam of light consists of sinusoidally fluctuating electric and magnetic fields. The frequency of the fluctuation ranges from about 8×10^{14} Hz (cycles per second) for violet light to about 4×10^{14} Hz for red light. Outside this narrow range of radiation visible to the eye is the rest of the spectrum of electromagnetic radiation. There are few regions of this spectrum in the range from 10^6 to 10^{18} Hz that have not been used to obtain information of interest to chemists, and this book presents a survey of the whole spectrum, showing what kind of information can be obtained by the measurement of the intensity of absorption or emission of radiation in the various regions. The range of the spectrum, with the names of some of the regions, is shown in Fig. 1.1.

The frequency ν of electromagnetic radiation is related to the wavelength λ by

$$c = \nu\lambda \tag{1.1}$$

where c is the velocity of propagation. In a transparent medium the velocity is less than in a vacuum, the reduction in velocity being given by the refractive index. The refractive index of the medium is equal to the ratio of the velocity in a vacuum to the velocity in the medium. As radiation enters a region of higher refractive index the wavelength is reduced, the frequency remaining constant. The refractive index of air is so near unity, being about 1.0028 for visible light in air at normal atmospheric conditions, that the effect on the wavelength in air may be ignored except for work of the highest accuracy. The orders of magnitude of the wavelengths of the radiation in the various regions of the spectrum are indicated in Fig. 1.1.

Atoms or molecules only absorb radiation of frequency ν if they can thereby gain a quantum of energy E, where

$$E = h\nu \tag{1.2}$$

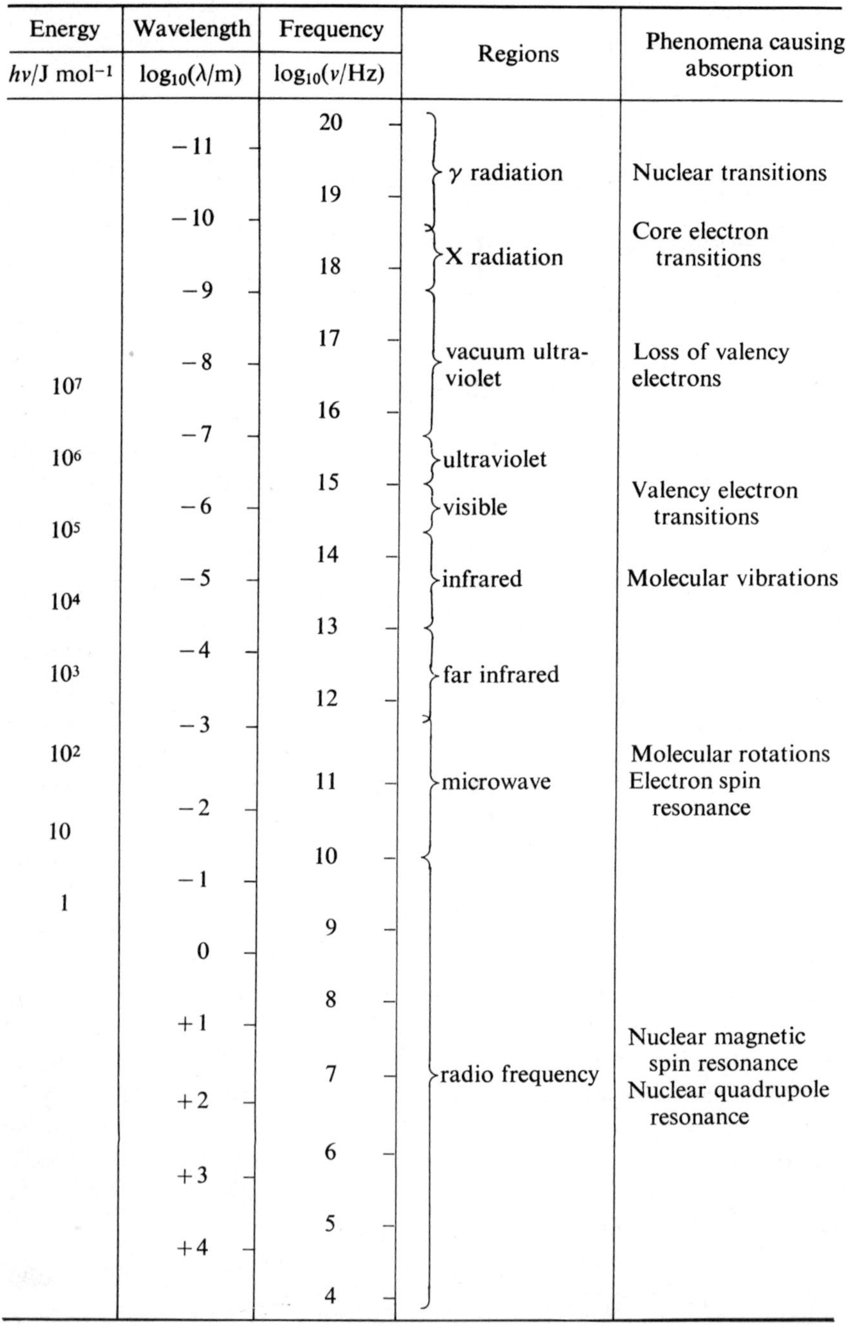

Energy	Wavelength	Frequency	Regions	Phenomena causing absorption
$h\nu$/J mol^{-1}	$\log_{10}(\lambda/\text{m})$	$\log_{10}(\nu/\text{Hz})$		
		20		
	−11			
			γ radiation	Nuclear transitions
		19		
	−10			
			X radiation	Core electron transitions
		18		
	−9			
		17		
	−8		vacuum ultra-violet	Loss of valency electrons
10^7				
		16		
	−7			
10^6			ultraviolet	
		15		Valency electron transitions
	−6		visible	
10^5				
		14		
	−5		infrared	Molecular vibrations
10^4				
		13		
	−4			
10^3			far infrared	
		12		
	−3			
10^2				Molecular rotations
		11	microwave	Electron spin resonance
	−2			
10				
		10		
	−1			
1				
		9		
	0			
		8		
	+1			
		7	radio frequency	Nuclear magnetic spin resonance Nuclear quadrupole resonance
	+2			
		6		
	+3			
		5		
	+4			
		4		

FIG. 1.1 The electromagnetic spectrum.

h being Planck's constant. The gain or loss of energy between two quantum states is termed a *transition*. There are many different ways in which a molecule may store energy, each corresponding to a different quantum size, so that the transitions for each correspond to a different absorption frequency. For example, a molecule may store energy by virtue of its rotation. The quantum size for rotational energy is found to correspond to absorption around 10^9 Hz, the microwave region. The same molecule may also store energy by virtue of the vibration of its bonds. The quantum size for vibration is found to correspond to absorption around 10^{13} Hz, the infrared region. The molecule may also store energy by excitation of its electrons from full to empty molecular orbitals, which causes absorption in the region of 10^{15} Hz, in the ultraviolet. Any one chemical species is likely to absorb radiation in several regions of the spectrum, each region providing information of a different type about the molecules of the species. Each chapter in this book is concerned with a particular region, showing how the absorption spectrum is measured and what information can be obtained from it.

It sometimes happens that there are several quantum states of exactly the same energy. This phenomenon is known as *degeneracy* and the energy levels are said to be degenerate. The student will already have encountered degenerate energy levels in his study of atomic structure. In the absence of applied electrostatic or magnetic fields, p orbitals of the hydrogen atom with a given main quantum number are three-fold degenerate, and the corresponding d orbitals are five-fold degenerate. The application of an electrostatic or magnetic field removes or lifts the degeneracy, causing the orbitals to have different energies. Degeneracy has great importance in spectroscopy. Other things being equal, the intensity of an absorption line depends on the population of the lower energy level, and this in turn partly depends on its degeneracy. Spectroscopic studies of energy levels for which the degeneracy has been lifted by the application of electrostatic or magnetic fields provide valuable information about atomic and molecular structure.

1.B Units

Various units are in common use to measure the frequency corresponding to a transition, the particular unit preferred depending on the region of the spectrum. The fundamental unit of frequency, the cycle per second, or Hz, is too small for convenience in most of the regions of interest. Even at the low-frequency end of the spectrum, in the microwave region, frequencies are expressed in GHz. The trouble is that the second is, for these purposes, a very large unit of time. A time unit of more convenient size is the time required by light to travel 1 cm in a vacuum, i.e. approximately 33.4 ps. The frequency expressed in terms of cycles per one of these units is termed the *wavenumber* and is assigned the unit cm^{-1}. Students who have been carefully trained in the use of units find that the expression of a frequency in units of reciprocal distance is disconcerting, and it is for this reason that the term wavenumber rather than frequency is used for this quantity. Some authors call the cm^{-1} unit the kayser, but this usage is not universal. The wavenumber is readily calculated from the corresponding wavelength, since it is numerically equal to the reciprocal of the wavelength in cm units. Frequency and wavenumber units have two common advantages over wavelength for the purposes

TABLE 1.1

Frequency and energy units used in spectroscopy

1 Hz ≡	1 cm^{-1} ≡	1 J mol^{-1} ≡	1 eV ≡	
1	2.9979×10^{10}	2.5053×10^{9}	2.4182×10^{14}	Hz
3.3356×10^{-11}	1	8.3567×10^{-2}	8.0663×10^{3}	cm^{-1}
3.9915×10^{-10}	11.9660	1	9.6522×10^{4}	J mol^{-1}
4.1353×10^{-15}	1.2397×10^{-4}	1.0360×10^{-5}	1	eV

of characterisation of a transition: they are both independent of the transmission medium, and they are both proportional to the energy of the transition by virtue of Eqn. 1.2. Wavenumber units are most commonly used in the infrared. For example, general-purpose infrared spectrometers are calibrated in wavenumber units, and cover the range 4000–650 cm^{-1}. Towards the high-frequency regions of the spectrum, wavenumbers become too large for convenience. A common unit is the electron-volt, or eV, the quantity of energy acquired by an electron falling through a potential difference of 1 volt. It is especially appropriate in these regions since the transitions for which this unit is employed are those in which electrons move from one orbital to another in the potential field of the nuclei. By virtue of Eqn. 1.2, an energy unit may be used as a frequency unit. Conversely, a frequency unit may be used as an energy unit. The student will encounter graphs in which the energy axis is marked off in units of cm^{-1}. The wavenumber for a transition may be directly read off such a graph. The electron-volt is related to a single molecule. In order to compare energies calculated by spectroscopic means with those from thermodynamic measurements it is often convenient to express the energy of a transition in terms of joules per mole. Conversion factors for these units are given in Table 1.1.

1.C The Principles of Spectroscopy

1.C.1 Instrumentation

The absorption spectrum of a chemical species is a display of the fractional amount of radiation absorbed at each frequency (or wavelength) as a function of frequency (or wavelength). The apparatus required to produce such a spectrum is made up of several components, the function of which is common to all regions of the spectrum, although its nature and design is that appropriate to the region of interest.

There must first of all be a source of radiation. Some sources are *monochromatic*, i.e. they emit radiation over a narrow frequency range, but if, as is often the case, the source emits radiation over a wide range of frequencies, there must be a monochromator to select a component of the radiation with a narrow frequency range. The sample whose spectrum is to be obtained must be confined in a sample cell in the path of the radiation, the window of the cell being transparent to radiation in the desired frequency range. After the radiation has passed through the sample, its intensity must be monitored, usually by a detector which generates an electrical signal proportional to the intensity of the radiation incident upon it. The signal from the detector is usually displayed on the vertical axis of a graph. The graph and the monochromator are linked in such a way that as the frequency is scanned it is displayed as the horizontal axis of the graph.

Emission spectra are also observed. The sample is induced to emit radiation by being supplied with energy in some fashion. Whereas in absorption spectroscopy the monochromator is often placed before the sample, to protect it against excessive radiation from the source, the monochromator in emission spectroscopy must of course be between the sample and the detector.

Various terms are in use to describe the apparatus used in spectroscopy. These are often used indiscriminately but a strict distinction may be made as follows. A *spectroscope* is a device whereby a spectrum is made visible to the eye, the eye taking the role of the detector. A *spectrograph* resembles a spectroscope, with a camera in place of the eye. The spectrum thus appears as a photographic negative; the more intense the blackening of the film at a position corresponding to some frequency, the more intense the radiation at that frequency. A *spectrometer* is a device whereby the frequency at which absorption occurs can be measured. A *spectrophotometer* resembles a spectrometer but also measures the intensity of absorption as well as the frequency at which absorption occurs. Since modern spectrometers are spectrophotometric in function, these two latter terms are often not distinguished. The term spectrophotometer is generally reserved for an instrument for which at least part of the frequency region scanned is in the visible. The technique is usually termed *spectroscopy* even though the eye is not used as detector; for example, one talks about microwave spectroscopy rather than microwave spectrometry. The term spectrometry is usually met as part of the term *mass spectrometry*. This is a technique whereby ionised molecules follow a curved path through crossed electric and magnetic fields, and arrive at a detector only if their mass bears the correct relationship to the field intensities. As the field intensity is varied, the detector current varies so as to show the relative abundance of the ions of various masses present in the sample chamber. The display of detector current as a function of ionic mass resembles a spectrum, so the technique is termed mass spectrometry, although it has nothing to do with the electromagnetic spectrum.

1.C.2 Noise

Every measurement has an uncertainty associated with it, which, for electrical measurements, is termed *noise*. Careful measurements on any supposedly constant voltage or current show that small rapid fluctuations occur. The longer the response time of the measuring device, the less these fluctuations are observed. They are, as it were, "damped", and the signal is less noisy. There are many types of noise, some of which are due to faulty construction or design, but some are due to the fundamental laws of the quantisation of matter and energy. One fundamental type of noise is *thermal*, or *Johnson noise*. This is due to the thermal agitation of the electrons in a resistor. Owing to this motion, currents flow at random to and fro across the resistor. The amount of noise may be measured as the root-mean-square of the voltage fluctuations appearing across the ends of the resistor. The higher the resistance the greater the voltage fluctuation for a given current fluctuation, and the higher the temperature the greater the thermal agitation. The fluctuations are at all frequencies, so the total noise measured depends on the range of frequencies monitored by the detector. This range is termed the *bandwidth*. The root-mean-square voltage of Johnson noise is given by

$$V_{\mathrm{rms}} = (4RkTB)^{\frac{1}{2}} \tag{1.3}$$

where R is the resistance in ohms, k is Boltzmann's constant in $\mathrm{J\,K^{-1}}$, T is the absolute temperature, and B is the bandwidth in Hz. Another fundamental type of noise is *shot noise*, which is due to the particulate nature of the electron. An electric current consists of a stream of electrons. If a small current is observed over a short period of time, there is an appreciable uncertainty in the number of electrons passing during that period. To take an extreme example, consider a current whose average value is 10 electrons per second. The number of electrons passing a given point in any interval of 1 second's duration is not always 10, but might be 9 for one interval and 11 for the next, corresponding to 10% noise. The magnitude of the shot noise current is related to the charge on the electron, the total current, and the time over which the observation is made, which in turn is related to the bandwidth of the detector. The root-mean-square shot noise current is given by

$$I_{rms} = (2eIB)^{\frac{1}{2}} \tag{1.4}$$

where I is the total current in amperes and e is the charge on the electron in coulombs.

Both Johnson noise and shot noise are examples of *white noise*, that is to say that the noise power for a given bandwidth interval is independent of the bandwidth. Reduction of the bandwidth thus gives a proportionate reduction in noise power. The effect of reducing the bandwidth, i.e. increasing the response time of the detector system from 1 s to 25 s, is shown in Fig. 1.2. The traces are of *baseline*, i.e. there is no absorption of radiation in the region scanned. It can be seen that, as predicted by theory, the amplitude of the noise voltage is reduced by a factor of 5. It can also be seen that the

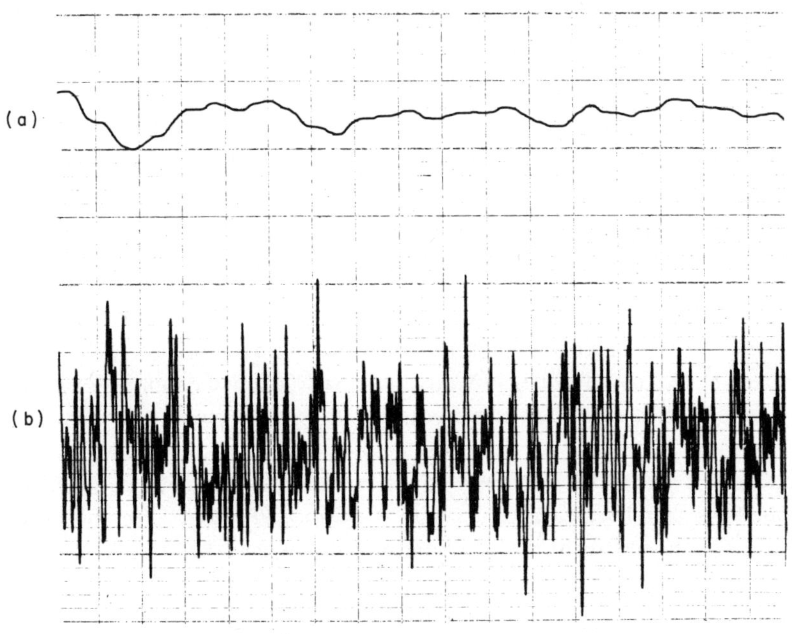

FIG. 1.2 The effect of electronic bandwidth on noise. Chart speed, 1 division in 10 s; time constant (a) 25 s, (b) 1 s.

increase in response time causes the trace to be less "spiky". Indeed, an experienced observer can estimate the response time from the spikiness. This has an important consequence for the scanning of spectra containing narrow peaks. If the frequency scanning rate is too high, and the detector system response time too long, the detector system may not have time to give a faithful display of a narrow peak, and the top of the peak may be truncated. Some spectrometers have a "fidelity meter" which warns if such truncation is likely. In case of doubt, the spectrum should be scanned again at a lower speed, or with a shorter response time.

Another type of noise, *flicker noise*, is generated by solid-state devices such as transistors and photoconductive detectors. Flicker noise is not white, since the noise power for a given interval of bandwidth increases with decreasing frequency. In many spectrometers the signal is *modulated*, i.e. caused to fluctuate at some predetermined frequency within a narrow range. The bandwidth is then reduced to the range of the modulation frequency, and the noise is much reduced. A signal observed by a detector system of response time 0.1 s will be subject to the same component of white noise as a signal modulated at 1 kHz with a bandwidth of 10 Hz. However, the latter will be much less subject to flicker noise, so modulation is a favoured technique for noise reduction.

In all discussions of the noise problem it must be remembered that it is not so much the absolute magnitude of the noise that is a matter of concern, but rather the *signal-to-noise ratio*. This term may be used to mean either the ratio of the total signal observed if no sample is present to the noise, or the ratio of the height of a particular peak in the spectrum to the noise.

1.C.3 Resolving power

An important limitation to the accuracy of measurement of the frequency for a transition is the *resolving power*, or *resolution*, of the apparatus. This is a measure of the narrowness of the range of frequencies of the radiation incident in the sample. It is obvious that the frequency cannot be measured to a greater precision than this frequency range. For some monochromatic sources, e.g. the klystron used in the microwave region and the laser used in the visible region, the frequency range is extremely narrow, so absorption frequencies can be measured to high precision. However, the intensity of radiation obtained from the combination of a wide-range frequency source and a monochromator is inversely proportional to the narrowness of the frequency range selected. The more intense the radiation the greater the signal-to-noise ratio, since the signal is proportional to the detector current, whereas, as shown in Eqn. 1.4, the shot noise is only proportional to the square-root of the detector current. It is thus necessary to use a compromise value of the frequency range, depending on whether precision in intensity or frequency is more important. Most spectrophotometers have controls that may be adjusted for optimum conditions.

The narrowness of the frequency range determines the minimum frequency difference between two absorption peaks, or *lines* as narrow peaks are sometimes called, below which the two cannot be separated, and appear as a single peak. If two lines, of similar frequency ν, separated by $\Delta\nu$, can just be clearly seen to be two lines rather than one, the resolving power is $\nu/\Delta\nu$. The resolution is $\Delta\nu$, or $\Delta\lambda$. Resolving power, unlike resolution, is dimensionless. There is a certain vagueness in this definition, since whether or not two lines are "clearly seen" is a matter of individual judgement, and varies with the signal-to-noise ratio. A more precise definition, due to Rayleigh, is applicable to a monochromator in which a narrow section of a wide range of frequencies is selected by a slit. The slit causes a diffraction pattern in the emergent beam. According to the Rayleigh criterion, two lines are resolved when the maximum of the diffraction pattern of one lies in the first zero of the other, and vice versa.

2 The Microwave Region

2.A The Rotation of a Diatomic Molecule

2.A.1 Quantisation of rotational energy

A diatomic molecule may be pictured as two masses, m_1 and m_2, held at a distance r apart. The rotation of this model rotor can be described in terms of the moment of inertia, I, and the angular rate of rotation, ω radian s^{-1} (i.e. the molecule rotates $2\pi\omega$ times each second). According to classical mechanics,

$$I = \mu r^2 \tag{2.1}$$

where μ, the *reduced mass*, is given by

$$\mu = m_1 m_2/(m_1 + m_2) \tag{2.2}$$

The angular momentum of the rotor is then $I\omega$, and the kinetic energy is $\frac{1}{2}I\omega^2$. According to classical mechanics, the momentum and kinetic energy can have any value, but according to quantum mechanics they are both quantised. The quantum sizes are such that quantisation effects are insignificant for rotors of macroscopic size, but dominate the behaviour of rotors of molecular size. Solution of the Schrödinger equation for the simple rigid (i.e. constant r) rotor shows that the angular momentum P and kinetic energy E_{rot} are given by

$$\begin{aligned} P &= [J(J+1)]^{\frac{1}{2}}\hbar \\ E_{rot} &= J(J+1)\hbar^2/2I \end{aligned} \tag{2.3}$$

where $\hbar$ is defined as $h/2\pi$, and J is the *rotational quantum number*, which may have value 0 or any integer. As is found for electrons in atoms, not only is the total angular momentum quantised, but also the resolved part of the momentum along some defined axis, which has the value $M\hbar$, where M is the *magnetic quantum number*. The value of M can be any integer in the range $J < M < -J$, i.e. $(2J+1)$ values in all (counting 0 as an integer). In the absence of a magnetic or electrostatic field, the lines of force of which define the axis along which the angular momentum is resolved, these $(2J+1)$ energy levels are degenerate.

The transition from a state of quantum number J to a state of quantum number $(J+1)$ requires energy ΔE_{rot}, where

$$\Delta E_{rot} = [(J+1)(J+2)-(J+1)J]\hbar^2/2I \tag{2.4}$$

Therefore,

$$\Delta E_{rot} = (J+1)\hbar^2/I \tag{2.5}$$

This transition can thus be accomplished by the absorption of a quantum of electromagnetic radiation of frequency ν, where

$$\nu = (J+1)h/4\pi^2 I \tag{2.6}$$

The quantity $h/8\pi^2 I$ is usually written as B, so that

$$\nu = 2B(J+1) \tag{2.7}$$

If h and I are in SI units, i.e. J s and kg m^2 respectively, then ν is in units s^{-1}, i.e. cycles per second or Hz. It is often convenient for ν to be in units cm^{-1}, and so a different rotational constant $\tilde{B}$ is defined as B/c where c is the velocity of light in cm s^{-1}. Then

$$\tilde{\nu} = 2\tilde{B}(J+1) \text{ cm}^{-1} \tag{2.8}$$

It is important to note this distinction between B and $\tilde{B}$, which are occasionally confused in the literature.

2.A.2 Selection rules

It does not necessarily follow that any diatomic molecule can acquire rotational energy from electromagnetic radiation. Two criteria, known as selection rules, must be satisfied. The first selection rule states that the molecule must have a permanent dipole moment, which rules out homonuclear molecules such as H_2, N_2, and Cl_2. This rule can be understood by consideration of the interaction between the rotating dipole, e.g. HCl, and the electromagnetic radiation. As the dipole rotates, the electric field due to it changes in direction, so that the resolved part of the field in any one direction fluctuates in a sinusoidal manner. This sinusoidally fluctuating electric field interacts with the sinusoidally fluctuating electric field of the electromagnetic radiation in such a way that energy can be transferred from one to the other if the frequencies are correctly related. If the molecule has no dipole moment, this interaction cannot occur. The transfer of energy from the radiation field to the molecule causes absorption of radiation; the transfer of energy from the molecule to the radiation field causes emission.

The second selection rule is that $\Delta J = +1$ for absorption and $\Delta J = -1$ for emission. This can be proved by the methods of quantum mechanics. A simple explanation of this selection rule has been given by Moynihan (1969). By contrast with the convention employed in other branches of spectroscopy, a transition is described by the lower level written before the upper level, so that the permitted transition from $J = 0$ to $J = 1$ is written as $J = 0 \rightarrow 1$. The $J = 0 \rightarrow 1$ transition causes a line in the absorption spectrum at $2B$ Hz, the transition $J = 1 \rightarrow 2$ a line at $6B$ Hz, the transition $J = 2 \rightarrow 3$ a line at $12B$ Hz, and so on. Lines due to such forbidden transitions as $J = 0 \rightarrow 2$ are not seen, in general. In some instances the selection rules do not apply strictly, and lines due to forbidden transitions may be seen, although they are much less intense than lines due to permitted transitions.

2.A.3 Frequency region of the spectrum for rotational excitation

The region of the spectrum in which absorption due to rotational excitation may be seen can be found by evaluation of B, by substituting numerical values into Eqns. 2.1, 2.2, 2.6, and 2.7. The masses of the atoms, m_1 and m_2, in kg, can be obtained by dividing the gram-atomic weights by $10^3 N_0$ (N_0 being the Avogadro number). The interatomic distance r is usually calculated from B, rather than the other way round, but for the sake of the argument a typical value for r of 100 pm may be assumed on the basis of X-ray data. Then, for the $J = 0 \rightarrow 1$ transition of CO, ν is found to be 1.4×10^{11} Hz. This corresponds to radiation of wavelength about 2 mm. More massive molecules, having higher moments of inertia, absorb radiation of longer wavelengths. Thus absorption due to rotational excitation is observed in the microwave region of the spectrum. The common frequency unit used in microwave spectroscopy is the MHz ($= 10^6$ Hz), so that the $J = 0 \rightarrow 1$ transition for CO is expected around 140 000 MHz. The observed value (Rosenblum *et al.*, 1958) for $^{12}C^{16}O$ is $115\,271.204 \pm 0.05$ MHz.

2.A.4 Line widths and intensities

In a gas at room temperature and pressure the mean time between collisions is of the order of 10^{-10} s. Energy can be gained or lost at each collision, so that, for the transition from one energy state to another, each state only has a specified energy for 10^{-10} s. According to the Heisenberg Uncertainty Principle,

$$\Delta E \times \Delta t \approx h \tag{2.9}$$

If Δt is 10^{-10} s, then $\Delta E \approx 6.6 \times 10^{-24}$ J. For radiation of frequency 10^{11} Hz, the quantum size $h\nu$ is only 6.6×10^{-23} J, so the uncertainty in the energy levels is of the order of 1% of the total. This causes the absorption line to have a width of the order of 10^9 Hz. This phenomenon is known as *pressure broadening*. As the potential resolving power of microwave spectroscopy apparatus is of the order of 10^5 Hz, the reduction of pressure broadening is obviously of prime importance. This is achieved by using gas at low pressure as the sample. Typically, pressures in the range 1.3–13 Pa (10^2–10^{-1} mm Hg) are used, which reduce the pressure broadening to 10^5 Hz or so.

The use of gases at low pressures as sample introduces the possibility of another problem, *power saturation*. As each molecule in a lower energy level has an equal chance of excitation, the intensity of absorption depends on the number of molecules in the lower level. The absorption of energy from the incident radiation raises molecules to an upper level. For absorption to be a steady-state process, the molecules in the upper level must lose energy by some mechanism so as to re-populate the lower level. For gases, energy is lost by collision of molecules with the walls of the container and with each other, so rotational energy can be converted into translational and vibrational energy. At best, this is not a very efficient mechanism, and the efficiency is much reduced at low pressures where collisions are comparatively infrequent. Thus, if the power input from the radiation is too high, molecules will be raised to the upper level at a greater rate than they can fall back to the lower level. The distribution of molecules among the rotational energy levels will become non-Boltzmann, i.e. the population in the lower level will be markedly less than that in the absence of radiation. The fraction of microwave power absorbed will thus decrease.

The intensity of absorption is measured by the *absorption coefficient* α, defined as the rate of loss of microwave power P as the microwaves are propagated along distance x cm:

$$\mathrm{d}P/\mathrm{d}x = \alpha P \tag{2.10}$$

α has units cm^{-1}. Note that this unit is a true unit of reciprocal length, not to be confused with the unit of wavenumber. The calculation of α from molecular parameters is possible, but mathematically involved. An approximate equation for the maximum absorption coefficient of a gas at 133 Pa (1 mm Hg) and room temperature is

$$\alpha \approx 7 \times 10^{-36} \mu \nu^3 \tag{2.11}$$

where μ is the dipole moment in debye units (see Section 2.E.1 for definition). Two important considerations for microwave spectroscopy are apparent from this equation. The intensity of absorption increases rapidly with frequency, so that, other things being equal, it will be advantageous to work at as high frequencies (i.e. short wavelengths) as possible. Secondly, whatever the frequency chosen, α is going to be small. Dipole moments are of the order of 1 debye, so α is of the order of 10^{-5} cm^{-1} at 10^{10} Hz.

The spacing between rotational energy levels is comparatively small. A frequency of 10^{11} Hz corresponds to a quantum of energy of magnitude 6.6×10^{-23} J, which is less than the value of kT at room temperature, 4×10^{-21} J. Statistical mechanics shows that the proportion of molecules of energy ε_i in a system at temperature T is given by

$$n_i \propto g_i \exp(-\varepsilon_i/kT) \tag{2.12}$$

where g_i is the statistical weighting, multiplicity, or degeneracy, of the energy level ε_i. Thus, at room temperature there is a considerable population of molecules in upper rotational levels. In fact, because of the degeneracy factor, which is $(2J+1)$ for rotational energy, there tend to be more molecules in some upper rotational levels than in the zeroth level. Substituting values for g_i and ε_i into Eqn. 2.12 gives

$$n_i \propto (2J+1) \exp[(J+1)2hB/kT] \tag{2.13}$$

Differentiation of Eqn. 2.13 gives the value of J for which n_i is a maximum, J_{max}. Transitions from this level are more numerous than from any other, so the transition $J = J_{max} \rightarrow (J_{max} + 1)$ gives the most intense line in the spectrum.

$$J_{max} = \text{nearest integer to } (kT/2hB)^{\frac{1}{2}} - \tfrac{1}{2} \tag{2.14}$$

For CO, substitution of numerical values gives $J_{max} = 7$. The more massive the molecule and the greater the moment of inertia, the less the value of B and the greater the value of J_{max}.

2.B Apparatus for Microwave Spectroscopy

Radiation in the microwave region of the electromagnetic spectrum presents special problems in instrumentation, since the wavelengths are of the same order of magnitude as laboratory apparatus. If microwaves are handled by optical techniques, diffraction effects are very pronounced. Microwave energy can also be thought of as carried by high-frequency alternating current, and can be transmitted along a coaxial cable, but the frequency is so high that techniques for handling electrical power at lower frequencies are not applicable. The most satisfactory way to convey microwave energy is along a waveguide, an accurately made hollow metal tube, usually of rectangular cross-section. The cross-sectional dimensions are of similar magnitude to the wavelength of the radiation to be conveyed. Microwave radiation cannot be conveyed along a waveguide whose cross-sectional dimensions are much smaller than the microwave wavelength. The radiation is propagated along the inner surface of the guide, which must be of high electrical conductivity.

Until about 1970 the klystron valve was the source of microwave radiation used for spectroscopy. A klystron delivers several mW of microwave power at a constant frequency, the frequency being controlled by the dimensions of a resonating cavity. The frequency is altered by mechanically altering the cavity dimensions. It is only possible to produce a variation of $\pm 10\%$ or so in the frequency in this manner, so the coverage of a wide range of the microwave spectrum requires several klystrons. Modern commercial microwave spectrometers use a backward-wave oscillator, in which the frequency is controlled by an applied voltage, and can be varied by a factor of up to 2 by varying the applied voltage. One backward-wave oscillator can thus be used for the whole frequency range for which one particular size of waveguide is used. No monochromator is needed for either klystron or backward-wave oscillator, since these sources are intrinsically monochromatic to a very high degree.

The sample, a gas or vapour at low pressure, is held in a length of waveguide, closed at each end by a mica or teflon septum (i.e. window) which is transparent to microwave radiation. The radiation transmitted through the cell is detected by a probe connected to a rectifying crystal, which converts the microwave radiation into an electric current, of magnitude proportional to the incident microwave power. In principle this current could be plotted against frequency to give the microwave spectrum, but a more elaborate detection technique is required. As has been shown in Section 2.A.4, absorption coefficients are in general very low, and the need to work at low pressures to minimise linewidths makes the problem still more acute. Large sample cells, up to 10 m long, coiled for convenience, are used to increase the amount of radiation absorbed, but even so only a few parts per million of the microwave power incident on the sample is absorbed. Obviously this quantity cannot be measured by direct observation of the microwave power before and after the sample cell, and comparison of the readings.

The solution to the signal detection problem is to use a modulation technique. The signal is modulated at a known frequency, and the modulated signal is then amplified using an amplifier tuned or "locked-in" to the modulating frequency. The effective bandwidth for the purposes of noise calculation is then the bandwidth of the amplifier, which can be very low, say a few Hz. The modulation can be applied in two ways. The microwave source can be modulated by the application of a sinusoidal voltage. The frequency of the radiation emitted by both klystrons and backward-wave oscillators is voltage-dependent, although less so for klystrons. The modulation frequency is chosen at some frequency high enough to eliminate flicker noise, but less than the linewidth, which would otherwise be increased by the modulation. A value of 5×10^4 Hz is typical. This is sufficient to produce a large change in the amount of radiation absorbed by the sample, so that the signal at the detector varies sinusoidally at the modulation frequency. This signal is amplified, rectified, and displayed. The other type of modulation commonly employed utilises the Stark effect. In Section 2.A.1 it was shown that each rotational level, quantum number J, is $(2J+1)$-degenerate. This degeneracy is lifted by the application of an electrostatic field. As a consequence, a transition that gives a single line in the absence of an electrostatic field gives a multiplet of lines when the field is applied. The absorption coefficient at the frequency of field-free maximum absorption is thus markedly reduced and the application of a varying electrostatic field to the sample modulates the detector signal. The modulation may be sinusoidal, but it is preferable to employ square-wave modulation. In square-wave modulation, the electrostatic field is alternately applied for a short time, say 10 μs, then removed for an equal duration. The signal can then be processed in such a way as to give not only the absorption intensity but also the magnitude of Stark splitting. The electrostatic field is in the range 10 to 2000 V cm^{-1}, the value chosen depending on the magnitude of the Stark splitting factor for the particular sample in question. Stark splitting is discussed further in Section 2.E.1. Values of α as low as 4×10^{-11} cm^{-1} may be observed using Stark effect modulation. However, for work at the highest frequencies in the millimetre wavelength region, Stark effect modulation is experimentally difficult, and source modulation is employed.

Microwave frequencies can be measured to great accuracy. A piezoelectric quartz crystal can be made to oscillate at some high frequency, say 10 MHz, stable to a few Hz. The crystal is calibrated by counting the number of oscillations over some period of time. The crystal oscillations drive a chain of frequency multipliers, so that oscillations in the microwave frequency range, stable and of known frequency to a few tens of kHz, are obtained. The output consists of a set of oscillations, spaced at 10 MHz intervals, spanning the frequency range of the klystron or backward-wave oscillator. A small fraction of the output of the klystron or backward-wave oscillator is mixed with this set of standard frequencies, and a frequency marker "pip" is produced in the output whenever the frequency of the microwave source is equal to each standard frequency during a scan. An accurate value of the frequency of absorption is obtained by interpolation between the "pips". The accuracy of microwave frequency measurements is such that they are used to define the unit of time. One second is defined as the time required for 9 192 631 770.00 cycles for a specified transition between two close electronic energy levels in the caesium atom.

Microwave spectra can be displayed in either of two ways. If a klystron is employed as source, it is usual for only one frequency to be measured at a time. Coincidence between klystron frequency and calibration frequency at maximum absorption is observed on an oscilloscope. Data so obtained are presented as a table of frequencies, indexed for the particular transition involved, as in Table 2.2. If a backward-wave oscillator is used as source, a wide frequency band can be scanned, and the data are presented as a graphical display of absorption as a function of frequency.

A schematic block diagram of a typical microwave spectrometer, employing Stark modulation, is shown in Fig. 2.1.

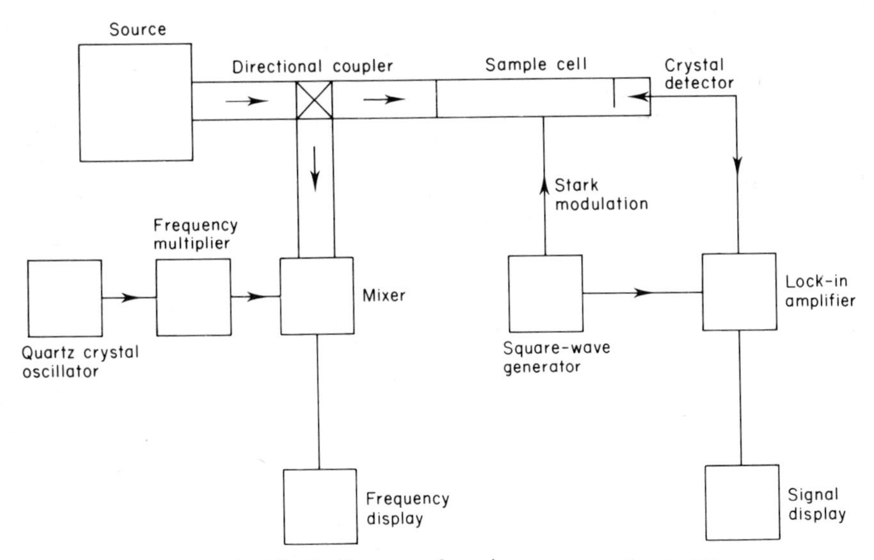

FIG. 2.1 Block diagram of a microwave spectrometer.

2.C The Rotation of Polyatomic Molecules

2.C.1 The classification of rotors

Any three-dimensional object, such as a molecule, has in general three moments of inertia, I_A, I_B, I_C. These are the moments of inertia about three mutually perpendicular axes. By convention, these three axes cross at the centre of gravity of the molecule, and are oriented so as to give maximum or minimum values for the moments of inertia. Axes so chosen are called *principal axes*. For symmetrical molecules, the principal axes coincide with axes of symmetry. Also by convention, $I_A < I_B < I_C$. For discussion of

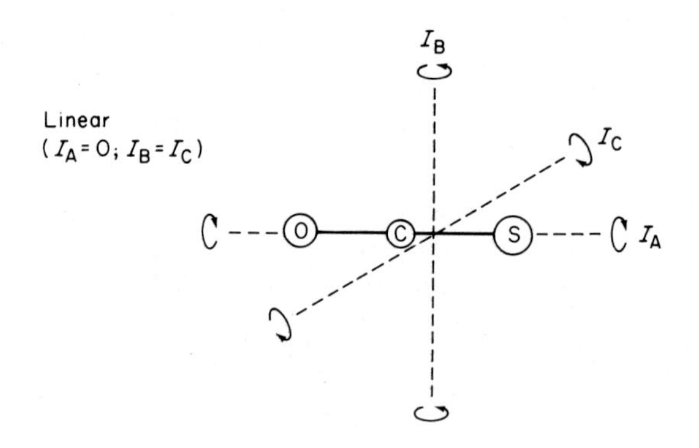

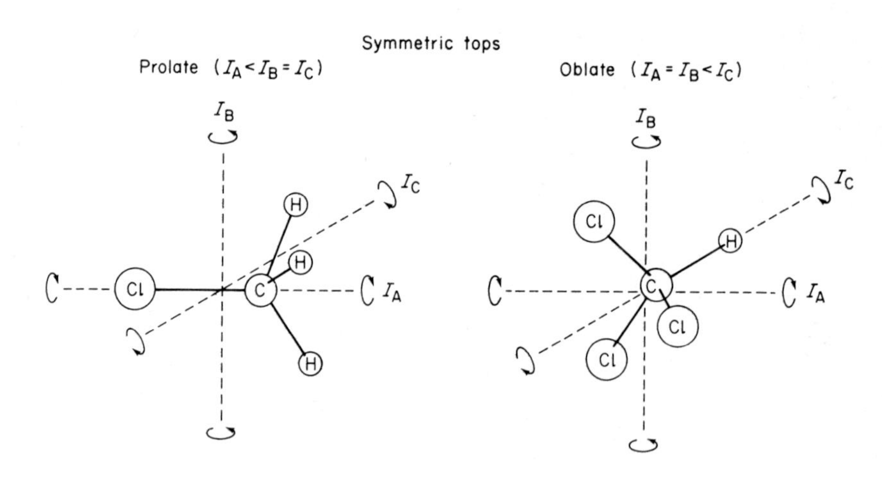

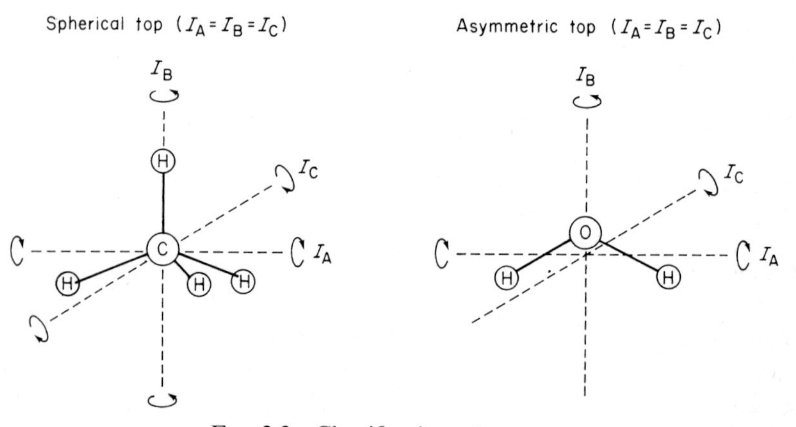

FIG. 2.2 Classification of rotors.

their rotation, polyatomic molecules are classified, according to the relative magnitudes of I_A, I_B, and I_C, into *linear molecules*, *spherical tops*, *symmetric tops*, and *asymmetric tops*. Examples are shown in Fig. 2.2.

2.C.2 Linear molecules

For linear molecules, $I_B = I_C$ and $I_A = 0$. The rotation about the axis of the molecule, for which I_A is zero, or, more properly, negligibly small, does not contribute to the rotational spectrum for two reasons. Firstly, a very small value of I_A means a very large value of B_A, corresponding to absorption at frequencies well above the microwave range. Secondly, the charge distribution about the axis is generally symmetrical, so that the molecule does not have a dipole moment perpendicular to its axis, which is a prerequisite for rotational excitation. This is strictly true for diatomic molecules only. Polyatomic molecules can bend so as to be no longer linear, and this can give rise to microwave absorption, as discussed in Section 2.D.5.

Diatomic molecules are a special type of linear molecule, and the treatment in Section 2.A is in general valid. The major difference for polyatomic molecules is that the moment of inertia is not dependent on one bond length only, but on $(n-1)$ bond lengths for an n-atomic molecule. An algebraic theorem states that it is necessary to have $(n-1)$ simultaneous equations for the elucidation of $(n-1)$ unknowns, so $(n-1)$ values of B are required to evaluate the $(n-1)$ bond lengths in an n-atomic molecule. This can be achieved by isotopic substitution. If it is assumed that bond lengths are unaltered by replacing ^{16}O, say, by ^{18}O, then the measurement of the two values of B for $^{16}O^{12}C^{32}S$ and $^{18}O^{12}C^{32}S$ give two values of I, from which the O–C and C–S bond lengths can be calculated. Conversely, for a diatomic molecule, isotopic substitution can be used to calculate the relative masses of the isotopically substituted atom. For example, the mass ratio $^{39}K/^{41}K$ can be found by measurements of B for $^{39}K^{127}I$ and $^{41}K^{127}I$. The approximation that the bond lengths are unchanged by isotopic substitution ignores the effect of zero-point energy on the bond length, as discussed in Section 2.D, and a small correction, which can be calculated, can be applied to account for this effect.

2.C.3 Spherical tops

Spherical tops are molecules for which $I_A = I_B = I_C$, e.g. methane. Such molecules are usually highly symmetric, and have no dipole moment, so they do not give rotational spectra. It is possible for a large polyatomic molecule to have $I_A = I_B = I_C$ just by chance, and also have a dipole moment, but this is too rare to be worth further discussion here.

2.C.4 Symmetric tops

Symmetric tops are molecules for which two moments of inertia are equal, but not equal to the third, e.g. $CHCl_3$ and CH_3Cl. These molecules have one axis of symmetry, e.g. the C–H bond in $CHCl_3$ and the C–Cl bond in CH_3Cl, about which they rotate like a child's spinning top (hence the name). Symmetric tops are further classified as *oblate* ($I_A = I_B < I_C$) or *prolate* ($I_A < I_B = I_C$). (A sphere is made oblate by compressing it along a diameter, to give a shape like a bun or a discus; a sphere is made prolate by extending it along a diameter, to give a shape like an egg or a cigar. $CHCl_3$ is thus oblate, and CH_3Cl prolate.) The existence of two different moments of inertia, each of which can contribute to the rotational energy of the molecule, implies the existence of two quantum numbers for the symmetric top. Rather than assign one quantum number to each moment of inertia, it turns out to be mathematically more convenient to retain the quantum number J for the total angular momentum, and define a new quantum number K for the angular momentum about the axis of symmetry of the top. K can have any integral value between $+J$ and $-J$, including zero, i.e. $(2J+1)$ values in all. K cannot be greater than J, since J is a measure of the total angular momentum. Negative values of K can be thought of as referring to rotation in the opposite sense. The rotational energy for a prolate symmetric top is then given by

$$E_{rot} = hJ(J+1)B + hK^2(A-B) \tag{2.15}$$

where

$$A = h/8\pi^2 I_A \quad \text{and} \quad B = h/8\pi^2 I_B \tag{2.16}$$

The equations for an oblate symmetric top are obtained by replacing B by C. As in Section 2.A.1, if h is in units J s, I_A and I_B in units kg m^2, then E_{rot} is in units J, and A and B are in units s^{-1}, i.e. Hz. The corresponding quantities in units cm^{-1} would be written $\tilde{A}$ and $\tilde{B}$. The dependence of the energy on K^2 shows that, as one would expect, the sense of the rotation about the axis of symmetry has no effect on the energy. The selection rules are that $\Delta J = \pm 1$ and $\Delta K = 0$. The selection rule $\Delta K = 0$ is due to the symmetry of the molecule about the axis of rotation, so that there is no dipole moment perpendicular to this axis. The microwave spectrum of a symmetric top thus closely resembles that of a linear molecule, and only one moment of inertia, I_B, which is equal to I_C, can be evaluated. Measurements of moments of inertia for isotopically substituted molecules give values for bond lengths and interbond angles. The total number of isotopically substituted molecules required is equal to the sum of the numbers of bond lengths and interbond angles to be determined. The symmetry of the molecule is invoked to reduce the number of unknowns; for example, only one C–Cl bond length is calculated for $CHCl_3$. Often more than the minimum number of types of isotopically substituted molecule are studied as a cross-check. It is not always necessary to synthesise isotopically substituted molecules especially for the measurement. The sensitivity of microwave techniques is such that for many molecules the natural abundance of isotopes is sufficient to provide the data required. The argument so far has assumed rigid molecules and fixed bond lengths. The effects of bond elasticity are different for linear molecules and symmetric tops, as shown in Section 2.D.

2.C.5 Asymmetric tops

Asymmetric tops are molecules for which $I_A \neq I_B \neq I_C$. These include the vast majority of molecules, e.g. H_2O, O_3, and NO_2. Unfortunately their lack of symmetry makes a general treatment extremely difficult. The usual procedure is to consider an unsymmetric molecule to be a slightly distorted

TABLE 2.1

Bond lengths and angles from microwave spectroscopy

HCN (*linear molecule*)

Values calculated from B for:
$^{1}H^{12}C^{14}N$, $^{1}H^{13}C^{14}N$, $^{2}H^{12}C^{14}N$, $^{2}H^{13}C^{14}N$, $^{1}H^{15}C^{14}N$
(i.e. three more than minimum necessary) (Winnewisser *et al.*, 1971)

C–H 106.549 ± 0.024 pm
C–N 115.321 ± 0.005 pm

CHCl$_3$ (*symmetric top*)

Values calculated from:
$^{12}C^{1}H^{35}Cl_3$, $^{12}C^{2}H^{35}Cl_3$, $^{13}C^{1}H^{35}Cl_3$, $^{12}C^{1}H^{37}Cl_3$, $^{12}C^{1}H^{35}Cl_2{}^{37}Cl$
(i.e. two more than minimum necessary) (Jen and Lide, 1962)

C–H 110.0 ± 0.4 pm	Cl–C–Cl 111.3° ± 0.2°
C–Cl 175.8 ± 0.1 pm	

Pyridine (*asymmetric top*)

Values for 11 isotopically substituted molecules (i.e. one more than minimum necessary) (Bak *et al.*, 1958)

N–C(2) 134.0 pm	C(6)–N–C(2) 116.8°
C(2)–C(3) 139.5 pm	N–C(2)–C(3) 123.9°
C(3)–C(4) 139.4 pm	C(2)–C(3)–C(4) 118.5°
C(2)–H(2) 108.4 pm	C(3)–C(4)–C(5) 118.3°
C(3)–H(3) 108.0 pm	N–C(2)–H(2) 115.9°
C(4)–H(4) 107.7 pm	C(4)–C(3)–H(3) 121.3°
Error ± 0.1 pm	Error ± 0.2°

version of either a prolate or an oblate symmetric top, and apply perturbation techniques to the relevant equations. Despite the mathematical complexity required, bond lengths and angles have been evaluated for many molecules, even as complex as fluorobenzene and pyridine. The bond lengths and interbond angles for some simple molecules are given in Table 2.1, to show the type of information obtained.

2.D The Effects of Bond Elasticity

2.D.1 The simple oscillator

Chemical bonds are not rigid, but elastic so that atoms can vibrate within molecules. A simple model for the diatomic molecule as an oscillator is that of two masses, m_1 and m_2, joined by a spring which resists extension or contraction to an equal degree. If the spring obeys Hooke's Law, then the restoring force is proportional to the displacement from the equilibrium length, i.e.

$$\mu(\mathrm{d}^2x/\mathrm{d}t^2) = -kx \qquad (2.17)$$

where μ is the reduced mass as defined by Eqn. 2.2, $\mathrm{d}^2x/\mathrm{d}t^2$ is the acceleration of the masses at distance $(r_e \pm x)$ apart, r_e being the equilibrium distance, and k is the constant of proportionality, called the *force constant*. Integration of Eqn. 2.17 shows that the masses move in simple harmonic motion, of frequency ω, where

$$\omega = (k/\mu)^{\frac{1}{2}}/2\pi \qquad (2.18)$$

For ω to be in units Hz, with μ in units kg, k must be in units N m^{-1}. Tables of k are usually in units dyn cm^{-1} (10^7 dyn = 1 N), and μ must be in gram units if these values are to be used. It can be seen that ω is independent of the amplitude for simple harmonic motion. The energy of the oscillator is the sum of the kinetic energy of the two masses and the potential energy of the spring. This sum stays constant, although during each oscillation there is exchange between kinetic and potential energy terms. At maximum extension, x_{max}, the masses are at rest, so all the energy is in the potential energy form. Thus the energy of the harmonic oscillator may be calculated to be

$$E_{vib} = \tfrac{1}{2}kx_{max}^{\;2} \qquad (2.19)$$

Application of quantum mechanics to this classical model shows that the permitted vibrational energy levels of a diatomic molecule are given by

$$E_{vib} = (v + \tfrac{1}{2})h\omega \qquad (2.20)$$

where v is the *vibrational quantum number*, which can have any positive integral value from zero upwards. Comparison of Eqns. 2.19 and 2.20 shows that v gives a measure of the amplitude of the vibration. For $v = 0$, E_{vib} has the value $\frac{1}{2}h\omega$, so there is a certain amount of vibrational energy even at the lowest vibrational level. This is termed the *zero-point energy*. The existence of a finite amount of zero-point energy is a consequence of the Heisenberg Uncertainty Principle. If a diatomic molecule could exist with no such energy, both the bond length and bond energy would be simultaneously exactly defined.

The strength of a covalent chemical bond is due to the overlap of atomic orbitals. As the interatomic distance increases during a vibration, the amount of overlap decreases, so the bond becomes slightly weaker. This effect is hardly perceptible for low values of v, but at high values of v, corresponding to large amplitudes of oscillation, the restoring force is appreciably weakened at the maximum extension of the bond. This has the effect of reducing the frequency of oscillation, by virtue of Eqn. 2.18. The oscillation is no longer simple harmonic, and a correction term for *anharmonicity* must be applied. The frequency is now slightly dependent on the amplitude, and can be written as ω_v, where

$$\omega_v = \omega_e[1 - x_e(v + \tfrac{1}{2})] \tag{2.21}$$

Here ω_e is the limiting value of the vibration frequency for an infinitely small amplitude of vibration, and x_e is a constant, the *anharmonicity constant*, giving a measure of the change of bond strength with interatomic distance. Typically x_e is of the order of 10^{-2}. ω_e is not an actual vibration frequency of the molecule, since even at the zeroth vibrational level the molecule is vibrating with finite amplitude. The minimum observable frequency is that for the zeroth level, ω_0. A value can be obtained for ω_e by extrapolation of a set of values for ω_v. The vibrational energy corrected for anharmonicity is then

$$E_{vib} = (v + \tfrac{1}{2})h\omega_e - (v + \tfrac{1}{2})^2 h\omega_e x_e \tag{2.22}$$

2.D.2 The effect of vibration on moments of inertia

Bond vibration frequencies are hundreds or thousands of times higher than rotational frequencies, so a molecule vibrates many times in the course of a single rotation. The moment of inertia is the average for all the values of the bond lengths during the vibrations. Even a symmetrical, i.e. harmonic, vibration, in which the bond shortens by exactly as much as it lengthens, causes a change in the moment of inertia of a diatomic molecule by comparison with the value for a rigid rotor. This is because the moment of inertia is proportional to the square of the bond length. The average of the squares of a set of numbers is not equal to, but greater than, the square of the average. To take a simple numerical example, the average of the squares of 5 and 7 is 37, whereas the square of the average is 36. Thus the effective moment of inertia is increased by bond vibration, even at the zeroth vibrational level. It is necessary to distinguish between B_0, the rotational constant observed for the zeroth vibrational level, and B_e, the rotational constant that would be observed if the molecule were rigid. Higher vibrational levels are associated with further values of B, i.e. B_1, B_2, etc. where $B_e > B_0 > B_1 > B_2 \ldots$. It is possible to calculate B_e by extrapolation from the higher values of B. The value of B for the vibrational level v is given by

$$B_v = B_e - \alpha_e(v+\tfrac{1}{2}) + \gamma_e(v+\tfrac{1}{2})^2 \tag{2.23}$$

where the $\alpha_e(v+\frac{1}{2})$ term is the correction for harmonic oscillation, and the $\gamma_e(v+\frac{1}{2})^2$ term is a further small correction for anharmonicity. Typically α_e is of the order of 1% of B_e, and positive, and γ_e is of the order of 0.1% of α.

Equations similar to Eqn. 2.23 can be produced for polyatomic molecules, but since polyatomic molecules have many modes of vibration, a contribution to the equations must be made for each mode.

2.D.3 Centrifugal distortion

Centrifugal force tends to stretch a rotating molecule, thus increasing the moment of inertia and decreasing B. The higher the rotational energy, the greater the effect, so one can write

$$B = B_e - D_j J(J+1) \tag{2.24}$$

where D_j is the *centrifugal stretching constant*. The value of D_j depends on the moment of inertia and the elasticity of the bond. The higher the moment of inertia, the less the angular velocity for any given energy, and hence the

TABLE 2.2

Microwave absorption spectrum of $^{39}K^{127}I$ (Honig *et al.*, 1954)

	$J = 4 \rightarrow 5$	$J = 5 \rightarrow 6$	$J = 6 \rightarrow 6$
$v = 0$	18 209.8	21 851.3	25 492.8
$v = 1$	18 129.6	21 755.2	25 380.7
$v = 2$	—	21 659.4	25 268.9
$v = 3$	—	21 563.9	25 157.0

Frequencies in MHz; errors 0.1 MHz.

TABLE 2.3

Data for diatomic molecules obtained by microwave spectroscopy (Honig *et al.*, 1954)

Molecule	B_e/MHz	α_e/MHz	γ_e/kHz	D_J/kHz	ω_e/cm^{-1}	r_e/pm	Electron diffn. values of r_e/pm	μ/debye
$^{6}Li^{81}Br$	16 650.0	168.6	650	—	—	217.04	—	6.19
$^{23}Na^{79}Br$	4534.5	28.30	85	7	235	250.20	261	—
$^{39}K^{35}Cl$	3856.4	23.68	50	—	—	266.66	276	10.48
$^{39}K^{127}I$	1825.0	8.034	12.21	1.03	165	304.78	319	11.05
$^{85}Rb^{35}Cl$	2627.4	13.60	21	—	—	278.68	286	—
$^{85}Rb^{79}Br$	1424.8	5.578	7.9	0.45	169	294.48	302	—
Error	±0.1 MHz	±0.005 MHz	±5%	±20%	±20%	±0.01	—	±0.03

less the centrifugal effect. Similarly, the stiffer the bond, the less the centrifugal effect. It can be shown that, for a diatomic molecule,

$$D_j = 4B_e^3/\omega_e^2 \qquad (2.25)$$

so a value of ω_e can be obtained from D_j. Typically, ω_e is of the order of 10^{14} Hz whereas B_e is of the order of 10^{11} Hz, so D_j is of the order of 10^5 Hz. Large centrifugal effects in the value of B are found for molecules of low moment of inertia, undergoing transitions at high J. Such transitions tend to occur at frequencies too high for the microwave region, and centrifugal stretching has only a very small effect on the microwave spectra of diatomic molecules.

Data illustrating the various effects of bond elasticity are comparatively rare for the microwave spectra of diatomic molecules. The best example is provided by the spectra of the alkali halides, in the vapour form at around 700 °C (Honig *et al.*, 1954). The high temperature and low vibrational frequencies result in several vibrational levels being accessible to study. The high dipole moments of these molecules aid the production of good microwave spectra. Typical microwave data for $^{39}K^{127}I$ are given in Table 2.2. These are fitted to the general expression for the frequency:

$$\nu = 2B_e(J+1) - 2\alpha_e(J+1)(v+\tfrac{1}{2}) + 2\gamma_e(J+1)(v+\tfrac{1}{2})^{\frac{1}{2}} - 4D_j(J+1)^3 \qquad (2.26)$$

Values of the parameters for best fit for some of the alkali halides are given in Table 2.3. The trends and relative magnitudes are as expected from the discussion above.

The inclusion of terms for bond elasticity in the spectrum of symmetric tops gives

$$\nu = 2B_e(J+1) - 2\Sigma[\alpha_i(J+1)(v+\tfrac{1}{2}) - \gamma_i(J+1)(v+\tfrac{1}{2})^2] - 4D_j(J+1)^3 - 2D_{jk}(J+1)K^2 \qquad (2.27)$$

where D_{jk} is the centrifugal stretching constant describing the "opening out" of the molecule as it rotates about its axis of symmetry. The magnitude of K now affects ν as a result of centrifugal stretching, so that the $J \rightarrow (J+1)$ transition gives rise to a multiplet of $(2J+1)$ lines, the spacing within the multiplet giving a value for D_{jk}. This splitting distinguishes the microwave spectra of symmetric tops from those of linear molecules.

2.D.4 Coriolis forces and *l*-type doubling

A polyatomic linear molecule executing a bending vibration is not constantly linear, and so has an internal angular momentum about the principal axis. At the extreme of an O–C–S bend, for example, the shape of the OCS molecule resembles that of H_2O, as illustrated in Fig. 2.2. This angular momentum is, of course, quantised, and has values $\hbar l$, where l can have one of the values

$$l = v, \quad (v-2), \quad (v-4), \ldots, (2-v), \quad -v$$

v being, as in Section 2.D.1, the vibrational quantum number for the bending mode. The rotational energy including a term for this rotation is then

$$E_{\text{rot}} = hB_{\text{v}}[J(J+1)-l^2] \tag{2.28}$$

Each energy level, except for $l = 0$, is doubly degenerate, since the energy depends on l^2, so the sense of the rotation does not affect the energy. The selection rule is $\Delta l = 0$, so this effect should not be observable in the spectrum. The situation is closely analogous to that for symmetric tops, where the selection rule $\Delta K = 0$ simplifies the spectrum. However, interaction between the bending vibration and other vibrations makes it possible for this rotation to affect the microwave spectrum.

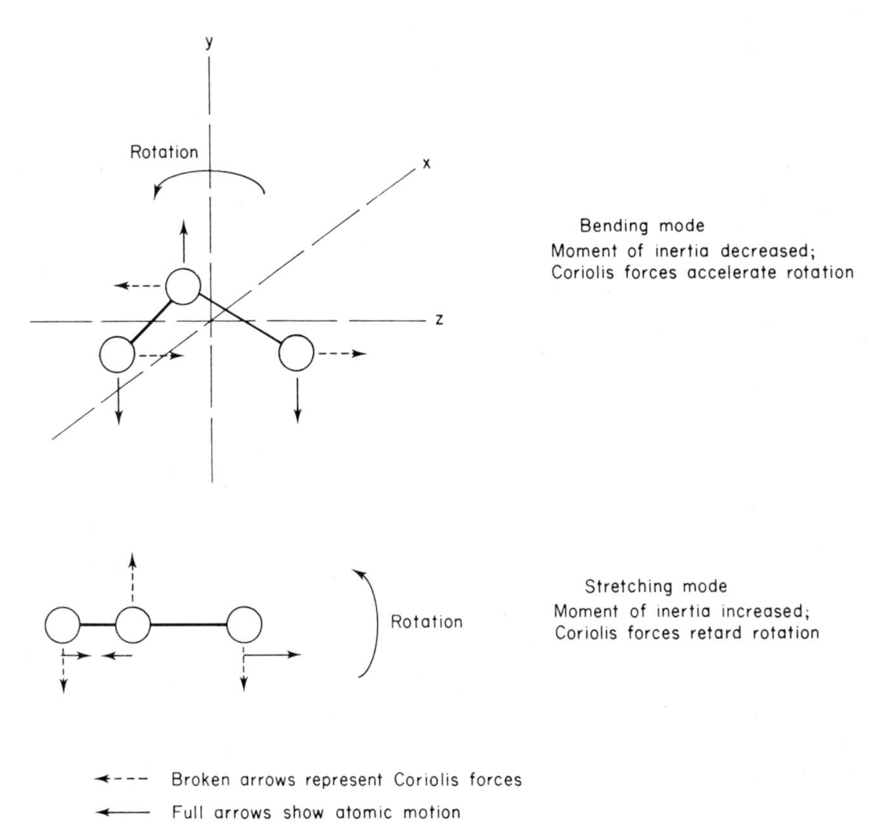

FIG. 2.3 Coriolis interaction.

The angular momentum of an isolated molecule is a constant, so a change of shape, due to a vibrational contraction, which decreases the moment of inertia must cause the angular velocity to increase in compensation, and vice versa for a vibrational expansion. It is as if tangential forces were periodically applied to the molecule to slow its rotation and then speed it up again. These forces are known as Coriolis forces. The Coriolis forces produced by one vibrational mode can act so as to increase the amplitude of vibration in another mode, as illustrated in Fig. 2.3. The bending mode is doubly degenerate. If the axis of the molecule is taken as the z axis, bending can be in the yz plane (as in Fig. 2.3) or in the xz plane. For rotation about the x axis (as in Fig. 2.3) the Coriolis interaction occurs with the vibration in the yz plane, and does not affect the vibration in the xz plane. The consequences are a lifting of the degeneracy of the bending modes, a splitting of the vibrational energy levels, and a doubling of the lines in the microwave spectrum. This is termed *l*-type doubling. The splitting is given by

$$\Delta\nu = q_l J(J+1) \tag{2.29}$$

where q_l is approximately $2.3B/\omega$, ω being the frequency of the bending vibration. For some molecules transitions between the two energy levels occur in the microwave region (Shulman and Townes, 1950); for example, for HCN, q_l is 223.6 MHz.

2.E The Effects of Nuclear Spin

2.E.1 The magnetic properties of the nucleus

The atomic nucleus is found to have magnetic as well as electrical properties. A body of macroscopic size carrying an electric charge acts as a magnet if it spins on its axis, and so by analogy the atomic nucleus is said to have *spin*, although whether or not the nucleus does really spin is a matter for metaphysical discussion. The angular momentum $\boldsymbol{I}$ of the nucleus is quantised, and given by

$$\boldsymbol{I} = \hbar[I(I+1)]^{\frac{1}{2}} \tag{2.30}$$

where I is the quantum number. The *spin quantum number*, I, is characteristic of the nucleus, and has different values for different isotopic species. Some examples are given in Table 2.4. The magnetic moment $\boldsymbol{\mu}$ of the nucleus is given by

$$\boldsymbol{\mu} = g_N \mu_N [I(I+1)]^{\frac{1}{2}} \tag{2.31}$$

where g_N is a dimensionless number, the *nuclear g-factor*, and μ_N is the *nuclear magneton*, defined by

$$\mu_N = e\hbar/2m_p \tag{2.32}$$

where e is the charge on the electron and m_p is the mass of the proton. If e is in units C and m_p in units kg, μ_N is in units A m^2, which are equivalent to units J T^{-1}. (Students unfamiliar with the SI units should refer to the Appendix.) The quantity $g_N\mu_N$ is sometimes denoted by γ and termed the *magnetogyric* or *gyromagnetic ratio*. Substitution of numerical values shows that the nuclear magneton has value 5.051×10^{-27} A m^2.

TABLE 2.4

Some nuclear properties (Shulman and Townes, 1950)

Nucleus	Spin	Magnetic moment, μ /nuclear magnetons	Quadrupole moment, Q, $\times 10^{28}$/m^2
^{1}H	1/2	2.79270	0
^{2}H	1	0.85738	+0.0028
^{12}C	0	0	0
^{13}C	1/2	0.70216	0
^{14}N	1	0.40357	+0.02
^{15}N	1/2	−0.28304	0
^{16}O	0	0	0
^{17}O	5/2	−1.8930	−0.0301
^{19}F	1/2	2.6273	0
^{23}Na	3/2	2.22161	+0.1
^{33}S	3/2	0.64274	+0.005
^{35}Cl	3/2	0.82089	−0.079
^{37}Cl	3/2	0.68329	−0.017
^{57}Co	7/2	4.6	+0.1
^{79}Br	3/2	2.0990	+0.332
^{81}Br	3/2	2.2626	+0.287
^{127}I	5/2	2.7939	−0.785

The spin of a nucleus in a molecule can couple with the rotation of the molecule to cause splitting of the rotational energy levels. The total angular momentum is given by a quantum number F, which is the sum of the nuclear spin quantum number I and the rotational quantum number J. The energy levels are $(2F+1)$-degenerate in the absence of a field. The effect is generally very small, in the range 1 to 100 kHz, except for molecules containing an odd number of electrons, the magnetic field of which affects the nuclei, e.g. NO, ClO, NO_2.

2.E.2 Nuclear quadrupoles

Atomic nuclei do not possess an electric dipole moment, but do possess an electric quadrupole moment if I is equal to or greater than unity. The *quadrupole moment* can be thought of as being due to distortion of the spherical shape of the nucleus. Prolate distortion deprives the equatorial region of positive charge, giving a positive quadrupole moment, whereas oblate distortion deprives the polar regions of positive charge, giving a negative quadrupole moment, as illustrated in Fig. 2.4.

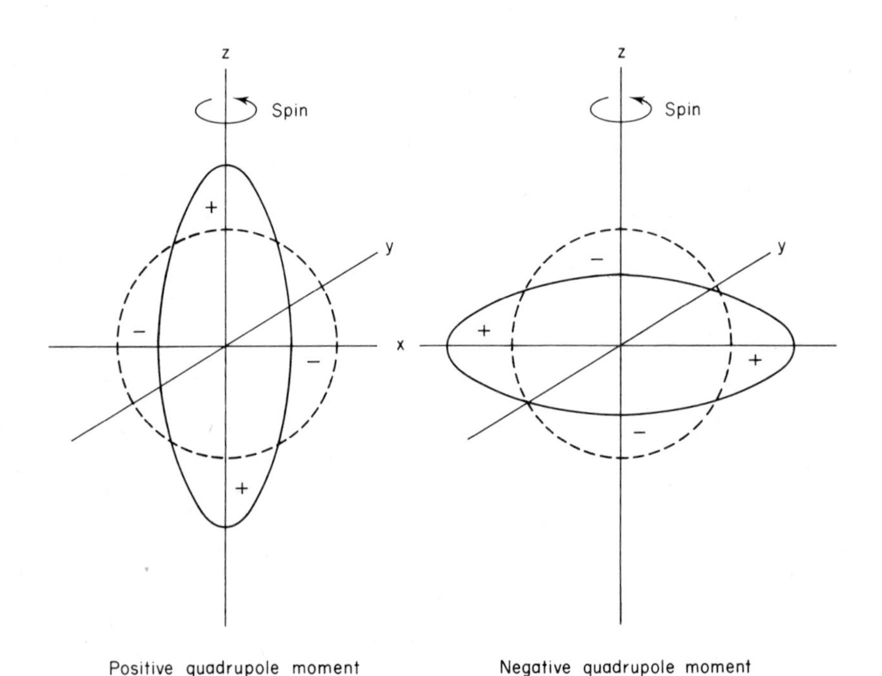

FIG. 2.4 Nuclear quadrupole moments.

Nuclear quadrupole moments are commonly expressed as eQ, where e is the charge on the electron and Q is of dimensions $(\text{length})^2$, usually quoted in units cm^2. Nuclear diameters are of the order of 10^{-12} cm, so Q is of the order 10^{-24} cm^2. The student should note that electric dipole moments are of dimensions (charge) × (length), and quadrupole moments are of dimensions $(\text{charge}) \times (\text{length})^2$. Values for some nuclei are given in Table 2.4. In a uniform electrostatic field, the torque on one of the two dipoles comprising the quadrupole exactly balances the torque on the other. There is a net torque only if the quadrupole is in a non-uniform field, varying along the spin (i.e.

z) axis, such that one part of the nucleus is subject to a different field intensity from another part. The energy of interaction, E_Q, is given by

$$E_Q = \tfrac{1}{4}eQ(\partial^2\phi/\partial z^2) \tag{2.33}$$

where ϕ is the electric potential, the first derivative of which is the field strength. If e is in units C, Q in units m^2, ϕ in units V, and z in units m, then E_Q is in units C V, i.e. J. Equation 2.33 may be compared with Eqn. 2.36 from which it can be seen that the energy of interaction between an electric dipole and an electric field is given by the product of the dipole moment and the field strength E. The quantity $(\partial^2\phi/\partial z^2)$ is usually written as q. E_Q is quantised, so the nucleus has various permitted orientations with respect to the direction of maximum field gradient.

An order-of-magnitude calculation for E_Q may be performed by considering the electric field to be due to a single electron distant 100 pm from the nucleus. The gradient of a field at distance r m from a point charge of magnitude e C is given by

$$q = e/2\pi\varepsilon_0 r^3 \text{ V m}^{-2} \tag{2.34}$$

Substitution of numerical values gives a value of 2.88×10^{21} V m^{-2} for q. If the quadrupole is ^{2}H, then E_Q is calculated to be 3.22×10^{-29} J, which corresponds to a frequency of 48.6 kHz. If the quadrupole is ^{127}I, then E_Q is calculated to be 9.05×10^{-27} J, corresponding to a frequency of 13.6 MHz.

The interaction between the quadrupole and the electrons around it provides a comparatively strong coupling between the nuclear spin and the rotation of the molecule. The total angular momentum has quantum number F, as in Section 2.E.1, which can have any value in the range $(J+I)$ to $(J-I)$. The various possible orientations of $\boldsymbol{J}$ and $\boldsymbol{I}$ correspond to various possible orientations of the quadrupole with respect to the field produced by the electrons, and cause a splitting of the rotational level into $(2I+1)$ components, or $(2J+1)$ if $I > J$. The calculation of the magnitude of the splitting is difficult, but it has been shown that

$$E_Q = eQq_J\frac{[\tfrac{3}{4}C(C+1)-I(I+1)J(J+1)]}{2J(2J+1)I(2I+1)} \tag{2.35}$$

where

$$C = F(F+1)-I(I+1)-J(J+1)$$

The term in square brackets is known as Casimir's function, after the original solver of the problem. q_J is the field gradient along the direction of rotation, i.e. the line of the vector $\boldsymbol{J}$. The direction of $\boldsymbol{J}$ depends on the rotational state of the molecule, and so q_J is similarly dependent. It is necessary to have a measure of field gradient that is independent of the rotational state, and this is provided by q_m, the field strength along a geometric axis of the molecule. q_m is related to q_J by a function involving the various quantum numbers. The fine structure of rotational transitions due to quadrupole interactions gives the value of eQq_m, henceforth written eQq, which is termed the *quadrupole coupling constant*. The magnitude of this constant varies widely, and splittings of up to several hundred MHz have been observed. A discussion of the significance and interpretation of values of the quadrupole coupling constant is given in Chapter 3, in which another method of obtaining these values is described.

2.F The Effect of Applied Fields

2.F.1 The first-order Stark effect

When an electrostatic field of strength E V m^{-1} is applied to a symmetric top the energy of interaction between the field and the dipole moment μ of the molecule is quantised in such a manner that the energy is given by

$$E_S = -\mu EMK/(J+1)J \tag{2.36}$$

where M and K are the magnetic and subsidiary quantum numbers introduced in Sections 2.A.1 and 2.C.4 respectively. Each rotational level is thus split by the applied field into $(2J+1)$ components. The splitting of spectroscopic energy levels by an applied electrostatic field was first observed by Stark, and the effect is named in his honour. If μ is in units C m, E_S is in units J. In the literature dipole moments are generally given in debye units. One debye is the dipole moment of two charges each of 1×10^{-18} e.s.u. separated by 1 Å, and is equal to $10^{-29}/3$ C m in the SI. (An explanation of these systems of units is given in the Appendix.) The dipole moment of a diatomic molecule consisting of a pair of univalent ions separated by 100 pm is 4.8 debye, showing that the debye is of convenient size. Substituting values of $\mu = 1$ debye and $J = K = 2$ into Eqn. 2.36 shows that a field of 100 V cm^{-1} produces a splitting of 34 MHz, a readily observable effect.

This first-order effect, so called because it is linearly proportional to the applied field, is not in general observed for linear molecules, since their dipole moments are necessarily perpendicular to their axes of rotation. An exception to this is found for l-type doubling spectra, for which rotation about the axis is significant.

2.F.2 The second-order Stark effect

The application of an electrostatic field to a molecule induces a dipole moment by polarising the electrons. The magnitude of the induced dipole is linearly proportional to the field strength. The induced dipole then interacts with the field, with an interaction energy proportional to the product of the induced dipole and the field strength, i.e. proportional to the square of the field strength. The quantum-mechanical treatment shows that the energy levels for a diatomic molecule are split by a term

$$E_s = \frac{\mu^2 E^2 [J(J+1) - 3M^2]}{2hBJ(J+1)(2J-1)(2J+3)} \tag{2.37}$$

The substitution of typical numerical values, $B = 5000$ MHz, $J = M = 2$, $\mu = 1$ debye, shows that a field strength of 1000 V cm^{-1} produces a splitting of 25 MHz. Second-order Stark effects are thus readily observable for diatomic molecules. Corresponding, but more complex, expressions can be derived for polyatomic molecules.

2.F.3 The application of the Stark effect

The multiplicity and relative intensities of the lines produced by the Stark effect depend on J. The Stark splitting pattern may thus be helpful in the interpretation of complex microwave spectra, since it immediately gives a value of J for each line observed.

The amplitude of the Stark splitting gives a value for the dipole moment of the molecule. Extremely accurate values are obtained, partly because of the intrinsic accuracy of microwave frequency measurements, and partly because the sample is a gas at low pressure, so intermolecular perturbations are negligible. Impure samples may be used, provided that the correct lines can be identified. Dipole moments can be measured both along and perpendicular to the molecular axis, and in excited as well as in ground vibrational states. An example of what can be done is the study by Lide and Mann (1957) of the dipole moments of propylene. The dipole moment along the C=C axis was found to be 0.360 ± 0.001 debye, and the dipole moment perpendicular to the C=C axis, due to the substituent methyl group, was found to be 0.05 ± 0.02 debye.

2.F.4 The Zeeman effect

The splitting of a spectroscopic energy level by the application of a magnetic field is called the Zeeman effect, after its discoverer. The magnetic field interacts with a rotating molecule by virtue of the magnetic moment of the molecule. A few molecules, e.g. NO, ClO, and NO_2, possess a large magnetic moment due to unpaired valency electrons, and hence show large Zeeman splittings. The resulting spectra are complex. The applied magnetic field also interacts with the magnetic moments of the nuclei in the molecule. The magnetic interaction energy can either add to or subtract from the field-free energy level, so that each absorption line splits to a doublet, symmetrically disposed about the field-free position. The frequency shift $\Delta\nu$ for the $J = 0 \rightarrow 1$ transition for an applied field of strength B teslas is given by

$$\Delta\nu = g\mu_N B/h \tag{2.38}$$

where μ_N is the nuclear magneton, as defined in Section 2.E.1, and g is a factor containing terms for both the nucleus causing the splitting and the molecule as a whole. For example, $\Delta\nu$ for $^{14}NH_3$ inversion (as described in Section 2.G.1) is 0.359 MHz for a field of 0.1 T (Jen, 1948). Substitution into Eqn. 2.33 shows that g has the value 0.44. Shifts of this magnitude are typical. More complex effects are observed for lines due to nuclear quadrupole interactions.

2.G Other Processes Causing Microwave Absorption

2.G.1 Inversion

The ammonia molecule is pyramidal, with the nitrogen atom lying outside the plane of the three hydrogen atoms, the H–N–H bond angle being 106°42′. The molecule may be thought of as resembling that of methane, with a lone pair of electrons in place of a hydrogen atom. One mode of vibration of the ammonia molecule is that in which the nitrogen atom moves to and fro through the plane of the hydrogen atoms along a line perpendicular to that plane. This motion is represented as (a) in Fig. 2.5. If the vibration is of sufficiently high energy, i.e. amplitude, the nitrogen atom will reach a position in the plane of the three hydrogen atoms, as shown in (b). This is at an energy maximum. The nitrogen atom can then go on through the plane, to give (c). This process is known as *inversion*. The potential energy diagram for the system is shown in Fig. 2.5. Vibrations (a) and (c) correspond to energy levels in the separate troughs, and the energy required for inversion to occur is that of the maximum between the troughs, V (if the system can be described by classical physics). If the molecule has energy much greater than V, it passes freely from state (a) to state (c).

The application of the principles of wave mechanics produces some modifications to the picture. The energy levels in each trough are affected by the proximity of the other trough, since the wave function for each energy level in one trough extends beyond the boundaries of the trough and into the region of the other trough. Thus the resultant wave function for each trough is the combination of the two wavefunctions which would describe the

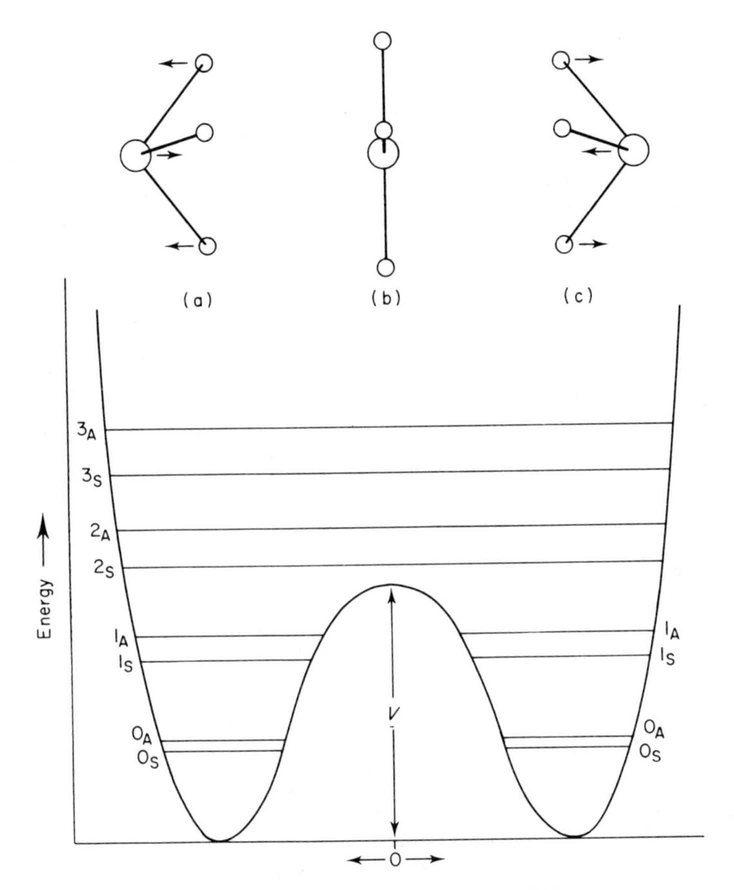

FIG. 2.5 The inversion of the ammonia molecule, NH_3.

system if the two troughs were completely independent. As for the combination of two atomic orbitals to make a molecular orbital, this combination can be in either the symmetric or the antisymmetric mode. Each vibrational level is thus split into two, symmetric S and antisymmetric A, the antisymmetric levels being of higher energy. The magnitude of the splitting varies inversely with the exponential of $V^{\frac{1}{2}}\mu^{\frac{1}{2}}$, where μ is the reduced mass of the system, and is zero if V is infinite. The splitting of the zeroth level for NH_3 is 1.576×10^{-23} J molecule^{-1}, i.e. 9.50 J mol^{-1}. Thus the transition $0_S \rightarrow 0_A$ gives rise to microwave absorption at 23 786 MHz. This transition occurs from the 0_S level of one trough to the 0_A level of the other, so the frequency is known as the inversion frequency. Because of the interaction between the two troughs, the molecule does not need to possess energy V for the inversion to occur, but can pass from one state to the other by quantum-mechanical *tunnelling*. There is only one other vibrational level within each trough, 1.99×10^{-20} J molecule^{-1} above the zeroth level, split by 7.15×10^{-22} J molecule^{-1}. The thinner barrier between the states at this higher energy causes the interaction, as measured by the splitting, to be greater. Since the magnitude of the splitting is very sensitive to V, a value of V, 4.116×10^{-20} J molecule^{-1}, or 2.48 kJ mol^{-1}, may be calculated from the inversion frequency (Costain and Sutherland, 1952). This value agrees well with that calculated from the bond-bending force constant, found from frequencies observed in the near infrared spectrum of the bond-bending vibration. This value of V shows that the inversion must be a non-classical, quantum-mechanical process. The frequency of the HNH bending mode, found from the near infrared spectrum, is 3.165×10^{13} Hz. The Boltzmann factor, $\exp(-\varepsilon_i/kT)$ is 5×10^{-5} for $\varepsilon_i = 4.116 \times 10^{-20}$ J and T = 300 K. If inversion occurred only when a molecule had acquired sufficient vibrational energy to surmount the barrier, only two vibrations in every million would result in inversion, so the inversion frequency would be 1500 MHz, much less than the observed frequency. Furthermore this classical frequency would be temperature-dependent, which the observed frequency is not.

The microwave absorption of ammonia is of great historical interest. The NH_3 molecule was the first to give a sharp microwave spectrum, the first to show nuclear quadrupole splitting, the power saturation effect, and the pressure-broadening effect, and the first to be used as an "atomic clock", i.e. frequency standard. The NH_3 molecule was the first to be used in experiments on amplification by stimulated emission of radiation. A beam of NH_3 molecules in the 0_A state was prepared by the application of an electrostatic field, and allowed to pass into a resonant cavity. The cavity was irradiated with low-intensity radiation at the inversion frequency, which stimulated the $0_A \rightarrow 0_S$ transition, so that the NH_3 molecules emitted intense radiation at the inversion frequency. The original radiation was thus amplified. The device is called a *maser*, the name being an acronym for *microwave amplification by stimulated emission of radiation*. The application of a similar principle to molecules in excited vibrational and electronic states has led to the development of the *laser*, the name being an acronym for *light amplification by stimulated emission of radiation*.

Inversion spectra have been observed only for NH_3 and amines. Calculations show that the inversion frequency of PH_3 is 140 kHz, whereas that for AsH_3 is 0.5 cycles per year. This dramatic decrease is due to increasing V and μ down the series. The inversion of PH_3 might just be observable, but certainly not the inversion of AsH_3. The high rate of inversion for nitrogen explains why compounds of the type $NR^1R^2R^3$ are not optically active, and it may be predicted that compounds of the type $AsR^1R^2R^3$ will be optically active.

2.G.2 Internal rotation

Substituent groups in organic molecules, linked to the rest of the molecule by single bonds, can rotate about that bond. Consider a molecule such as CH_3CCl_3. When the molecule has such a shape, or *conformation*, that the chlorine atoms are as close as possible to the hydrogen atoms, the molecule has a higher energy than when they are as far apart as possible, due to steric repulsion. This energy difference is denoted by V. Thus, as the CCl_3 group rotates relative to the CH_3 group, about the C–C bond, the energy of the molecule varies in a periodic manner, as shown in Fig. 2.6. As can be seen by comparison with Fig. 2.5, there is a similarity between this phenomenon of *internal rotation* and inversion. There are two limiting types of internal rotation: that for which the barrier is very low, and can readily be surmounted by thermal energies, and that for which the barrier is high, in which rotation is replaced by torsional oscillation, accompanied by quantum-mechanical tunnelling through the barrier. The high-barrier type is illustrated in Fig. 2.6, for a three-fold barrier, in which the rotating group is a symmetric top with a three-fold axis of symmetry, as in CH_3CCl_3. According to classical mechanics the internal molecular motion for energy less than V consists of a triply degenerate torsional oscillation (triply degenerate because there are three equivalent energy troughs). If the energy is greater than V, the molecule exhibits free internal rotation, doubly degenerate because the rotation can be in either sense (clockwise or counter-clockwise) referred to the rest of the molecule. The application of quantum mechanics to the problem shows that each torsional oscillation energy level is split into two, as for inversion. One level, the A level, corresponds to an undisturbed torsional level (the effects of the troughs on either side cancelling out), and the other level, the E level, is doubly degenerate, being perturbed by tunnelling through to the troughs in either side. At high energies the A levels associate in pairs to give the doubly degenerate free-rotation levels. In Fig. 2.6, both the torsional levels, $v = 0, 1, 2, 3, \ldots$, and the corresponding rotational levels, $m = 0, 1, 2, 3, \ldots$, are labelled, and it can be seen how the increasing splitting between E and A for the torsional levels corresponds to the development of degeneracy of the rotational levels. The lower the value of V, the lower the value of m at which the splitting of the rotational levels disappears. Thus an analysis of the microwave spectrum gives a value for V. Another method of evaluating V involves the measurement of the frequency for the $0_A \rightarrow 0_E$ transition, but this method is applicable only to high barriers, above about 3 kJ mol^{-1}.

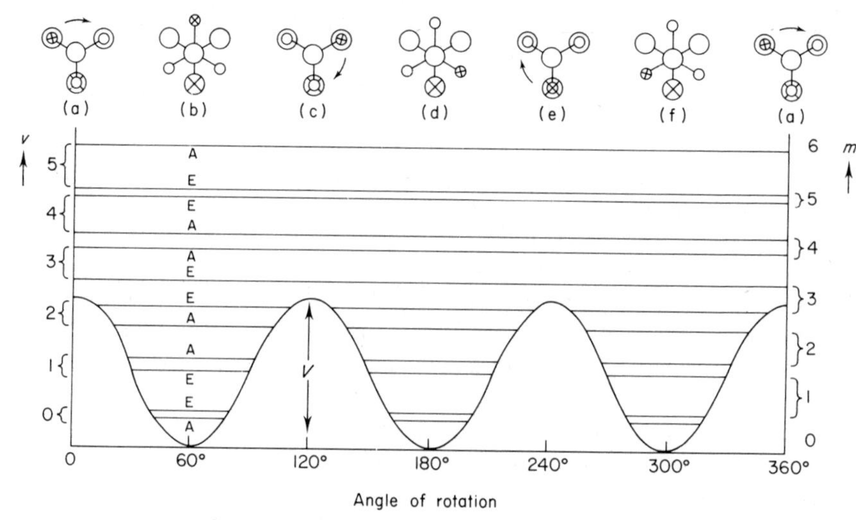

FIG. 2.6 Internal rotation in a molecule such as CH_3CCl_3.

There is a centrifugal interaction between torsional oscillation and molecular rotation, since the more energetic the torsional oscillation, the more the substituent atoms splay out from the axis of symmetry, and the greater the moment of inertia. This causes rotational lines to have *satellites*, i.e. weak adjacent lines, due to rotational excitation of molecules in excited torsional states. The relative intensities of the satellites, compared to that of the principal line, give an approximate, but easily calculated, value for V. By the Boltzmann distribution law, the ratio of the intensities of the lines for the zeroth and first torsional level is given by

$$I_1/I_0 = \exp(-h\nu/kT) \tag{2.39}$$

where ν is the torsional frequency of oscillation. If the torsional oscillation is assumed to be simple harmonic, then

$$\nu = n(V/2I_r)^{\frac{1}{2}}/2\pi \tag{2.40}$$

where I_r is the reduced moment of inertia of the molecule, and n is the number of equivalent energy troughs. If the molecule consists of a pair of mutually-rotating symmetric tops, e.g. CH_3CCl_3, of moments of inertia I_1 and I_2, then

$$I_r = I_1 I_2/(I_1 + I_2) \tag{2.41}$$

This method is necessarily less accurate than those using frequency data, since the error in intensity measurements is comparatively large, of the order of $\pm 10\%$. Some values of V calculated from microwave data by various methods are given in Table 2.5.

If the rotating part of the molecule is not a symmetric top, the potential energy troughs and maxima are of different heights during a complete rotation. Suppose that in Fig. 2.6 the atoms or groups marked by a cross are much bulkier than the others. If 1-iodopropane is taken as an example, the iodine atom and the methyl group would be the bulky substituents. Conformations (d) and (f), although still corresponding to energy minima, would be of higher energy than conformation (b). Conformations (d) and (f) are termed *gauche*, and conformation (b) is termed *trans*. *Gauche* and *trans* forms are examples of *conformational isomers*. Conformational isomers differ in their microwave spectra, so microwave spectroscopy is one of the few techniques available for their study.

TABLE 2.5

Barriers to internal rotation (Lide and Mann, 1957)

Substance	V/kJ mol^{-1}
Isobutane	9.3
Ethyl chloride	8.1
Propylene	4.74
Methanol	2.55
Acetone	1.87
Acetic acid	1.16
Nitromethane	0.014

2.G.3 The effects of electronic energy levels

The unpaired electron in atoms such as H, Na, Rb, and Cs has a spin which can couple with the nuclear magnetic moment in either of two ways, giving an energy gap which corresponds to the absorption and emission of radiation in the microwave region. The frequency of such a transition in the caesium atom is known with such accuracy that it provides the definition of the fundamental unit of time. An analogous transition in the hydrogen atom corresponds to a frequency of 1420.4 MHz, and radiation of this frequency can be detected in radiation from the galaxy, due to emission from interstellar hydrogen.

Although oxygen has no dipole moment, it does absorb microwave radiation in two regions, around wavelengths of 5 mm and 2.5 mm. The absorption is weak, the power being reduced by about 30% per kilometre of transmission, but is of considerable practical importance in the choice of microwave frequencies for telecommunications and radar. The absorption is due to interaction between the angular momentum of the unpaired electrons in the molecule and the rotational momentum of the molecule as a whole. There are two possible values for S, the net *spin momentum quantum number*, namely 0 (i.e. $\frac{1}{2}-\frac{1}{2}$) and +1 (i.e. $\frac{1}{2}+\frac{1}{2}$). The spin momentum of +1 can couple with or against the rotational momentum, so each rotational level is split into a triplet. The transition between the level for which $S = 0$ and either of the levels for which $S = 1$ can be excited by microwave radiation.

The interaction between the orbital momentum of an unpaired electron in a molecule and molecular rotation gives rise to a phenomenon known as Λ-type doubling. The interaction removes the degeneracy of the rotational energy levels about the y and z axes of the molecule, so each rotational line becomes a doublet. This Λ-type doubling is mathematically similar to the *l*-type doubling discussed in Section 2.C.4. Λ-type doubling is prominent in the spectra of diatomic free radicals, such as OH, SH, CN, and ClO. There are considerable experimental difficulties in the study of these species, since they are highly reactive and of short life. They are usually produced by irradiation of stable source molecules with high intensity microwaves, and allowed to pass continuously through the sample chamber, so that their concentration is maintained at a detectable level.

2.H Analytical Applications

2.H.1 Qualitative analysis

A typical modern microwave spectrometer with a backward-oscillator source can scan a range of 10 000 MHz or so with a resolution of 0.1 MHz, and can thus display a great deal of information about a molecule. As has been shown in the preceding sections, there are many causes for complexity in a microwave spectrum. The interpretation of the microwave spectrum of a polyatomic molecule is usually lengthy and difficult. However, a full interpretation is not needed if the spectrum is used to "fingerprint" a molecule. No two people have the same fingerprints, so, even though a fingerprint cannot be used to predict other features of its owner, e.g. height or age, it can give an unambiguous identification if the owner's fingerprints are on record. In just the same way, a microwave spectrum may give an unambiguous identification of an unknown substance if an adequate library of data is available. Manual processing of the data is tedious, but the problem lends itself to computerised data handling, and such systems are now available. Mixtures can also be analysed, the practical limit so far being five components. It is possible readily to distinguish conformational isomers.

2.H.2 Quantitative analysis

Quantitative analysis is possible, but is more difficult than qualitative analysis. The height of an absorption peak, or, to be more correct, the area under a peak, is proportional to the partial pressure of the substance in the sample cell, as long as power saturation is avoided. Peak areas can be measured by electronic integrating techniques. Sample handling presents some problems, since the sample is a gas at low pressure. Sample may be lost by adsorption on to the sample cell walls, and then perhaps desorb during the investigation of the next sample. This phenomenon is termed the *memory effect*.

REFERENCES

Bak, B., Hansen-Nygaard, L. and Rastrap Andersen, J. (1958) *J. Mol. Spectroscopy* **2**, 361.

Costain, C. C. (1958) *J. Chem. Phys.* **29**, 864.

Costain, C. C. and Sutherland, G. B. B. M. (1952) *J. Phys. Chem.* **56**, 321.

Honig, A., Mandel, M., Stitch, M. L. and Townes, C. H. (1954) *Phys. Rev.* **96**, 629.

Jen, C. K. (1948) *Phys. Rev.* **74**, 1396.

Jen, M. and Lide, D. R., Jr. (1962) *J. Chem. Phys.* **36**, 2525.

Lide, D. R., Jr. and Mann, D. E. (1957) *J. Chem. Phys.* **27**, 869.

Moynihan, C. T. (1969) *J. Chem. Educ.* **46**, 431.

Rosenblum, B., Nethercot, A. H., Jr. and Townes, C. H. (1958) *Phys. Rev.* **109**, 400.

Shulman, R. G. and Townes, C. H. (1950) *Phys. Rev.* **77**, 421.

Winnewisser, G., Maki, A. G. and Johnson, D. R. (1971) *J. Mol. Spectroscopy* **39**, 149.

BIBLIOGRAPHY

Carrington, A. (1974) "Microwave Spectroscopy of Free Radicals", Academic Press, New York and London.

Flygare, W. H. (1972) in "Techniques of Chemistry, Vol. I, Physical Methods of Chemistry, Part IIIA, Optical, Spectroscopic and Radioactivity Methods" (A. Weissberger and B. W. Rossiter, eds.), pp. 439–497, Wiley-Interscience, New York and London.

Gordy, W., Smith, W. V. and Trambarulo, R. F. (1953) "Microwave Spectroscopy", Chapman and Hall, London.

Kroto, H. W. (1975) "Molecular Rotation Spectroscopy", Wiley, New York and London.

Morino, Y. and Saito, S. (1972) in "Molecular Spectroscopy; Modern Research" (K. Narahari Rao and C. Weldon Mathews, eds.), pp. 9–78, Academic Press, New York and London.

Townes, C. H. and Schawlow, A. L. (1955) "Microwave Spectroscopy", McGraw-Hill, New York.

Wollrab, J. (1967) "Rotational Spectra and Molecular Structure", Academic Press, New York and London.

Sheridan, J. (1973) in "MTP International Review of Science, Physical Chemistry Series One, Vol. 12, Analytical Chemistry Part 1" (T. S. West, ed.), pp. 251–278, Butterworths, London.

3 Nuclear Quadrupole Resonance Spectroscopy

3.A Apparatus for Nuclear Quadrupole Spectroscopy

The magnitudes of the nuclear quadrupole coupling constant, eQq, discussed in Section 2.E.2 are such that transitions between nuclear quadrupole energy levels may be observed in the radiofrequency region of the spectrum, which is from about 100 kHz to several hundred MHz. These transitions can thus be studied using apparatus made up of standard radio components, such as capacitors and coils. The corresponding wavelengths are so much greater than the dimensions of the apparatus that the optical properties of the electromagnetic radiation can be ignored. The sample, a single crystal or several grams of microcrystalline powder in a thin glass vessel, is placed inside a coil of wire forming part of an oscillating circuit. The oscillation frequency is varied, by use of a variable capacitor, until the frequency is equal to the nuclear quadrupole transition frequency. Energy is then absorbed by the sample, so that extra energy is required to keep the circuit in oscillation. This extra energy requirement is monitored and displayed as the signal. The type of circuit termed a super-regenerative oscillator is most favoured. The transfer of energy from one oscillating system to another oscillating at the same frequency is known as *resonance*, so this technique is termed nuclear quadrupole resonance spectroscopy, or n.q.r. for short.

The energy levels for a nuclear quadrupole in the field of adjacent electrons are given by

$$E_{Q} = eQq[3M_{I}^{2} - I(I+1)/4I(2I-1)] \qquad (3.1)$$

where, as in Section 2.E.1, I is the nuclear spin quantum number, and M_I is the quantum number giving the orientation of the quadrupole relative to the field. M_I can have any integral value between $+I$ and $-I$. The coupling to the radiation field is by the magnetic dipole rather than the electric quadrupole, so the selection rule is $\Delta M_I = \pm 1$. Thus the absorption frequency is given by

$$h\nu = eQq[3(2M_{I}-1)/4I(2I-1)] \qquad (3.2)$$

M_I being taken as positive, and the larger value involved in the transition. Thus for I having the value 3/2, as for ^{35}Cl, there is one absorption frequency, given by

$$h\nu = \tfrac{1}{2}eQq \tag{3.3}$$

If I has the value 5/2, as for ^{127}I, there are two absorption frequencies, given by

$$h\nu_1 = (3/20)eQq; \qquad h\nu_2 = (3/10)eQq \tag{3.4}$$

For some nuclei, such as ^{127}I, a correction must be applied to the energies, to take account of lack of axial symmetry of the field, which is of the order of a few per cent for iodine compounds, but 15% for molecular iodine.

The sample must be a solid, so that the molecules have a fixed orientation in space. In a liquid, the molecules are in continuous rotation, so the angle between the applied electromagnetic field and the field around the nucleus due to adjacent electrons is continuously varying. This has the effect of broadening and lowering the absorption lines below the limits of detection. Molecular motion within the crystal will similarly tend to degrade the n.q.r. spectrum, so it is customary to cool the sample to 70 K in a bath of liquid nitrogen. There are often significant contributions to q from the field of the crystal, which may cause eQq to be different in the solid and gaseous states (the latter being determined by microwave spectroscopy). For example, for $H^{35}Cl$ the values of eQq found in the solid and gaseous states are 52.94 and 67.3 MHz respectively.

3.B The Interpretation of Quadrupole Coupling Constants

Values of Q may be determined independently from eQq measurements by a technique of nuclear physics, termed *coulombic excitation*. The sample is bombarded with heavy ions, e.g. ^{14}N or ^{40}A, which are scattered by the nuclei in the sample. The nuclei thereby acquire rotational energy, which is lost again by γ-radiation. The γ-ray spectrum gives a value of Q for the scattering nuclei. Hence quadrupole coupling constants, eQq, give values of q. For some species, e.g. HD, values of q can be calculated from first principles by the self-consistent field method of quantum mechanics, and used to evaluate Q from eQq. If a value of Q for one nucleus has been found, that for another isotopic nucleus can be readily calculated from eQq, since it may be assumed that q is unaffected by isotopic substitution. s orbitals are spherically symmetrical around the nucleus, so an electron in an s orbital does not produce the field gradient across the nucleus necessary for quadrupole coupling effects. The same argument applies to any closed shell, e.g. six electrons in the three p orbitals, or ten electrons in the five d orbitals. A single electron in a d orbital does produce a field gradient at the nucleus, but d orbitals penetrate to the nucleus much less than p orbitals, so the field gradient is much less. Quadrupole coupling effects due to d electrons are thus much more difficult to detect, but have been observed in cobalt complexes. Most research has concentrated on the effects of p electrons. The extent to which p electrons penetrate to the nucleus depends on the principal quantum number; the greater the principal quantum number, the more significant the penetration. Values of q are thus greater for atoms in the later rows of the Periodic Table. Values of Q tend to be larger for heavier nuclei, so eQq values tend to be much larger for heavy nuclei such as ^{127}I than for light nuclei such as ^{2}H, and the range of values found is extremely large. For example, eQq for $^{1}H^{2}H$ is 227 kHz, whereas eQq for $^{127}I^{35}Cl$ is 2930 MHz.

TABLE 3.1

Nuclear quadrupole coupling constants for ^{35}Cl

Molecule	eQq/MHz	$q \times 10^{-21}$/V m^{-2}	U_p
KCl	0.04	0.021	~0.00
TlCl	−15.8	7.95	0.14
SiF_3Cl	−43.0	22.1	0.39
CH_3Cl	−74.8	38.4	0.68
ICl	−82.5	42.4	0.75
FCl	−146.0	75.0	1.33
Atomic ^{35}Cl	−109.7	56.3	—

Experimental values of eQq for a series of ^{35}Cl compounds are given in Table 3.1. Values of q are calculated using the value -0.0802×10^{-28} m^2 for Q of ^{35}Cl. It can be seen that the charge distribution around the Cl^- ion is spherically symmetrical, as expected from elementary valency theory. In covalent compounds one of the p electrons of the filled 3p shell is partially removed from the environment of the nucleus to make the covalent bond, so q is greater. In the FCl molecule, the Cl atom is polarised so as to acquire a partial positive charge, and q is greater than for the free atom. A parameter U_p, the ratio of eQq for the molecule in question to eQq for the free atom, is defined to give a measure of the number of unbalanced p electrons affecting the nucleus. Unfortunately eQq is affected not only by the degree of ionic character of the bond, but also by hybridisation. The properties of the p orbitals may be modified by mixing, or hybridising, with s orbitals. It has been suggested that U_p may be approximately given by

$$U_p = (1-s)(1+i) \tag{3.5}$$

where s is a parameter measuring the extent of hybridisation with s orbitals and i is a parameter measuring ionic character. The relative contributions of s and i can only be estimated. It is probably better to use U_p values as indicative of trends rather than as possessing absolute significance.

BIBLIOGRAPHY

Schultz, H. D. (1970) in "Spectroscopy in Inorganic Chemistry", Vol. I (C. N. R. Rao and J. R. Ferraro, eds.), pp. 301–346, Academic Press, New York and London.

Semin, G. K., Babushkina, T. A. and Yakobson, T. A. (1975) "Nuclear Quadrupole Resonance in Chemistry", Wiley, New York.

Smith, J. A. S. (ed.) (1974) "Advances in Nuclear Quadrupole Resonance", Vol. I, Heyden, London and New York.

Smith, J. A. S. (1971) *J. Chem. Educ.* **48**, A39, A77, A147, A243.

4
The Infrared Region

4.A The Vibration of a Diatomic Molecule

4.A.1 Energy levels

The simple harmonic oscillator model for a diatomic molecule was discussed in Section 2.A.1, in which it was shown that the vibrational energy is given by

$$E_{vib} = (v+\tfrac{1}{2})h\omega \qquad (2.20)$$

where v is the vibrational quantum number. The simple harmonic oscillator model assumes that the restoring force is linearly proportional to the bond extension. This is a good approximation for small extensions, but at large extensions the bond weakens owing to reduction in atomic orbital overlap, so the restoring force is less than that predicted by the model. To take account of this anharmonicity, an extra term is added to the expression for E_{vib}, to give

$$E_{vib} = (v+\tfrac{1}{2})h\omega_e-(v+\tfrac{1}{2})^2h\omega_e x_e \qquad (2.22)$$

where ω_e is the theoretical limiting frequency for zero bond extension, and x_e is the anharmonicity constant.

The simple harmonic oscillator model corresponds to a *potential energy function* (i.e. the variation of bond energy U with internuclear distance r) of parabolic shape. This is obviously inadequate to describe the variation of U with r over a wide range of values of r. As the atoms in a diatomic molecule are pushed together, the energy rises very rapidly as work is done against the short-range but powerful nuclear repulsive forces. As the atoms are separated the energy also rises, but less steeply, since less work has to be done to stretch a bond already weakened by extension. A limiting value of the energy is reached at infinite separation, corresponding to the bond energy, i.e. the energy liberated when the two free atoms combine to form the diatomic molecule. A simple mathematical expression which describes such behaviour was given by Morse, and the potential energy function thus calculated is termed a *Morse potential*:

$$U = D_e\{1-\exp[-\beta(r-r_e)]\}^2 \tag{4.1}$$

D_e is the dissociation energy, i.e. the energy difference between the molecule in its hypothetical equilibrium state and the individual atoms at infinite separation. This is sometimes called the *spectroscopic dissociation energy*, as distinct from the *thermodynamic dissociation energy*, D_0, which is the energy difference between the molecule in its zeroth vibrational state and the atoms at infinite separation:

$$D_0 = D_e - \tfrac{1}{2}h\omega_e \tag{4.2}$$

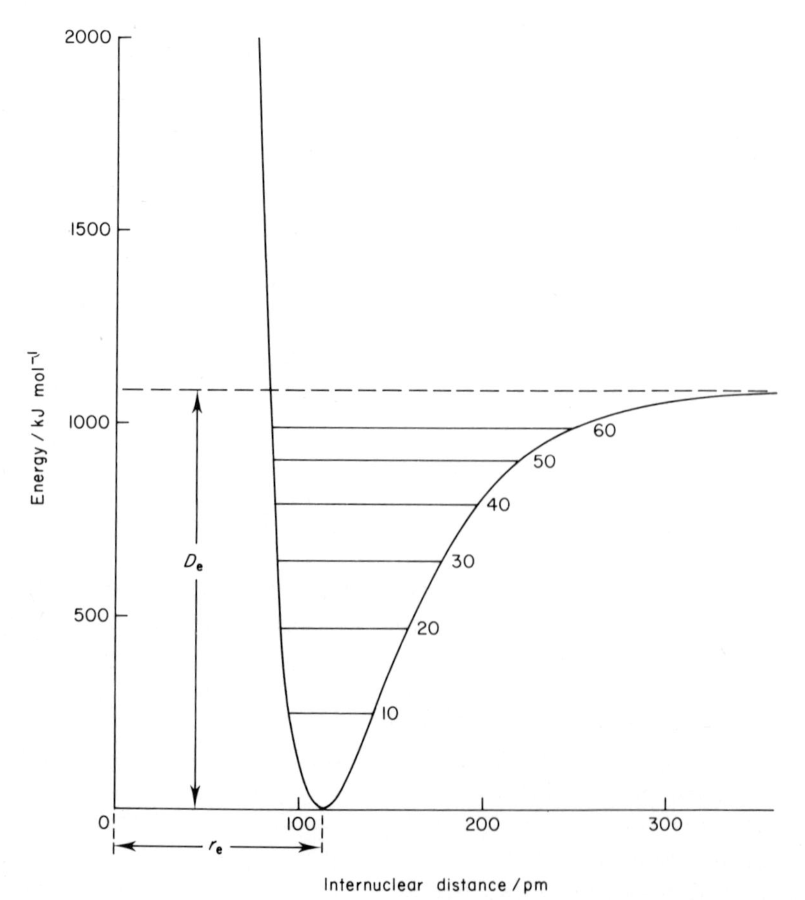

FIG. 4.1 The Morse potential for carbon monoxide, and upper vibrational levels.

The energy baseline, i.e. zero potential energy, is taken as the minimum energy for the system, where r is equal to r_e, the internuclear distance at equilibrium. β is a constant which controls the steepness of ascent from the energy minimum with increasing internuclear distance; the greater the value of β the steeper the ascent. β has the dimensions of reciprocal length, and is typically of the order of 2×10^{-2} pm^{-1}. β is related to the molecular parameters by

$$\beta = \pi\omega_e(2\mu/D_e)^{\frac{1}{2}} \tag{4.3}$$

where μ is the reduced mass, defined by Eqn. 2.2. The Morse potential for carbon monoxide is shown in Fig. 4.1, together with energy levels calculated

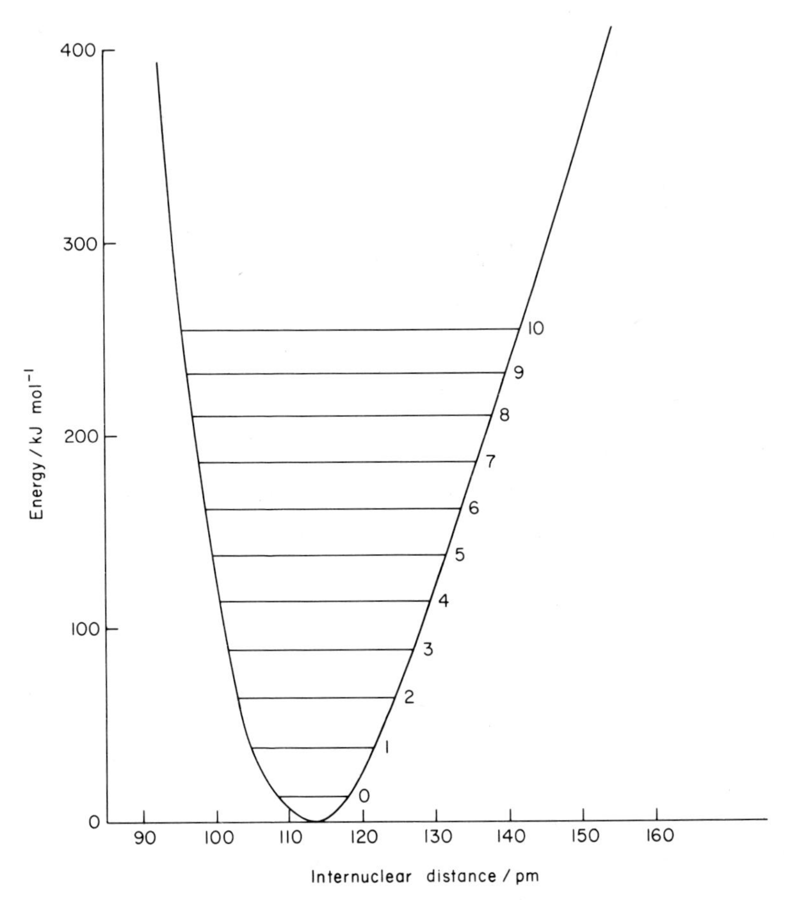

FIG. 4.2 The Morse potential for carbon monoxide, and lower vibrational levels.

by Eqn. 2.22. The bunching together of the upper levels, due to anharmonicity, can be clearly seen. Dissociation occurs for a value of v around 80. Equation 2.22 is not adequate to describe the behaviour of the molecule at vibrational energies close to D_e, since further terms in higher powers of $(v+\frac{1}{2})$ are required. Figure 4.2 shows the region of the Morse potential around the minimum. For diatomic molecules which dissociate into atoms, there is usually a finite number of vibrational levels below the dissociation limit. If no terms in addition to the $\omega_e x_e$ term are needed to describe the anharmonicity,

$$D_e = \omega_e/4x_e \tag{4.4}$$

On the other hand, there is usually an infinite number of vibrational levels below the dissociation limit for molecules which dissociate into ions.

4.A.2 Selection rules

Two conditions, or selection rules, must be satisfied if a diatomic molecule is to gain or lose vibrational energy by interaction with electromagnetic radiation. The first condition is that the vibration of the molecule causes a change in dipole moment, which is equivalent to saying that the molecule must possess a dipole moment. The sinusoidally varying electric field due to the vibrating molecule can then interact with the sinusoidally varying electric field due to the radiation. If the frequencies are equal, maximum transfer of energy can occur by resonance, as discussed in Section 2.A.2. Homonuclear molecules, e.g. H_2, O_2, and N_2, cannot acquire vibrational energy in this way.

The second condition is that $\Delta v = +1$ for absorption of radiation. The methods of quantum mechanics can be used to prove this selection rule rigorously for the harmonic oscillator model. For example, for carbon monoxide the energies of the zeroth and first vibrational levels are 12.88 and 38.63 kJ mol^{-1} respectively, so the transition from $v = 0$ to $v = 1$ requires 25.75 kJ mol^{-1}, corresponding to the absorption of radiation of frequency 6.46×10^{13} Hz, or wavenumber 2153 cm^{-1}. The transition from $v = 0$ to $v = 2$ is forbidden for the harmonic oscillator model, but may be observed as a weak absorption due to the slightly anharmonic nature of the vibration. This transition is called an *overtone*, the $v = 0$ to $v = 1$ transition being a *fundamental*, by analogy with musical notes. Excitation from the zeroth to the second vibrational level in carbon monoxide requires 51.1 kJ mol^{-1}, corresponding to absorption of radiation of frequency 1.282×10^{14} Hz, or wavenumber 4274 cm^{-1}. The overtone frequency is slightly less than twice the fundamental, owing to anharmonicity.

4.A.3 The frequency region of the spectrum for vibrational absorption

Absorption due to excitation of vibrational energy levels occurs mainly in the mid-infrared region of the spectrum, i.e. in the wavelength range 2.5–25 μm, corresponding to the frequency range 1.2×10^{14} to 1.2×10^{13} Hz and the wavenumber range 4000 to 400 cm^{-1}. The attribution of absorption in this region to vibrational excitation is based on the observations that homonuclear molecules are transparent in this region, and the absorption frequencies for different molecules follow the trends predicted on the harmonic oscillator model. Since

$$\omega = (k/\mu)^{\frac{1}{2}}/2\pi \tag{2.18}$$

the heavier the atoms and the weaker the bond, the lower the absorption frequency. Typically, values of k are in the range 100–1000 N m^{-1}.

The magnitude of the energy gap between the zeroth and first vibrational level for a frequency of 10^{14} Hz is 6.6×10^{-20} J per molecule, which is greater than the value of kT at room temperature, 4×10^{-21} J. The population of molecules in the first vibrational level, given by

$$n_i \propto g_i \exp(-\varepsilon_i/kT) \tag{2.12}$$

is thus less than that in the zeroth state by a factor of about $\exp(-16.5)$, i.e. 6×10^{-8}. Absorption due to transition from the first to the second vibrational level will thus be very weak, since so few molecules are available to make this transition, and will only be observed for low values of ω_e or at high temperatures. Groups of lines due to such transitions are therefore sometimes called *hot bands*.

4.B Apparatus for Infrared Spectroscopy

4.B.1 Introduction

The wavelengths of radiation in the infrared region are small compared with the dimensions of laboratory apparatus, so the design of an infrared spectrometer follows the principles of classical optics. An infrared spectrometer comprises a radiation source, monochromator, sample cell, and detector. The optical components cannot be made from glass, since even silica, the purest form of glass, is opaque to radiation of wavenumber less than 2500 cm^{-1}. Sodium chloride is the most commonly used material for prisms and windows, since it is transparent down to 600 cm^{-1}. For more specialised applications, potassium bromide, transparent to 350 cm^{-1}, or a mixed thallium iodide–bromide, transparent to 250 cm^{-1}, may be used. Concave mirrors are used, rather than lenses, to focus the radiation in order to minimise the absorption losses.

In common with other regions of the spectrum, the infrared is delimited in a rather arbitrary way, the limits being set by technical and economic considerations. The upper frequency region of the infrared may be taken as that at which the radiation becomes visible, corresponding to a wavenumber of about 14 000 cm^{-1}. Since only overtone transitions, which are more difficult to observe than fundamental transitions, cause absorption in the region 14 000 to 4000 cm^{-1}, most commercial infrared spectrometers are not designed to operate in this region, which may be termed the *near infrared.* The lower frequency limit to the operation of an infrared spectrometer is set by the material used for the optics. The less expensive spectrometers, with sodium chloride optics, thus go down only to 600 cm^{-1}, whereas the more expensive spectrometers go down to 250 cm^{-1}. Most fundamental vibrations of organic molecules occur in the 4000–600 cm^{-1} region, which is sometimes termed the *mid infrared.* The region between the mid infrared and the microwave region is termed the *far infrared.*

4.B.2 Radiation sources

Any object at a temperature above absolute zero emits electromagnetic radiation over a wide range of wavelengths. The variation of radiant energy with temperature and wavelength can be calculated on the basis of quantum theory, as shown by Planck, and the agreement between observed and calculated values provided the original proof of the quantum theory. Planck's equation is

$$E\delta\lambda = 2\pi c^2 h\lambda^{-5}\delta\lambda/[\exp(ch/\lambda kT)-1] \tag{4.5}$$

where $E\delta\lambda$ is the radiant energy emitted per square metre of surface from a body at temperature T K between wavelength λ and $(\lambda+\delta\lambda)$ metres. Plots of this function are shown in Fig. 4.3, and it can be seen that an object at a temperature in the range 1200–1800 °C is a useful source of radiation in the mid infrared region. The derivation of Eqn. 4.5 assumes that the radiant object is a perfect emitter. Since it can be shown on thermodynamic grounds that a perfect emitter must also be a perfect absorber of radiation (or else a system could be set up in which heat would flow spontaneously from a cold body to a hot one) the perfect emitter appears black when cold, and so is termed a *black body*.

Several sources which approximate to black bodies are in common use.

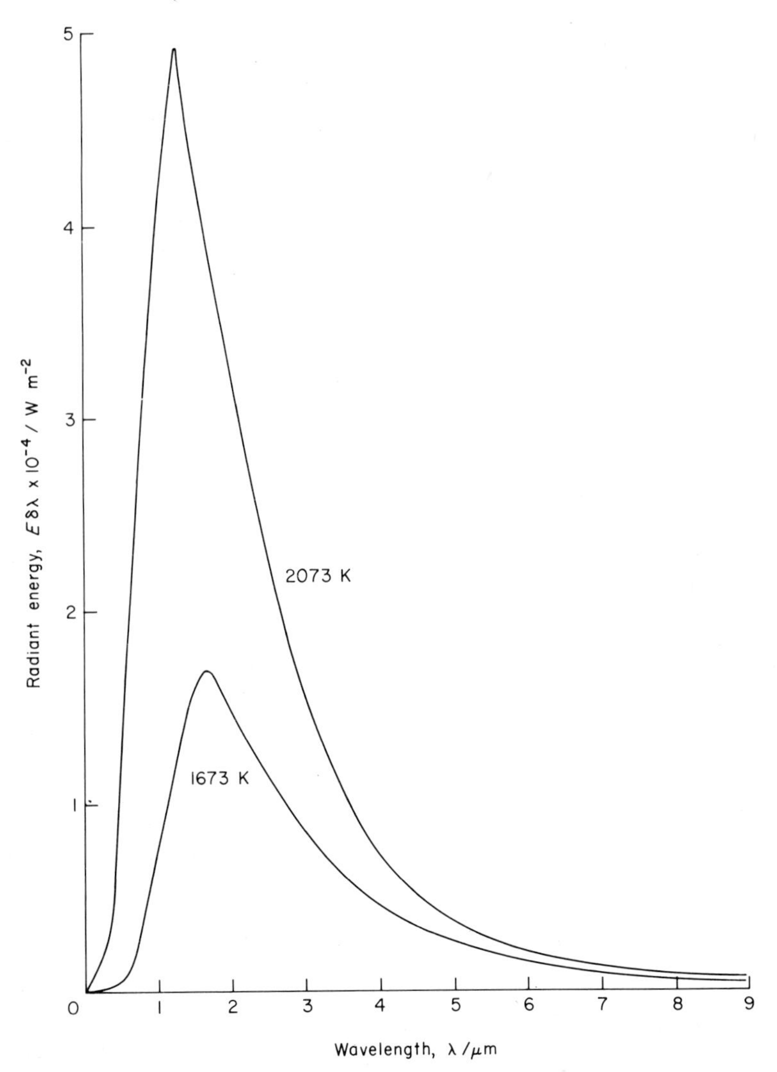

FIG. 4.3 Radiant energy from a black body.

The Nernst Glower is a rod of mixed rare-earth oxides, heated by the passage of an electric current to about 1900 °C. The Globar is a rod of silicon carbide, also electrically heated to about 1200 °C. As can be seen in Fig. 4.3, the amount of radiant energy falls off steeply with increasing wavelength. Although it is possible to increase the amount of energy in the longer wavelength region by increasing the temperature of the source, the amount emitted at shorter wavelengths would increase to a much greater extent, and this intense short wavelength radiation is liable to "leak" through the optics as stray light, and find its way into the detector, giving a spurious signal. It is customary to allow for the change in source intensity with wavelength with an automatically controlled slit, which opens up at longer wavelengths to allow a greater proportion of the radiation to reach the detector.

4.B.3 Monochromators

The radiant energy from a black body is spread over a wide range of wavelengths, so a monochromator is required to select a narrow wavelength band. Early monochromators depended on the variation of refractive index of transparent materials with wavelength, termed *dispersion*. The refractive index of a transparent material decreases with increasing wavelength, so the angle through which a beam of monochromatic radiation is deviated by passage through a prism is smaller the longer the wavelength of the radiation. If a beam of polychromatic radiation is sent through a prism, the wavelength of the radiation falling on some defined region on the far side of the prism depends on the orientation of the prism. The spectrum can thus be scanned by rotating the prism. Although comparatively simple to make, prism monochromators suffer from several disadvantages. The resolving power is proportional to the size of the prism. Large prisms require large pieces of optically perfect material for their manufacture, and the cost of optical working rises steeply with the size, since high-quality prisms cannot be mass-produced. The dispersion varies with the wavelength, so it is difficult to design an instrument with a linear wavelength or frequency scale. The optimum wavelength range for a prism is limited, so more than one prism may be required for complete coverage of the desired region. These disadvantages have led to the increasing popularity of the diffraction grating as a monochromator element.

A diffraction grating is a regular array of grooves, in a pattern repeating at distance d. Diffraction gratings may be of transmission or reflection type, the latter being used for the infrared. The grooves are ruled on an aluminium surface. Radiation of wavelength λ striking the grating at angle i to the grating normal is reflected at angle r with constructive interference if

$$m\lambda = d(\sin i + \sin r) \qquad (4.6)$$

Radiation of wavelength other than that satisfying Eqn. 4.6 is destroyed by destructive interference. The diffraction grating thus disperses polychromatic radiation into a spectrum. Several spectra may be produced by the grating, corresponding to various small integer values of m, which gives the *order* of the spectrum. The first-order spectrum ($m = 1$) is usually the brightest. It is often necessary to separate overlapping orders by the use of filters, which absorb radiation over a wide range of wavelengths. Primitive diffraction gratings consisted of alternate reflecting and non-reflecting lines. Modern diffraction gratings are *blazed*, i.e. the grooves are shaped in a sawtooth pattern as shown in Fig. 4.4. For light of a certain wavelength, given by

$$m\lambda = d \sin \alpha \tag{4.7}$$

the grating acts as a mirror, and reflects nearly all the radiation of that wavelength. A blazed grating concentrates the diffracted light into one order of spectrum, so radiation losses are minimised. The theoretical resolving power of a diffraction grating is given by mN, where N is the total number of grooves in the grating. Since d must be of the same order of magnitude as λ, to give convenient values of i and r, N can only be increased by increasing the total surface area of the grating. This is not a problem in the near and mid infrared

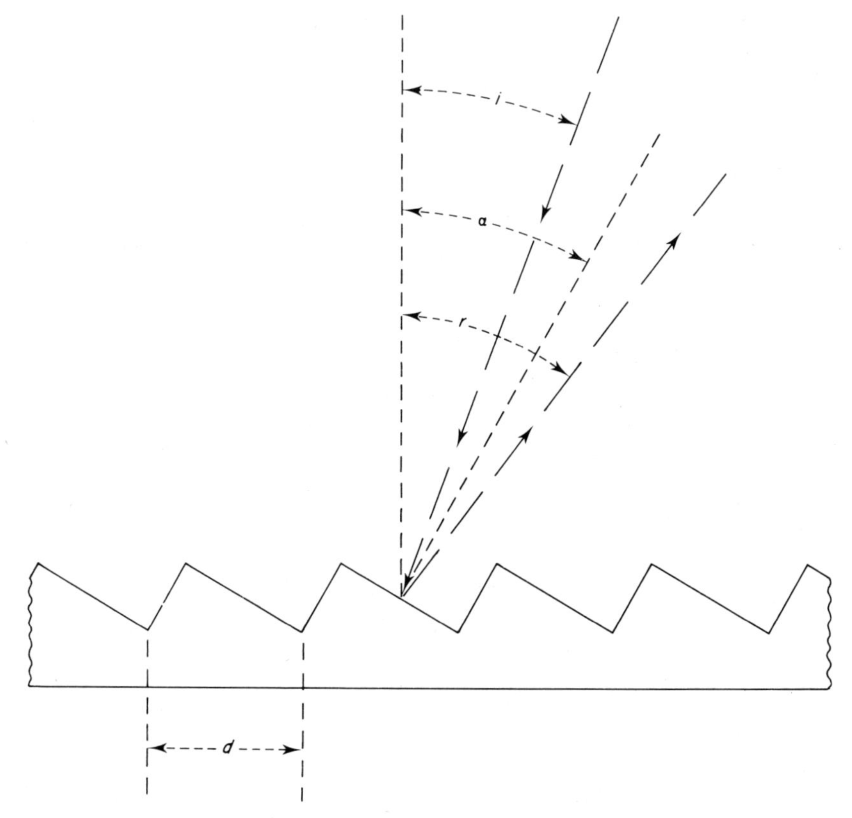

FIG. 4.4 The diffraction grating.

regions, but becomes a serious drawback to the use of diffraction gratings at low frequencies. Typically d is of the order of 10 μm for gratings used in the mid infrared region. In practice, the resolving power of a typical infrared spectrometer is of the order of 2 cm^{-1}. Gratings can be comparatively cheaply made by casting a plastic replica from a master, and aluminising the surface to increase its reflectance.

4.B.4 Sample cells and sample preparation

Infrared spectra may be run on gaseous, liquid, or solid samples. A typical gas sample cell consists of a glass tube, 4 cm in diameter and 10 cm long, fitted with detachable sodium chloride windows at each end, and two side arms for flushing out a previous sample and admitting a new one. For qualitative work, the spectrum of a pure organic liquid is obtained by smearing a thin film over the surface of a sodium chloride plate. Cells with sodium chloride plates separated by Teflon spaces are used for quantitative work, path lengths being in the range 0.02–1 mm. Solid samples are most often prepared by grinding with Nujol (a viscous paraffin oil) and smearing the paste on a sodium chloride plate. As well as causing the solid material to adhere to the plate, the Nujol much reduces scattering losses of radiation by reducing the difference in refractive index between the solid and the surrounding medium. If a region of interest in the spectrum coincides with the absorption of Nujol, a fluorocarbon oil can be used instead, or the sample can be ground up with potassium bromide and squeezed in a die at high pressure to make a transparent disc.

4.B.5 Detectors

There are two types of detector used for infrared radiation: those in which each quantum of radiant energy, or photon, is capable of exciting an electron to an upper energy level, thereby creating a measurable electric current, and those in which the energy of the radiation is used to heat the detector element, the temperature of which is measured to give the output signal.

Quantum effect detectors are made from semiconductor materials, e.g. indium antimonide, lead sulphide, or germanium. In the electronic structure of crystals of these materials, full sets of molecular orbitals, or *bands*, are separated from empty bands by small energy gaps. A photon of infrared radiation is sufficiently energetic to excite an electron from a full to an empty band, in which it is free to move, thereby increasing the conductivity of the crystal. The current passing through the crystal on application of an e.m.f. provides the signal output. An electron can also jump from the lower, full, band to the upper band by thermal excitation. If the semiconductor material is chosen so as to be sensitive to the low-frequency region of the infrared, the energy gap is so small that thermal excitation at room temperature is sufficiently probable to cause an appreciable conductivity in the absence of incident radiation, giving rise to a *dark current*. The probability that an electron has sufficient energy to jump a gap of ΔE joules is proportional to $\exp(-\Delta E/kT)$. The value of kT at room temperature, 4×10^{-21} J, is equal to the energy of a quantum of radiation of frequency 6×10^{12} Hz, i.e. of wavenumber 200 cm^{-1}. A steady dark current could be allowed for, but the statistical nature of the excitation process means that the current is subject to fluctuations, which appear as noise on the signal output. This noise may be reduced by cooling the detector in liquid nitrogen, which drastically reduces the number of electrons of thermal energy sufficient to jump the gap.

Thermal detectors include the *thermocouple*, the most widely used detector of infrared radiation. A small piece of blackened gold foil acts as target for the radiation. The temperature of the foil is monitored by a thermocouple, a junction of wires of dissimilar metals, which produces an e.m.f. proportional to the temperature. The e.m.f. is amplified and displayed as the signal. Another type of thermal detector, the *bolometer*, consists of a metal filament or semiconductor crystal whose electrical resistance varies with temperature. The *Golay cell* consists of a capsule of gas, one wall of the capsule being a window for incident radiation, and the opposite wall being a flexible diaphragm, silvered on the outside. Incident infrared radiation heats up the gas, causing its pressure to rise, so the diaphragm bulges outwards. Light from a lamp of constant brightness is reflected from the silvered face of the diaphragm on to a photocell, a device that produces a voltage proportional to the incident light intensity. The deformation of the diaphragm changes the intensity of the light incident on the photocell, so the electrical output of the photocell is controlled by the intensity of infrared radiation incident on the gas. Rather surprisingly, this device is both sensitive and reliable. The most modern type of thermal detector is based on the *pyroelectric effect*. Some crystals, e.g. barium titanate, develop an e.m.f. across their faces when heated, which may be amplified to give the output signal.

4.B.6 Double-beam operation

For quantitative work it is necessary to know the ratio of the intensity of the radiation transmitted by the sample to the intensity transmitted in the absence of sample, the 100% transmittance value. The radiation source is frequency-dependent, as are some detectors, so a 100% transmittance value is required at each wavelength. The black-body source does not emit much radiation in the infrared, so signal levels are low, and subject to noise. The path length from source to detector is so long that carbon dioxide and water vapour in the atmosphere, although only present in low concentrations, give rise to absorption peaks which are superimposed on the desired spectrum. These problems are solved by the double-beam design of spectrometer. A block diagram of a typical double-beam spectrophotometer is shown in Fig. 4.5. Radiation from the source is split into two beams, the sample beam and the reference beam. Each beam is sent through an attenuator, a comb-shaped shutter which progressively cuts off the radiation as it is slid into its path. A rotating mirror, or chopper, sends the sample beam and reference beam alternately through the monochromator and thence to the detector, at an alternation frequency that is not a simple multiple of mains frequency, say 11 or 13 Hz. The detector must have a response fast enough to follow the alternations in radiant power. The noise is thus greatly reduced by the narrow bandwidth restriction, and pickup from the mains is eliminated. The alternating voltage caused by the sample and reference beams being of different intensities is amplified by an amplifier tuned to the chopper frequency. The attenuator in the reference beam is driven by a motor which is controlled by the amplified detector signal. If there is a detectable signal, i.e. if the beams are not of equal intensities, the motor drives the attenuator so as to equalise the intensities. The extent by which the attenuator must be driven is monitored by a pen which moves up and down a chart to record the per cent transmission. Another motor rotates the prism or grating in the monochromator so as to scan the spectrum, simultaneously moving the chart across under the pen. The chart thus records the per cent transmission as a function of wavelength

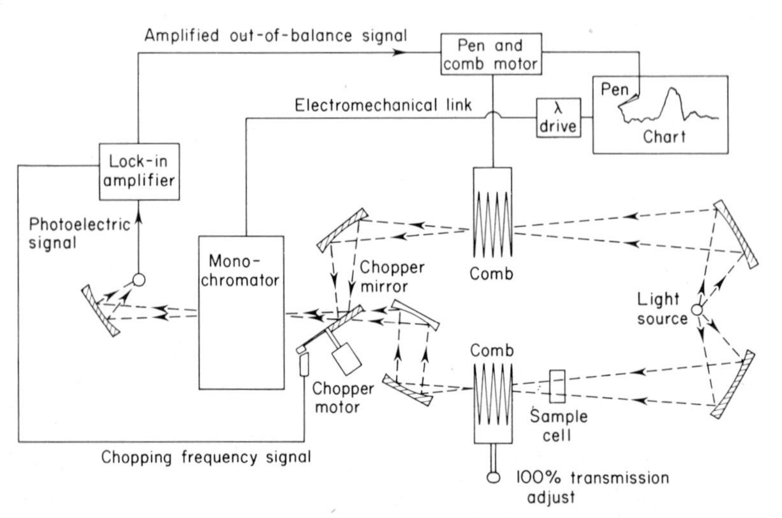

FIG. 4.5 Block diagram of a scanning double-beam spectrophotometer.

or frequency. Absorption due to atmospheric carbon dioxide and water vapour is cancelled out. Absorption due to solvent may be similarly cancelled out by placing a cell containing pure solvent in the reference beam. A possible disadvantage of this procedure is that in regions of high solvent absorption both sample and reference beams are extinguished. The recorder pen, receiving no signal, drifts aimlessly and misleadingly.

4.C The Vibrations of Polyatomic Molecules

4.C.1 Normal modes of vibration

An isolated atom has three *degrees of freedom*, i.e. it can store energy in three ways, the kinetic energy of motion along the x, y, and z axes of a coordinate frame. An assembly of N atoms thus has $3N$ degrees of freedom. It can be shown that the total number of degrees of freedom is unaltered by any linkages between the atoms, so an N-atomic molecule has $3N$ degrees of freedom. Of these, three are translational, i.e. those for motion along the x, y, and z axes of the coordinate frame, and three are rotational, i.e. those for rotation about the A, B, and C axes of the molecule. The remaining $(3N-6)$ must then refer to energy storage by vibrations of the molecule. Linear molecules do not acquire energy by rotation about the A axis, as discussed in Section 2.C.2, so they can store vibrational energy in $(3N-5)$ ways. Mathematical analysis shows that each way in which vibrational energy can be stored is a *normal mode of vibration* of the molecule. A normal mode of vibration is a molecular motion in which all the atoms move in phase with the same frequency. The actual vibration pattern of the molecule can be expressed as the sum of the vibrations in the normal modes. The total vibrational energy of the molecule is the sum of the vibrational energies stored in each mode, being $\sum(v_i+\frac{1}{2})h\omega_i$, where ω_i is the vibration frequency of the ith mode and v_i is the relevant quantum number. Some normal modes of vibration of simple molecules are illustrated in Fig. 4.6. Normal modes can be classified as *stretching* or *bending*, according to whether it is mainly bond lengths or bond angles that alter during the vibration. Normal modes of symmetric molecules can be classified as *symmetric* or *antisymmetric*, depending on whether the vibration would be in phase with, or 180° out of phase with, a similar molecule turned through 180° about its axis of symmetry. Vibrations can also be classified as *parallel* or *perpendicular*, depending on whether the dipole moment change is parallel to the principal axis of the molecule, or perpendicular to it. Examples of these classifications are shown in Fig. 4.6.

It can be seen that, as the number of normal modes of vibration of non-linear N-atomic molecules is $(3N-6)$, the complexity of the description of the vibrations of the molecule rises rapidly with N. Fortunately a branch of

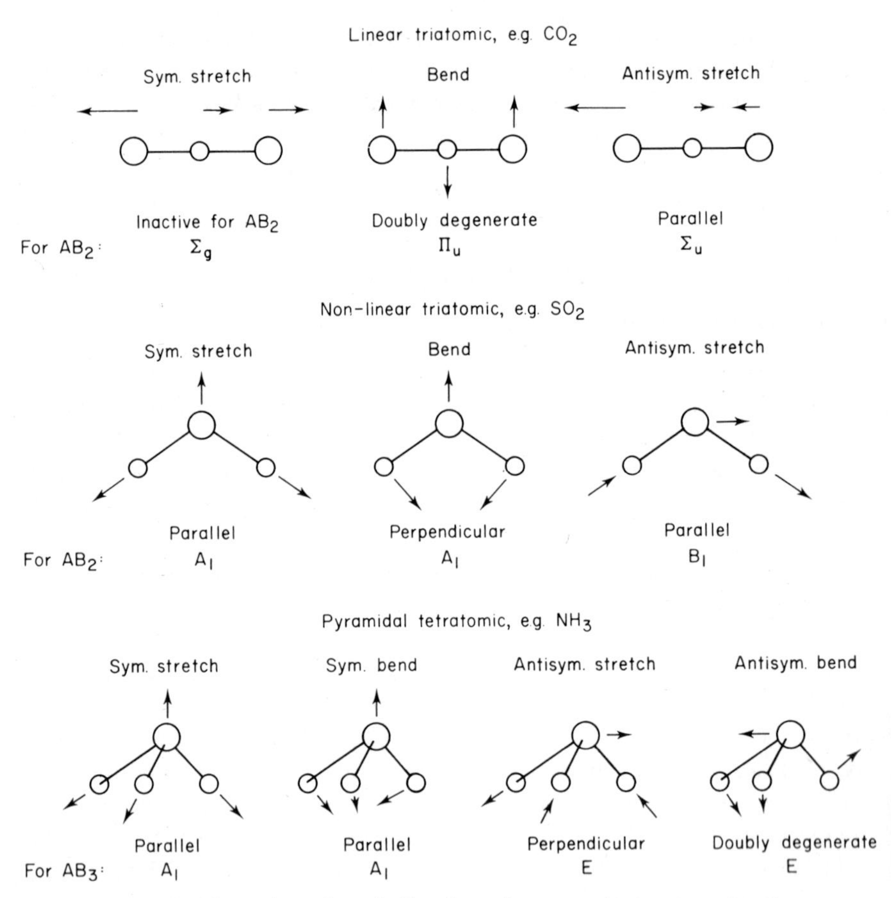

FIG. 4.6 Normal modes of vibration of some polyatomic molecules.

mathematics, group theory, which provides a technique for analysing symmetry properties, has proved extremely useful in tackling the problem of the normal modes of vibration of polyatomic molecules. Group theory is outside the scope of this book, but must be studied by those wishing to obtain a deeper insight into spectroscopy.

The modern convention for labelling the normal modes of vibration follows the principles of group theory. The motions of the atoms in each mode have a certain symmetry, and the mode is labelled according to the type of this symmetry:

> A: Symmetric with respect to rotation about the principal axis of symmetry, e.g. the symmetric stretching and bending modes of SO_2 and NH_3, as shown in Fig. 4.6.

B: Antisymmetric with respect to rotation about the principal axis of symmetry, e.g. the antisymmetric stretching mode of SO_2.
E: Doubly degenerate vibrations, e.g. the antisymmetric stretching and bending modes of NH_3.
F: Triply degenerate vibrations.
g and u (subscript): Symmetric or antisymmetric respectively with respect to a centre of symmetry.
1 and 2 (subscript): Symmetric or antisymmetric respectively with respect to an axis of symmetry other than the principal axis.
′ and ″ (superscript): Symmetric or antisymmetric respectively with respect to a plane of symmetry.

These terms are not used for linear molecules, for which the symbols below are used:

Σ^+ and Σ^-: Symmetric and antisymmetric respectively with respect to a plane of symmetry through the molecular axis.
Π, Δ, Φ: Degenerate vibrations of increasing order of degeneracy.

When listing the modes of vibration of a particular molecule it is customary to arrange the modes in order of decreasing symmetry, and arrange the modes of the same symmetry in order of decreasing frequency; the whole sequence is then labelled $\nu_1, \nu_2, \nu_3 \ldots$. Thus for CO_2, the symmetric stretching vibration at 1388 cm^{-1} is labelled ν_1, the bending vibration at 667 cm^{-1} is labelled ν_2, and the antisymmetric stretching vibration at 2349 cm^{-1} is labelled ν_3. Bending modes are sometimes denoted by δ, and parallel and perpendicular modes are often denoted by the symbols $\parallel$ and $\perp$ respectively.

4.C.2 Selection rules

The selection rule for a molecule to absorb radiation equal in frequency to a normal mode is that the vibration must cause a change in the dipole moment of the molecule. Such a mode is termed *infrared active*. It is possible for such a change to occur in a molecule with no permanent dipole moment. For example, the bending and antisymmetric stretching modes of vibration of CO_2 cause the molecule to have a dipole moment except when in the equilibrium configuration, so these modes are infrared active. The symmetric vibration is infrared inactive. As for diatomic molecules, the selection rule $\Delta v = +1$ holds, but overtone bands, e.g. $\Delta v = 2, 3$, etc., are also observed with much reduced intensity.

4.C.3 Combination bands

The absorption of infrared radiation may cause the simultaneous excitation of more than one vibrational mode, with the result that two or even three vibrational quantum numbers may change during the excitation. Sets of lines due to such transitions are termed *combination bands*, and may be either *sum*, i.e. the molecule acquires energy in two modes, or *difference*, i.e. the molecule acquires energy in one mode, but loses a smaller amount in another mode. Some of the vibrational transitions of SO_2 are listed in Table 4.1. It can be seen that overtone and combination bands are much less intense than fundamental bands. As will be discussed in Section 4.D, vibrational excitation is often accompanied by changes in the rotational energy of the molecule, so each vibrational transition gives a set of lines, or band, each line corresponding to a particular rotational transition. The frequencies listed in Table 4.1 are not observed infrared frequencies, but those calculated for zero rotational energy in both lower and upper vibrational levels. The selection rules for combination bands are rather complex and subtle, and are found by consideration of the overall change in symmetry of the molecule during the vibration. For example, the fundamental ν_1 of CO_2 is not observed in the infrared, but the combination $\nu_1 + \nu_2$ is observed.

TABLE 4.1

Infrared absorption of SO_2 (Shelton *et al.*, 1953)

Wavenumber/cm^{-1}	Rel. intensity	Assignment	Lower level v_1	Lower level v_2	Lower level v_3	Upper level v_1	Upper level v_2	Upper level v_3
517.69	455	ν_2	0	0	0	0	1	0
844.93	0.55	$\nu_3 - \nu_2$	0	1	0	0	0	1
1151.38	565	ν_1	0	0	0	1	0	0
1361.78	1000	ν_3	0	0	0	0	0	1
1535.06	0.1	$3\nu_2$	0	0	0	0	3	0
1665.07	0.1	$\nu_1 + \nu_2$	0	0	0	1	1	0
1875.55	6.0	$\nu_2 + \nu_3$	0	0	0	0	1	1
2295.88	5.5	$2\nu_1$	0	0	0	2	2	0
2499.55	20.0	$\nu_1 + \nu_3$	0	0	0	1	0	1
2715.46	0.2	$2\nu_3$	0	0	0	0	0	2
2808.32	0.8	$2\nu_1 + \nu_2$	0	0	0	2	1	0
3011.25	0.02	$\nu_1 + \nu_2 + \nu_3$	0	0	0	1	1	1
3431.19	0.01	$3\nu_1$	0	0	0	3	0	0
3629.61	0.8	$2\nu_1 + \nu_3$	0	0	0	2	0	1
4054.26	0.03	$3\nu_3$	0	0	0	0	0	3
4751.23	0.006	$3\nu_1 + \nu_3$	0	0	0	3	0	1
5165.64	0.002	$\nu_1 + 3\nu_3$	0	0	0	1	0	3

SO_2 is a non-linear molecule.
ν_1 is the symmetric stretching mode, ν_2 is the bending mode, and ν_3 is the antisymmetric stretching mode.

4.C.4 Fermi resonance

It may happen that two vibrational transitions, of the same symmetry type, are *accidentally degenerate*, i.e. very similar in magnitude. Accidental degeneracy must be distinguished from true degeneracy, as found for, say, the two bending modes of CO_2, for which the frequencies are necessarily equal. Two accidentally degenerate transitions interact in such a way that two new transitions are created, one of higher energy and one of lower energy than the original pair. This interaction is termed *Fermi resonance*. Typically, one of the original transitions is of overtone or combination type and would give rise to much weaker infrared absorption than the other, which is a fundamental. The effect of Fermi resonance is to produce two new transitions which give rise to similar intensities of absorption. For example, the symmetric C–Cl stretching of the CCl_4 molecule, denoted by ν_1, of wavenumber 458 cm^{-1}, which is infrared inactive, can be combined with the antisymmetric Cl–C–Cl bending mode, denoted by ν_4, of wavenumber 319 cm^{-1}, to give a combination transition, $\nu_1 + \nu_4$, at 777 cm^{-1}, which is infrared active. This is similar in frequency to the antisymmetric C–Cl stretching mode, ν_3, at 773 cm^{-1}, which is also infrared active. The observed spectrum shows vibrational transitions at 762 and 785 cm^{-1}, the higher frequency component being four times more intense (Plyler and Benedict, 1951).

4.C.5 Force constants

The force constant, k, calculated from Eqn. 2.18 gives a measure of the stiffness of the bond in a diatomic molecule. The calculation of k from ω is trivial for a diatomic molecule, but the calculation becomes very much more complex for polyatomic molecules. Each normal mode of vibration may involve the stretching and bending of many bonds, and it is difficult to disentangle the contribution of each bond to the overall vibration frequency. Consider the bent triatomic molecule XYZ. According to the *central force field model*, the potential energy, U, is given by

$$U = \tfrac{1}{2}k_{xy}(\delta r_{xy})^2 + \tfrac{1}{2}k_{yz}(\delta r_{yz})^2 + \tfrac{1}{2}k_{zx}(\delta r_{zx})^2 \tag{4.8}$$

where

$$\delta r_{xy} = r_{xy} - r^{\circ}_{xy}$$

r_{xy} is the actual distance between atoms x and y, r°_{xy} is the distance at equilibrium, and k_{xy} is the force constant. This model does not take molecular structure into account. The *valence-bond force field model* is more favoured by chemists. The potential energy of the molecule is given by this model to be

$$U = \tfrac{1}{2}k_{xy}(\delta r_{xy})^2 + \tfrac{1}{2}k_{yz}(\delta r_{yz})^2 + \tfrac{1}{2}kr^{\circ}_{xy}r^{\circ}_{yz}(\delta\alpha)^2 \tag{4.9}$$

where α is the bond angle, X–Y–Z. The $r^{\circ}_{xy}r^{\circ}_{yz}$ factor is included so that k_{α} may have the same units as k_{xy}. Equations of these types may be written for more complex molecules. A set of $(3N-6)$ equations of motion involving one of these potential energy functions can be solved to give the $(3N-6)$ frequencies of the normal modes. Given experimental values of these frequencies, it is possible, with considerable mathematical labour, to work back to find the values of the force constants. Some values found are listed in Table 4.2. It can be seen that in general bonds resist stretching more than bending, and that the stronger a bond the stiffer it is.

TABLE 4.2
Force constants for bonds in diatomic and polyatomic molecules

Vibration	Molecule	Force constant $\times 10^{-2}$/N m^{-1}	Ref.
C=O stretch	CO	19.02	*a*
N=O stretch	NO	15.94	*a*
H—F stretch	HF	9.66	*a*
H—I stretch	HI	3.14	*a*
C≡N stretch	C_2N_2	16.1	*b*
C—C≡N bend	C_2N_2	0.185	*b*
C—H stretch	C_2H_6	5.35	*c*
H—C—C bend	C_2H_6	0.51	*c*
C—C stretch	C_2H_6	4.57	*c*
C=C stretch	C_2H_4	9.57	*d*
C=C stretch	C_6H_6	7.62	*e*
C—H bend	C_6H_6	0.833	*e*
C≡C stretch	C_2H_2	15.8	*d*

a, Herzberg (1950). *b*, Schultz and Eggers (1958). *c*, Hansen and Dennison (1952). *d*, Herzberg (1945). *e*, Whiffen (1955).

4.D The Rotational Fine Structure of Vibrational Transitions

4.D.1 Diatomic molecules

A molecule can store energy in both vibrational and rotational modes, so the energy of a diatomic molecule may be represented, to a first approximation, by the sum of the energies expressed in Eqns. 2.3 and 2.20. This separation of the two energy terms is one aspect of the *Born–Oppenheimer approximation*, which is discussed further in Section 6.A.1.

$$E = h\omega(v+\tfrac{1}{2})+BJ(J+1) \tag{4.10}$$

The absorption of infrared radiation can cause changes in both the vibrational and rotational energies. The selection rules for a diatomic molecule are: $\Delta v = \pm 1$; $\Delta J = \pm 1$. As has been shown in Section 4.A.3, the magnitudes of the quanta are such that only the vibrational level $v = 0$ is populated to any significant extent at 300 K, whereas, as shown in Section 2.A.4, several rotational levels are significantly populated. Thus, although the transition $v = 0$ to $v = 1$ is the only important vibrational transition giving rise to infrared absorption, rotational transitions of both $\Delta J = +1$ and $\Delta J = -1$ may be observed in the infrared spectrum. For the transition from vibrational quantum number 0 and rotational quantum number J, to vibrational quantum number 1 and rotational quantum number $(J+1)$, absorption occurs at frequency ν_R, where

$$\nu_R = \omega + B_1(J+1)(J+2) - B_0(J)(J+1) \tag{4.11}$$

As explained in Section 2.D.2, B_1, the rotational constant for the upper vibrational state, is slightly less than B_0, the rotational constant for the ground vibrational state. Equation 4.11 may be rearranged to give

$$\nu_R = \omega + 2B_1 + (3B_1 - B_0)J - (B_0 - B_1)J^2 \tag{4.12}$$

Since $(B_0 - B_1)$ is typically of the order of 1 per cent of B_0, it is possible to make the approximation:

$$\nu_R = \omega + 2B_0 + 2B_0 J \tag{4.13}$$

Transitions of this type thus give rise to a set of lines in the infrared absorption spectrum, spaced at intervals of approximately $2B_0$, the first being at a frequency $2B_0$ greater than that of the vibration of the molecule. This set, or series, is termed the *R branch*. The resolution of many infrared spectrometers is inadequate to resolve these lines, which thus appear as a broad band. An individual line is labelled according to the rotational quantum number of the ground vibrational state, so the first line is R(0). The relative intensities of the lines depend on the relative population level of the rotational level of the ground vibrational state, which, as discussed in Section 2.A.4, depends on the temperature, the magnitude of the rotational quantum, and the multiplicity of the rotational level, which is $(2J+1)$. Typically the intensities increase up to lines R(5) to R(10), and then decrease as the energy factor comes to predominate, until the lines are imperceptible around R(30) to R(40). The inequality of B_0 and B_1 causes the lines to be more closely spaced at higher values of J. At high values of J the lines crowd together, and a limiting value of J is reached above which the frequency starts to decrease, giving rise to a *band head*. Band heads are usually not seen in the infrared spectrum, but are often seen in the visible and ultraviolet region, as will be shown in Chapter 6.

Another effect observable at high values of J is centrifugal distortion, as discussed in Section 2.D.3.

For the transition from vibrational quantum number 0 and rotational quantum number J, to vibrational quantum number 1 and rotational quantum number $(J-1)$, absorption occurs at frequency ν_P, where

$$\nu_P = \omega + B_1(J-1)(J) - B_0(J)(J+1) \tag{4.14}$$

Equation 4.14 may be rearranged to give

$$\nu_P = \omega - (B_1 + B_0)J - (B_0 - B_1)J^2 \tag{4.15}$$

If B_1 is taken as equal to B_0, the approximation gives

$$\nu_P = \omega - 2B_0J \tag{4.16}$$

Transitions of this type thus give rise to a set of lines, the *P branch*, spaced at intervals of approximately $2B_0$. There is no P(0) line, since the upper level cannot have a rotational quantum number of -1. The first line in the series is thus P(1), of frequency $2B_0$ less than the molecular vibration frequency. The inequality of B_1 and B_0 causes the spacing of the lines in the P branch to increase with increasing J.

The origin of P and R branches is illustrated in Fig. 4.7. Wavenumbers are used rather than frequencies, so that the energy axis can be plotted in units of cm^{-1}. The rotational constant is thus taken as $\tilde{B}$ rather than B. In the interests of clarity the relative difference between $\tilde{B}_1$ and $\tilde{B}_0$ is taken an order of magnitude greater than that typically found, and the ratio between vibrational and rotational quantum sizes is taken an order of magnitude less.

For lines due to transitions starting from the same rotational level, J, Eqns. 4.12 and 4.15 may be combined to give

$$\nu_R - \nu_P = 2B_1(2J+1) \tag{4.17}$$

from which B_1 can be evaluated. Similarly, for lines due to transitions to the same rotational level, i.e. for the R component starting from level J and for the P component starting from level $(J+2)$,

$$\nu_R - \nu_P = 2B_0(2J+3) \tag{4.18}$$

from which B_0 may be evaluated.

For molecules with an odd number of electrons, e.g. NO and free radicals, the selection rule $\Delta J = 0, \pm 1$ holds. Transitions for which $\Delta J = 0$ cause absorption at frequency ν_Q, where

$$\nu_Q = \omega + (B_1 - B_0)(J)(J+1) \tag{4.19}$$

Since B_1 is similar to B_0, this set of lines, termed the *Q branch*, is closely

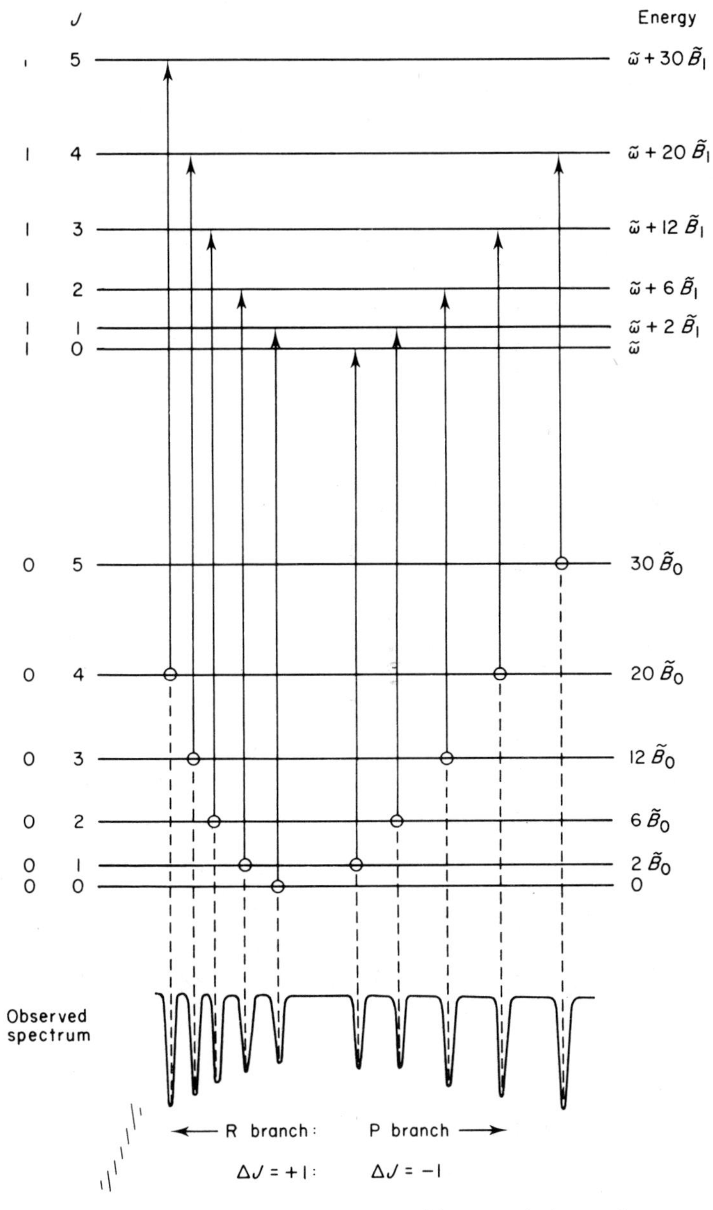

FIG. 4.7 Vibration–rotation energy levels, transitions, and observed spectrum for a diatomic molecule.

packed up to the molecular vibration frequency. The student may find it a help to his memory to note that, after the transition, the molecule is R*icher* in rotational energy in the R branch, P*oorer* in the P branch, and *e*Q*ual* in the Q branch.

The fine structure of the P and R branches is only seen for samples in the gas phase. Intermolecular interactions blur the energy levels of molecules in liquid and solid phases, by effects similar to the pressure broadening described in Section 2.A.4. The lines in each branch coalesce to give a broad peak.

4.D.2 Linear molecules

The rotational fine structure associated with parallel modes of vibration of linear molecules is of the same type as for diatomic molecules, i.e. there is an R branch ($\Delta J = +1$) and a P branch ($\Delta J = -1$) for each vibrational transition. For perpendicular modes the selection rule $\Delta J = 0$ is also valid, so P, Q, and R branches are observed. Figure 4.8 shows these features in the spectrum of HCN.

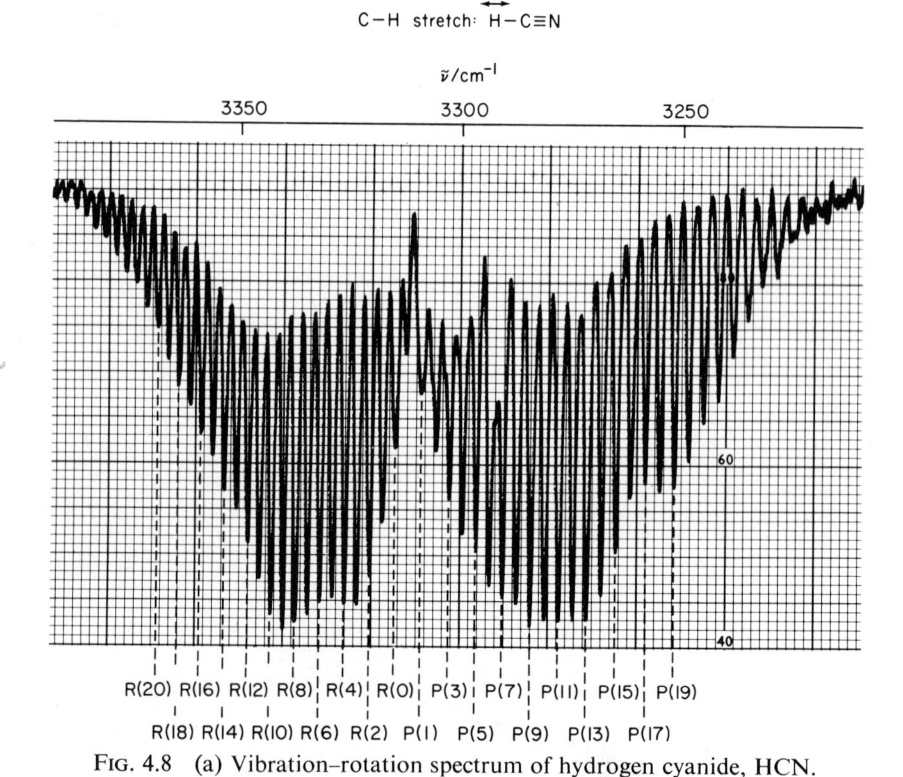

FIG. 4.8 (a) Vibration–rotation spectrum of hydrogen cyanide, HCN.

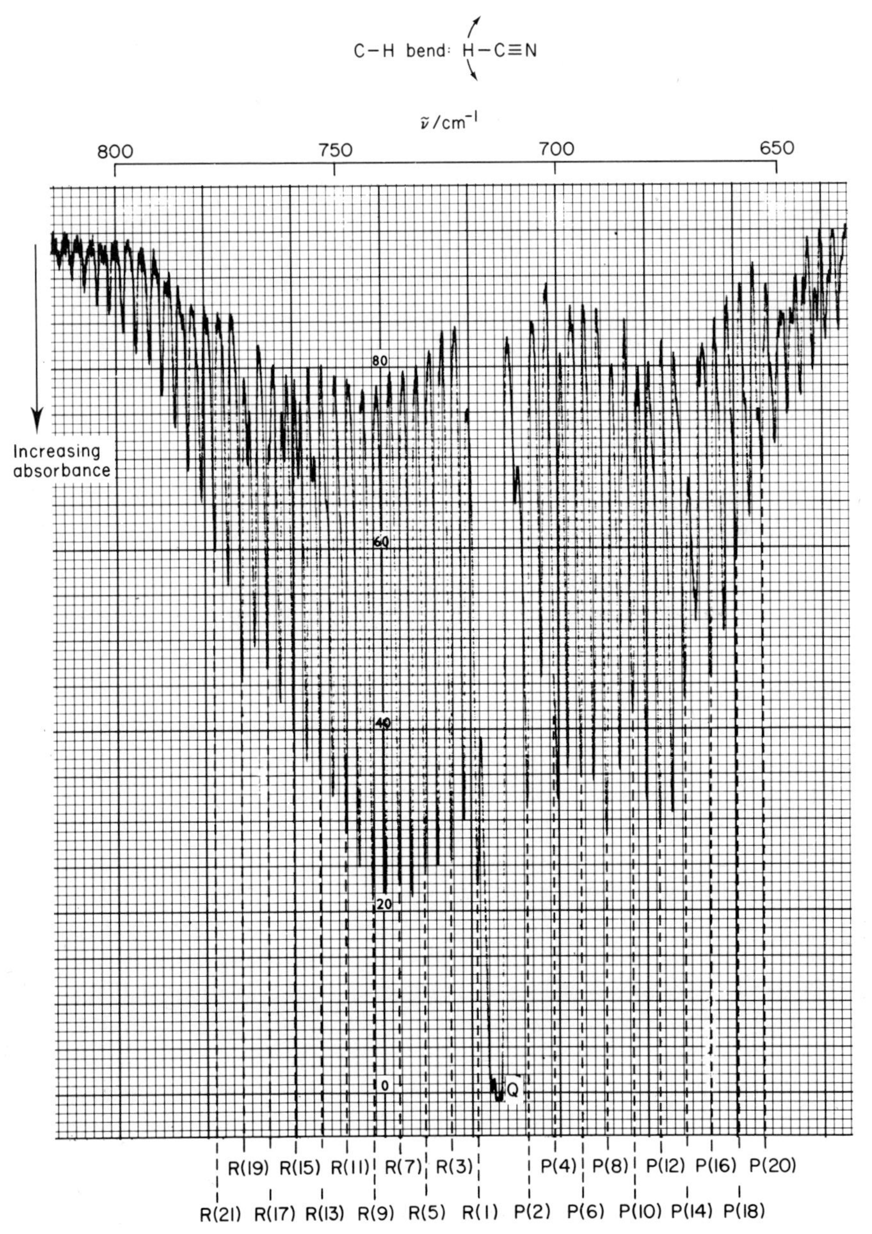

FIG. 4.8 (b) Vibration–rotation spectrum of hydrogen cyanide, HCN.

A subtle effect of the wave properties of atomic particles is observed in the infrared spectra of symmetrical linear molecules, such as CO_2 and C_2H_2. CO_2 is an example of a linear molecule with a centre of symmetry, each half having a total nuclear spin quantum number that is a whole integer (see Section 2.E.1). For such molecules it is a consequence of the Pauli exclusion principle that only rotational levels for which J is even are populated. The lines R(1), R(3), . . . and P(1), P(3), . . . are thus absent from the infrared spectrum. The spacing between the lines in the P and R branches is approximately $4B_0$, and the gap between the branches, for the parallel vibration, is $6B_0$. C_2H_2 is an example of a linear molecule with a centre of symmetry, each half having a total spin quantum number that is a half-integer. For such molecules, rotational levels for which J is odd have a multiplicity three times that of the levels for which J is even. Lines R(1), R(3), . . . and P(1), P(3), . . . are thus more intense than their adjacent neighbours, for which J is even, and the spectrum shows a pattern of alternating intensities.

4.D.3 Symmetric tops and spherical tops

As discussed in Section 2.C.4, symmetric tops are characterised by two moments of inertia, and hence by two rotational constants, B and A, and two rotational quantum numbers, J and K. For the rotational fine structure of parallel vibration bands, the selection rules are: $\Delta J = \pm 1$, $\Delta K = 0$, if $K = 0$; and $\Delta J = 0, \pm 1$, $\Delta K = 0$, if $K \neq 0$. The spectrum resembles that for a perpendicular vibration of a linear molecule, except that each observed line is in fact a cluster of lines, each component corresponding to one value of K. The resolving power of the spectrometer is usually insufficient to resolve the component lines of the cluster. The spacing between the clusters gives a value for B, by the application of Eqn. 2.15.

For the rotational fine structure of a perpendicular vibration band, the selection rules are: $\Delta J = 0, \pm 1$, $\Delta K = \pm 1$. Each change in K corresponds to a set of P, Q, and R branches. The lines in the various P and R branches overlap, so the fine structure cannot usually be resolved. The Q branches are more intense than the P and R branches, since many lines are superimposed, and stand out clearly from the background of unresolved P and R branches. If it is assumed that $B_0 = B_1$, the frequencies of the Q branches, ν_Q, can be calculated.

For $\Delta K = +1$:

$$\nu_Q = \omega + (A - B)[(K+1)^2 - K^2] \tag{4.20}$$

Rearranging:

$$\nu_Q = \omega + (A - B) + 2(A - B)K \tag{4.21}$$

Similarly, for $\Delta K = -1$:

$$\nu_Q = \omega + (A - B) - 2(A - B)K \tag{4.22}$$

The Q branches are thus spaced at intervals of $2(A - B)$, so $(A - B)$ can be evaluated. Thus information from the parallel and perpendicular bands may be combined to give values for A and B, and hence values for the two moments of inertia. Data for isotopically substituted molecules may be used to calculate bond lengths and angles, as described in Section 2.C.4. The accuracy of these data is, however, much lower than that from microwave spectroscopy.

Spherical tops may be considered as special cases of symmetric tops, with A equal to B. Only antisymmetric vibrational modes are infrared active. It is thus possible to obtain bond lengths for molecules, such as methane, that have not pure rotational spectra.

4.E Analytical Applications

4.E.1 Qualitative analysis

The discussion in the preceding section shows that the infrared spectrum of a polyatomic molecule is very complex. However, even if a detailed assignment cannot be made, much useful information may be obtained by the study of *group frequencies*. A group frequency is associated with a particular functional group and is common to all molecules containing that group. In Section 4.C, it was stated that it was necessary to consider the vibration of the molecule as a whole, but in certain circumstances, to a first approximation, it is possible to consider a particular bond as vibrating independently of the rest of the molecule. For example, the mass of the H atom is so much less than the mass of the adjacent C atom in an organic molecule that this adjacent C atom behaves as if it were the whole remainder of the molecule, and the C–H vibration can be considered as separate from the other vibrations of the molecule. In other words, the C–H vibrations are not *coupled* to the others. The C=O bond is much stiffer than adjacent C–C bonds, so its vibrations are weakly coupled to the rest of the molecule, and C=O group frequencies are well defined. A list of some group frequencies is given in Table 4.3. This list is for illustration only, and comprehensive lists are given in standard texts. Group frequencies are affected by substitution in other parts of the molecule. For example, although most ketones have a C=O stretching vibration in the region 1700–1720 cm^{-1}, the wavenumber may be as high as 1780 cm^{-1} if the ketone has a strained ring, e.g. cyclobutanone, or as low as 1600 cm^{-1} if the C=O is conjugated to a C=C bond. The deduction of the structure of a

TABLE 4.3

Some group frequencies

Group vibration	Wavenumber region (cm^{-1})
O—H stretch	3600
O—H stretch (hydrogen bonded)	3300–2500
N—H stretch	3400
Aromatic C—H stretch	3100–3000
Aliphatic C—H stretch	3000–2800
C≡C stretch	2100
C=O stretch	1700
NH_2 bend	1600
Aromatic C—C stretch	1600–1450 (several bands)
C=C stretch	1600
N=O stretch	1550
S=O stretch	1400
Aliphatic C—H bend	1400
C—N stretch	1000–800
Aromatic out-of-plane C—H bend	800–700 (only one band for mono-substituted benzenes)
C—Cl stretch	600

(1) C—H bands are usually much narrower than other bands.
(2) Stretching vibrations of groups of type AB_2 give two bands of similar intensity (i.e. due to symmetric and antisymmetric modes) separated by a few hundred wavenumbers.
(3) Adjacent electronegative groups tend to raise the group frequency; adjacent electropositive groups tend to lower it.

molecule from its infrared spectrum is as much an art as a science, since evidence from all regions must be combined and evaluated. Conclusive identification is only achieved if the spectrum of the unknown can be shown to be identical with that of a known reference compound. Vibrations of the C–C bonds in the "skeleton" of an organic molecule are closely coupled, and so are not good group frequencies. These vibrations usually occur in the 1200–760 cm^{-1} region, which is sometimes known as the fingerprint region, since absorption in this region is more useful for comparison with spectra of reference compounds (compare Section 2.H.1). Figure 4.9 shows a spectrum which may be used to illustrate the type of reasoning employed in the identification of an unknown substance. An unknown substance, X, is an organic liquid, and the spectrum is that of a thin film. X is aliphatic, rather than aromatic, since there is no absorption in the 700–800 cm^{-1} attributable to out-of-plane aromatic C–H bending. The C–H stretch region is below

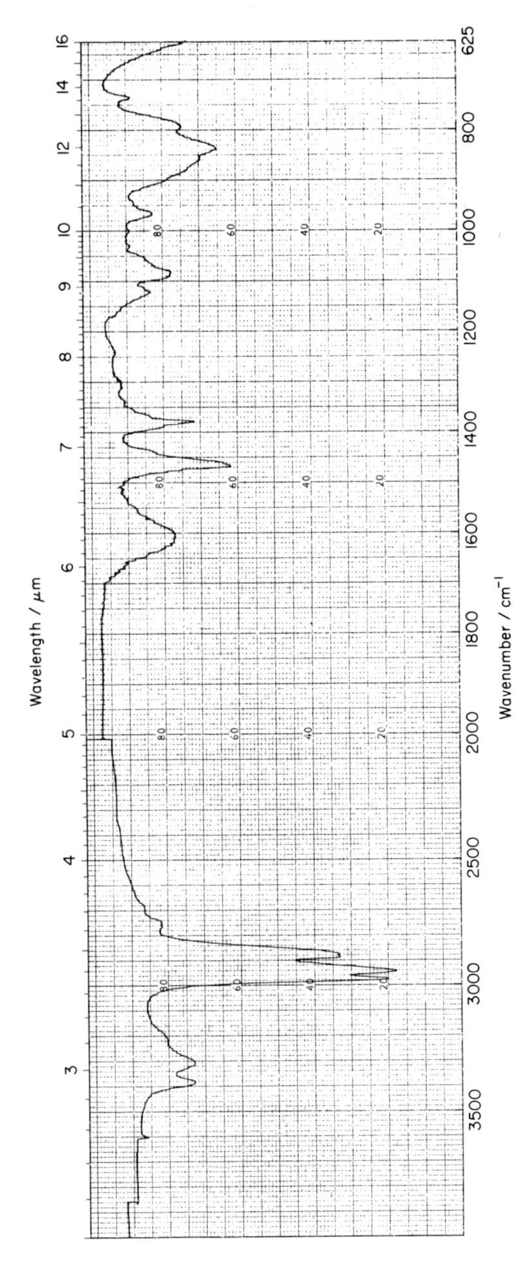

FIG. 4.9 Infrared spectrum of n-butylamine (as a liquid film).

$3000\ cm^{-1}$, rather than above $3000\ cm^{-1}$ where aromatic C–H stretching is seen. The twin peak at 3300–3400 cm^{-1} may be attributed to NH_2 stretching. The identification as a primary aliphatic amine is confirmed by the absorption at 800–900 cm^{-1}, due to C–N stretching, and that at 1600 cm^{-1} due to NH_2 bending. Final identification would require comparison with a reference spectrum. X is n-butylamine.

4.E.2 Quantitative analysis: the Beer–Lambert Law

Consider a sample cell, containing some substance capable of absorbing electromagnetic radiation, placed in the path of radiation of some given frequency. It is reasonable to suppose that all molecules of the same type absorb radiation of that frequency to the same extent. Thus doubling the concentration should double the amount of light absorbed. Similarly, each layer of the sample contains the same number of molecules, and should absorb the same proportion of the incident radiation. The actual amount of radiation absorbed by each layer is not the same, since absorption by layers nearer the source reduces the amount of radiation incident on layers further from the source. Let each layer be of some arbitrary thickness l_0, there being n layers, so the total path length, l, through the sample is equal to nl_0. [This derivation is based on that given by Loudon (1964).] In general, the intensity of light emerging from the ith layer is I_i, and is a fraction f of the intensity of radiation incident on that layer, which is the radiation emerging from the $(i-1)$th layer. Thus $I_i = fI_{(i-1)}$, and the intensity of radiation emerging from the nth layer, i.e. the intensity of radiation transmitted by the sample as a whole, is given by

$$I_n = fI_{(n-1)} = f^2 I_{(n-2)} = \ldots = f^n I_0 \tag{4.23}$$

I_0 being the intensity of the radiation incident on the sample. Hence

$$\log_{10}(I_0/I_n) = n \log_{10} f \tag{4.24}$$

Doubling the concentration c and halving the path length produces no change in the number of molecules in the path of the radiation, so for a fixed path length the variation of n with concentration can be expressed by $n = c/c_0$, where c_0 is a standard concentration. The combination of the variation of n with c and l gives

$$n = cl/c_0 l_0 \tag{4.25}$$

Substituting into Eqn. 4.24 gives

$$\log_{10}(I_0/I_n) = (cl/c_0 l_0) \log_{10} f \tag{4.26}$$

which may be written as

$$A = \varepsilon c l \qquad (4.27)$$

in which A, the *absorbance*, is defined as $\log_{10}(I_0/I_n)$ and ε, the *extinction coefficient*, is a constant, being equal to $(\log_{10} f)/c_0 l_0$. It is usual to refer ε to a standard concentration c_0 of 1 mol l^{-1} and standard length l_0 of 1 cm. The quantity I_n/I_0 is termed the transmittance, T. Equation 4.27 is the Beer–Lambert Law, the variation of A with concentration at constant path length being Beer's Law.

Beer's Law is only valid for monochromatic radiation, since the detector response is proportional to the intensity of radiation transmitted by the sample. For polychromatic radiation, a variation in the concentration of sample causes a variation in the sum of the absorbances at each wavelength, whereas the detector response is proportional to the sum of the intensities of the radiation transmitted at each wavelength. The non-linearity of the detector response to changing concentration is due to the mathematical fact that $\log(a+b)$ is not equal to $\log a + \log b$. A simple numerical example shows this. Suppose the sample is exposed to radiation of equal intensities at two different wavelengths. At one wavelength the absorbance is 1, corresponding to 10% transmission, and at the other the absorbance is 0.1, corresponding to 79% transmission. The absorbance measured by the detector is, by Eqn. 4.26, $\log_{10}[200/(10+79)]$, i.e. 0.35. If the sample concentration is halved, the absorbances at the two wavelengths fall to 0.5 and 0.05 respectively, giving an absorbance measured by the detector of 0.22, rather than the 0.175 required for a linear response. Stray light, i.e. radiation that reaches the detector without having passed through monochromator or sample, has a similar effect, and so must be eliminated as far as possible.

The sharp peaks observed in the spectra of gases mean that the extinction coefficient is likely to vary over the minimum range of wavelengths selected by the monochromator, so quantitative measurements are generally made on solutions, for which the absorption peaks are much broader. The linear variation of absorbance with path length is not subject to the restriction of monochromatic radiation (Strong, 1965).

Molecular interaction is another cause of variation of extinction coefficient with concentration. For example, in dilute solutions of phenols in non-polar solvents there is absorption in the region of 3500 cm^{-1} due to the O–H stretch. In concentrated solutions of phenols, hydrogen-bonding between the molecules occurs, so the relative absorption at 3500 cm^{-1} due to free O–H decreases.

The technique for the measurement of absorbance depends on whether a single-beam or double-beam spectrophotometer is being used. For single-beam operation, I_n and I_0 are measured by sliding the sample cell and a similar cell containing solvent successively into the beam. In double-beam operation, I_n and I_0 are measured simultaneously, the reference beam passing through the blank cell. The logarithmic conversion is performed automatically on most modern spectrophotometers. Once the extinction coefficient for a species has been measured, it is a simple matter to measure the concentration of any solution containing only that species. It is wise to check the validity of Beer's Law over the concentration range of interest. More than one species in a solution can be analysed if absorbances are measured at as many wavelengths as there are components. Ideally the absorbance of each species should be measured at a wavelength at which none of the others absorbs, but this is rarely possible in the infrared region. As long as each species has a different extinction coefficient at the wavelengths selected, the problem is soluble. For example, for a solution containing two species, i and j, whose absorbance is measured at two wavelengths, 1 and 2:

$$\begin{aligned} A_1 &= \varepsilon_{i1}c_i + \varepsilon_{j1}c_j \\ A_2 &= \varepsilon_{i2}c_i + \varepsilon_{j2}c_j \end{aligned} \tag{4.28}$$

in which only c_i and c_j are unknown. This technique can be used, with decreasing accuracy, for mixtures of up to about five species, a computer being recommended to ease the arithmetical labour.

Extinction coefficients in the infrared are typically in the range 10–100 $\mathrm{mol^{-1}\ cm^{-1}}$. Unfortunately the values obtained depend on the instrument used, and even differ between two different examples of the same model. This is probably due to non-linear photometric response. A linear Beer's Law plot does not necessarily mean that the photometric system gives a signal exactly proportional to the intensity of incident radiation. Consider an instrument which indicates a transmittance T' for a sample whose true transmittance is T. It is a common fault in infrared spectrophotometers that

$$T' = T^p \quad \text{where} \quad p \neq 1$$

Substitution into Eqn. 4.26 shows that the observed absorbance, A, is given by

$$A = -\log T' = -p \log T = p\varepsilon cl \tag{4.29}$$

Thus the non-linear photometric response does not affect the linearity of the Beer's Law plot of A against c, but does produce an incorrect value of ε.

REFERENCES

Hansen, G. E. and Dennison, D. M. (1952) *J. Chem. Phys.* **20**, 313.
Loudon, F. C. (1964) *J. Chem. Educ.* **41**, 391.

Plyler, E. K. and Benedict, W. S. (1951) *J. Res. Nat. Bur. Stand.* **47**, 202.
Schultz, J. W. and Eggers, D. F., Jr. (1958) *J. Mol. Spectroscopy* **2**, 113.
Shelton, R. D., Nielsen, A. H. and Fletcher, W. H. (1953) *J. Chem. Phys.* **21**, 2178.
Strong, F. C. (1965) *J. Chem. Educ.* **42**, 342.
Whiffen, D. H. (1955) *Phil. Trans. Roy. Soc.* *A***248**, 131.

BIBLIOGRAPHY

Alpert, N. L., Keiser, W. E. and Szymanski, H. A. (1970) "Theory and Practice of Infrared Spectroscopy", Plenum Press, New York and London.

Anderson, D. H. and Woodall, N. B. (1972) in "Techniques of Chemistry, Vol. I, Physical Methods of Chemistry, Part IIIB, Optical, Spectroscopic and Radioactivity Methods" (A. Weissberger and B. W. Rossiter, eds.), pp. 2–84, Wiley-Interscience, New York and London.

Bellamy, L. J. (1975) "The Infrared Spectra of Complex Molecules", Chapman and Hall, London.

Brittain, E. F. H., George, W. D. and Wells, C. H. J. (1970) "Introduction to Modern Spectroscopy; Theory and Experiment", Academic Press, London and New York.

Colthup, N. B., Daly, L. H. and Wiberley, S. E. (1975) "Introduction to Infrared and Raman Spectroscopy", Academic Press, New York.

Eglinton, G. (1970) in "An Introduction to Spectroscopic Methods for the Identification of Organic Compounds, Vol. I" (F. Scheinmann, ed.), pp. 123–143, Pergamon, Oxford.

Flett, M. St. C. (1970) in "An Introduction to Spectroscopic Methods for the Identification of Organic Compounds, Vol. I" (F. Scheinmann, ed.), pp. 109–122, Pergamon, Oxford.

Gans, P. (1975) "Vibrating Molecules", Chapman and Hall, London.

Herzberg, G. (1945) "Molecular Spectra and Molecular Structure, Vol. II, Infrared and Raman Spectra of Polyatomic Molecules", Van Nostrand, Reinhold, New York.

Herzberg, G. (1950) "Molecular Spectra and Molecular Structure, Vol. I, Spectra of Diatomic Molecules", Van Nostrand, Reinhold, New York.

Ross, S. D. (1972) "Inorganic Infrared and Raman Spectra", McGraw-Hill, London and New York.

Stewart, J. E. (1970) "Infrared Spectroscopy", Dekker, New York.

5
The Far Infrared Region

5.A The Problems Involved in Experimental Observations in the Far Infrared Region

5.A.1 The use of infrared techniques

The lower the frequency at which one attempts to observe infrared spectra using the techniques described in Section 4.B, the greater the difficulties that arise. The radiant power output from a black body decreases with decreasing frequency, as shown in Fig. 4.3. Although the radiant power in the low frequency region can be increased by increasing the source temperature, the radiant power at high frequencies is increased to a much greater extent, and may find its way to the detector as stray light. The Globar and Nernst Glower have low emission in the far infrared, since their radiative coatings are transparent in this region, and by Kirchhoff's law a body must absorb strongly at a given frequency if it is to radiate strongly at that frequency. A satisfactory source of radiation in the far infrared is the mercury arc lamp in a quartz envelope. The mercury arc heats the quartz, which, since it is opaque down to 100 cm^{-1}, radiates as a black body down to that wavenumber. The mercury arc is a *plasma*, i.e. an electrically neutral but highly ionised gas, containing a mixture of atoms, ions, and free electrons. Collisions between these various species cause the emission of radiation of intensity in the far infrared corresponding to a black body at 10^4 K. Below 100 cm^{-1}, this radiation is transmitted by the quartz envelope, and is of adequate intensity down to 2 cm^{-1}.

A diffraction grating must be used as the dispersive element, since no suitable prism material is known for use below 180 cm^{-1}. As discussed in Section 4.B.3, a diffraction grating has grooves shaped so as to concentrate the radiant energy into one order, and gives a maximum intensity of diffracted radiation at one wavelength, the blaze wavelength. In practice, a grating is useful over a range 2/3 to 3/2 of the blaze wavelength. In the far infrared this range of wavelengths does not cover a wide wavenumber range. For example, if the blaze wavelength is 300 μm, the grating is useful in the wavelength range 200–450 μm, corresponding to a wavenumber range of only 50–22 cm^{-1}. A commercial instrument, the Perkin-Elmer 301, uses six diffraction gratings to cover the range 665–14 cm^{-1}, each of which must be brought into operation at the appropriate point during a scan of the spectrum. Overlapping orders of spectra present a problem. If the grating is set so as to send radiation of wavenumber 50 cm^{-1} to the detector, radiation of wavenumber 100 cm^{-1}, 150 cm^{-1}, etc. will also arrive at the detector. The intensities of these higher orders is reduced by the blaze of the grating, but still significant by virtue of the increased source energy at higher wavenumbers. These higher orders must be absorbed by suitable filters. Absorption filters of black polyethylene or quartz may be used. Filters of a more elaborate design are made by electroforming metal grid and mesh patterns, the transmission region depending on the relationships between the grid spacing and the wavelength of the radiation. It has been shown in Section 4.B.3 that the resolving power of a grating is proportional to the total number of lines, so, if the number of lines per mm is small, to cope with radiation of long wavelength, the grating must be large. Gratings of size about 7 cm × 7 cm are typical.

The detector may be of thermal or quantum type, as discussed in Section 4.B.5. Both thermocouples and Golay cells are in common use, thermocouples being more often used in commercial apparatus. Quantum effect detectors are more sensitive and have a much faster response, but suffer from the disadvantage that they must be cooled in liquid helium, or else the effects of intrinsic thermal excitation swamp those of excitation by incident radiation.

Stray light is a great hazard. The maximum radiant intensity from a body at 300 K occurs at 1030 cm^{-1}, so the whole apparatus is acting as a source of far infrared radiation.

The one aspect of experimental technique in which far infrared spectroscopy is easier than that in the mid infrared region is that robust and inert window materials are readily available. Polyethylene is a useful window material, since a thin film absorbs little radiation below 700 cm^{-1}, and it has the advantage of absorbing heavily in the mid infrared region, thereby reducing stray light. Quartz may be used below about 200 cm^{-1}.

5.A.2 The use of microwave techniques

Another approach to the far infrared can be made by increasing the frequency used in microwave spectroscopy. The wavelength of the radiation emitted by a klystron depends on its geometric size, and there is a lower limit to the size of klystron that it is possible to make, corresponding to radiation of a few mm wavelength. It is possible, however, to use a silicon crystal to amplify the radiation in a non-linear fashion, so that higher harmonics of the fundamental frequency can be produced. The resulting polychromatic radiation is sent down a waveguide so narrow that the fundamental is rapidly attenuated, and only the high-frequency harmonics are propagated. Harmonics up to the twelfth have been used. Frequencies up to 806 MHz, i.e. wavenumbers up to 27 cm^{-1}, have been produced in this way. The radiation is not confined in the waveguide, but is directed by a tapered horn, and focused by a lens through the sample. A quantum-effect detector, an indium antimonide crystal cooled to 1.6 K is used, since it is more sensitive than the point-contact crystal diode that is satisfactory at lower frequencies. This technique is best used, as discussed in Chapter 2, for high-resolution work in gases.

Measurements made by microwave techniques are referred ultimately to a time standard, since in effect the frequency for a transition is measured by counting the number of cycles in a second. Measurements made by infrared techniques are referred ultimately to a distance standard, the spacing of lines in the diffraction grating, which can in principle be referred to the standard metre. Measurements on the same transition by the two different methods thus can be combined to give a value for the velocity of light (Gordy, 1971). For example, B_0 for $H^{35}Cl$ has been measured by the microwave technique to be 312.9913 ± 0.00026 GHz, and $\tilde{B}_0$ by the infrared technique to be 10.440254 ± 0.000010 cm^{-1}. Hence the velocity of light is calculated to be $(2.997928 \pm 0.000004) \times 10^{10}$ cm s^{-1}, in good agreement with the accepted value. Unfortunately, subsequent work showed there to be a consistent error of 4 MHz in the measurements.

5.B Interferometry and Fourier Transform Spectroscopy

The discussion in Section 5.A shows that extension of either the infrared technique or the microwave technique into the far infrared presents difficulties. Fortunately another, totally different, technique is available which is ideally suited to the region, called interferometry. There are several types of interferometers, of which only the simplest, the Michelson pattern, is described here. The Michelson interferometer achieved fame as the apparatus used in the Michelson–Morley experiment which disproved the existence of the ether and laid the experimental foundations for Einstein's theory of relativity, since it proved that the velocity of light is independent of the velocity of the observer. A schematic diagram of the Michelson interferometer is shown in Fig. 5.1. Radiation from the source falls on a beam splitter, a slab of material which reflects approximately half the radiation, beam A, perpendicular to the

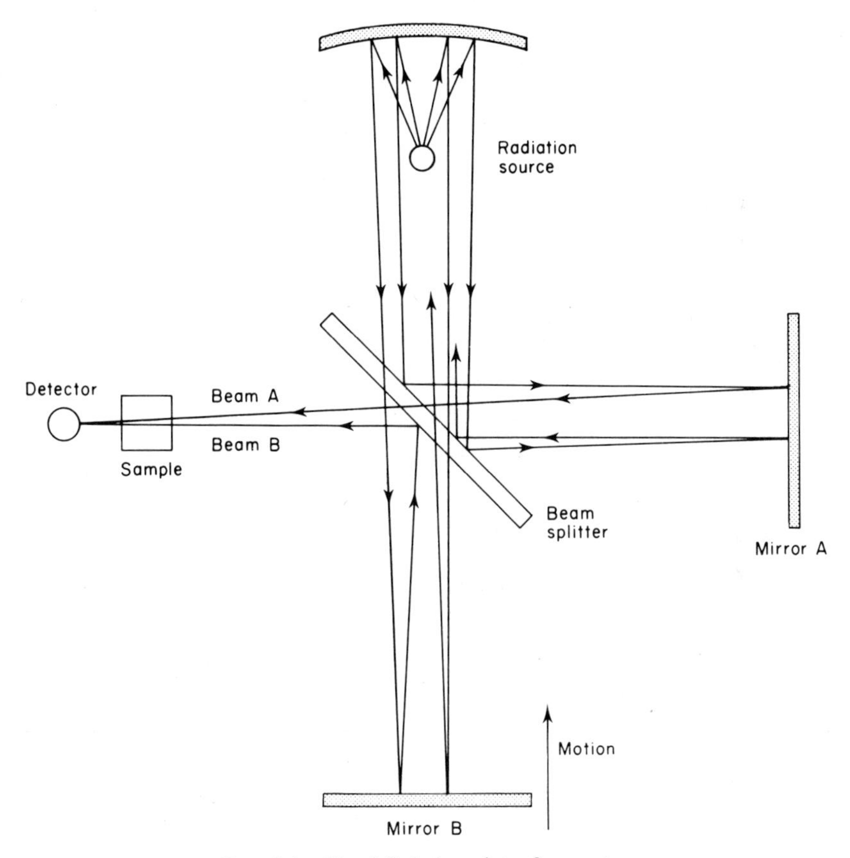

FIG. 5.1 The Michelson interferometer.

incident radiation and allows the rest, beam B, to pass through. Beams A and B are both reflected back to the beam splitter by mirrors A and B respectively. Approximately half the radiation in beam A passes through the beam splitter, and approximately half the radiation in beam B is reflected perpendicular to the incident beam B, so that beams A and B are recombined. The combined beams pass through the sample cell on to the detector. Consider monochromatic radiation of wavelength λ. If the path lengths of beams A and B are equal, or differ by $n\lambda$, where n is an integer, the two beams are in phase where they are recombined, and constructive interference occurs. If the path lengths differ by $\frac{1}{2}\lambda$, or $(n+\frac{1}{2})\lambda$, the two beams are exactly out of phase, and destructive interference occurs. The path length for beam B can be altered by moving mirror B towards or away from the beam splitter. A plot of the intensity of radiation, of wavelength 200 μm, against the path difference is given in Fig. 5.2. If the absorption spectrum of the sample is as shown in Fig.

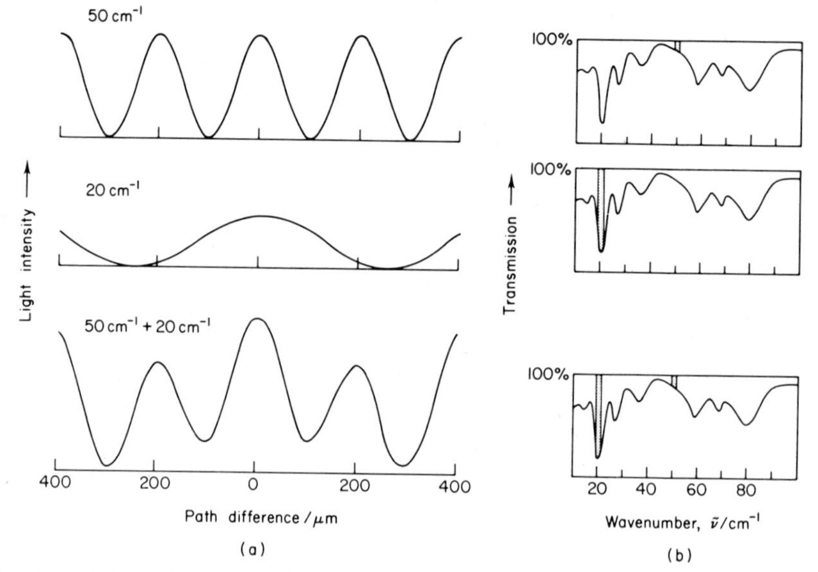

FIG. 5.2 (a) Interferometer signals at various frequencies, and (b) the corresponding features in a conventional spectrum.

5.2, then the intensity of the radiation at 200 μm, i.e. 50 cm^{-1}, is not much reduced by passage through the sample. Also in Fig. 5.2 is a plot of the variation of radiant intensity with path difference for monochromatic radiation of wavelength 500 μm. The spectrum of the sample shows that absorption at this wavelength is quite strong, so the intensity of the radiation of wavelength 500 μm incident on the detector is comparatively low. If the source emits radiation at equal intensities at 200 and 500 μm, the net variation at the detector as the mirror is moved is as shown in Fig. 5.2. It can be seen that, if the source emits radiation over a wide range of wavelengths, the detector signal will vary in a complex manner as mirror B is moved. This variation displayed as a graph is termed an *interferogram*. An example is given in Fig. 5.3. The salient feature of the interferogram is the large peak at zero path difference, since constructive interference occurs there at all wavelengths. The wiggles at the sides of the main peak contain all the information necessary to plot an absorption spectrum, although not in a form that can be easily interpreted. The conversion of an interferogram into an absorption spectrum is performed by application of a mathematical technique called Fourier transformation. This type of spectroscopy is thus sometimes called Fourier transform spectroscopy. Fourier transforms can be worked out by hand, but it is much more convenient to have a small computer permanently linked to the spectrometer. The data obtained as mirror B is moved over its full range

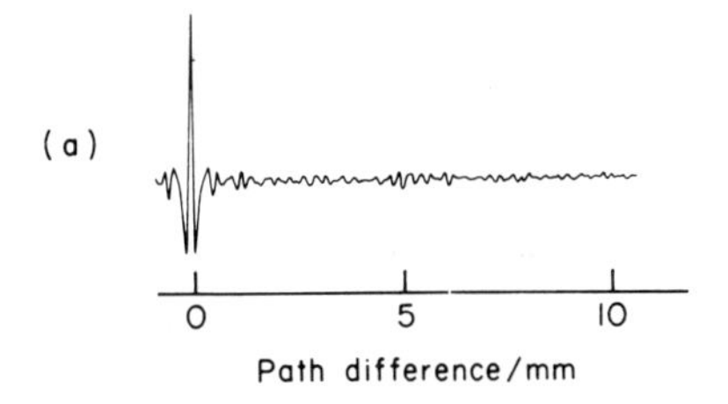

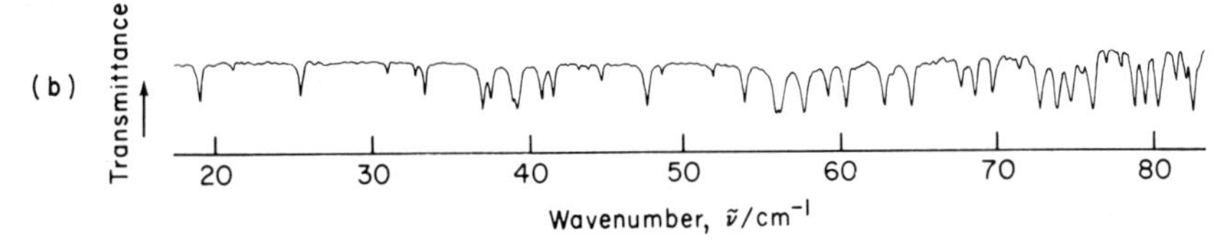

FIG. 5.3 Far infrared spectrum of water vapour (Richards, 1964): (a) interferogram of water vapour; (b) spectrum of water vapour obtained by Fourier transformation of (a).

are stored in the computer memory, and the results of the computation are presented by a graph-plotter. The advent of cheap small computers has made Fourier transform spectroscopy a worthwhile technique for the chemistry laboratory.

The resolution of the Michelson interferometer is approximately equal to the reciprocal of the traverse distance of the movable mirror. Thus, if the mirror moves through 10 cm, the resolution is 0.1 cm^{-1}, a typical value. This applies to *double-ended operation*, in which the zero path-difference position is at the middle of the traverse. In *single-ended operation* the zero path-difference position is at one end of the traverse, and the resolution is thereby improved by a factor of 2. Since the resolution is independent of frequency, the resolving power, $\Delta v/v$, is inversely proportional to frequency. The wide range of frequencies to be studied means that an apparatus with adequate resolving power in the high-frequency regions may not have adequate resolving power in the low-frequency regions.

The high-frequency limit of operation is set by the mechanical tolerances on the movable mirror mounting and drive. The adjustment of the mirror must be accurate to 1/10 of the shortest wavelength to be observed. A tolerance of 1 μm thus corresponds to an upper frequency limit of 1000 cm^{-1}, a typical value. Radiation from the source of frequencies higher than this limit must be filtered out to prevent degradation of the interferogram. Below about 20 cm^{-1} it is difficult to design an efficient beam splitter. The splitter usually consists of a plastic film, 0.1 mm thick Mylar sheet, which markedly loses efficiency below 20 cm^{-1}. An alternative form of interferometer, based on a *lamellar grating*, is more efficient in this region. A lamellar grating consists of two interleaved sets of thin plates, with the edges optically flat and polished. The path difference between beams A and B is produced by the reflections from the two sets of faces, and is altered by moving one set of plates through the other. The lamellar grating combines the functions of beam splitter and movable mirror. Despite the theoretical advantages of the lamellar grating interferometer over the Michelson interferometer, the difficulty of construction of the former causes the latter to be more common.

The interferometer is a single-beam instrument. Since water vapour absorbs extensively in the far infrared, it is necessary to purge the apparatus with dry air or, preferably, to evacuate it. The absorption bands of water may be put to good use for frequency calibration, since their frequencies are accurately known from calculations using data from microwave spectroscopy.

Interferometers are superior to grating spectrometers in two respects. The more important is termed the *multiplex* or *Fellgett advantage* (after the scientist who first gave a detailed discussion of this advantage). Consider the frequency range of interest to be sliced into n equal elements, where the width of each element is equal to the resolution of the spectrometer. There is a time t available for the recording of the spectrum in this range. A grating spectrometer records the spectrum by studying each element for time t/n, moving on to the next element at the end of this interval. An interferometer studies all the elements simultaneously, i.e. in a multiplex fashion, and so studies each element for time t. As shown in Section 1.C.2, the detector noise is inversely proportional to the square-root of the time during which the observation is made, so for equal signal-to-noise ratio and resolution the time required to obtain a spectrum is reduced by a factor of $\sqrt{n}$ for the interferometer compared with the grating spectrometer. For the same total observation time, the signal-to-noise ratio is greater by a factor of $\sqrt{n}$ for the interferometer, and, for the same observation time and signal-to-noise ratio, the resolution of the interferometer is greater by a factor of $\sqrt{n}$. The interferometer also has the advantage that the resolution can be increased by a factor of 2 simply by doubling the traverse distance of the movable mirror, thereby doubling the time required to obtain the same signal-to-noise ratio. To produce an equivalent increase in the resolution of a grating spectrometer, both the entrance and exit slits of the monochromator must be halved in width, reducing the intensity of the light, and hence the signal, by a factor of 4. The scanning time must therefore be increased by a factor of 16 to regain the original signal-to-noise ratio.

Interferometric spectroscopy may in principle be used in any part of the electromagnetic spectrum, but it is especially suitable for the far infrared;

this is a fortunate coincidence since more conventional techniques are not suited to this region. The construction of interferometers for use at shorter wavelengths requires finer mechanical tolerances, and a practical limit appears to be reached in the infrared, at about 10 000 cm^{-1}. Other things being equal, an interferometer can span a spectrum much more rapidly than a grating instrument, and there are special applications for which this speed is important. The traverse distance, and hence the overall size of the interferometer, limits the application of the interferometer to the microwave region. An instrument with a traverse distance of 1 m would only measure the frequency of a transition at 30 GHz to an accuracy of 300 MHz, which is several orders of magnitude less accurate than the value obtained by the conventional microwave techniques described in Chapter 2.

5.C Applications of Far Infrared Spectroscopy

5.C.1 Rotations and torsional modes

As discussed in Chapter 2, the transitions between rotational levels of low values of J tend to occur in the microwave region. Only if the rotational levels are highly excited do the transitions occur in the far infrared. These upper levels are markedly affected by centrifugal distortion, as discussed in Section 2.D.3. In general, the much greater accuracy of microwave data means that interferometry is not a preferred technique for obtaining values of moments of inertia and bond lengths for small polar molecules in the gas phase. The reverse procedure is used; interferometers are often calibrated by scanning the spectrum of a gas, usually water vapour, whose spectrum has been computed using microwave data extrapolated to high J values.

Far infrared techniques are useful for the study of restricted rotation, which has been discussed in Section 2.G.2. The transitions observed are those between different torsional levels of type A or type E (see Fig. 2.5), whereas in the microwave region the torsional levels manifest themselves by splitting the energy levels for rotation of the molecule as a whole. For example, the absorption of ethyl fluoride vapour at 242.7 cm^{-1} is attributed to the $0 \rightarrow 1$ transition for the A type levels of the methyl torsion. Sage and Klemperer (1963) used these data and calculated the barrier height to be 13.94 kJ mol^{-1}, in good agreement with the value of 13.87 kJ mol^{-1} obtained by microwave spectroscopy (Herschbach, 1956).

Restricted rotation can only occur in open-chain molecules. The analogous motion for closed-chain molecules is *ring-puckering*. For example, the cyclobutane ring tends to be planar to maximise the carbon–carbon bond angle, which even at its maximum value of 90° is still much less than the "natural" tetrahedral angle of 108°. Energy is required to bend, or pucker, the ring out of its planar configuration. Repulsion between substituents on the cyclobutane ring may increase the distance between bulky substituents, thereby reducing their steric repulsion energy. Cyclobutane itself is non-polar, so has no infrared spectrum, but another simple four-membered ring molecule, trimethylene oxide, has been much studied (Chan *et al.*, 1966). Trimethylene oxide,

$$\begin{array}{l} H_2C{-}CH_2 \\ \;\;\;| \quad\;\; | \\ H_2C{-}O \end{array}$$

shows a set of absorption peaks in the far infrared, which are listed in Table 5.1. The frequencies for the successive transitions increase, showing that the potential energy function is not that of the simple harmonic oscillator, i.e. the potential energy must be a function not of the square of the displacement from the mean position, but of the fourth or even higher power. There is also an irregularity in the low-frequency transitions, suggesting that the lower levels are perturbed by a small central hump in the potential function. The function

$$V = 0.08887z^4 - 8.0684z^2 \tag{5.1}$$

gives the potential energy V as a function of the displacement z, V being in units J and z in units pm. Equation 5.1 may be used to calculate a set of energy levels which agree well with those observed. A plot of this function, with some of the energy levels, is given in Fig. 5.4. The central potential hump affects the vibrational levels of even v more than those of odd v, and this effect is perceptible even though the zeroth level is above the top of the barrier. The lowest energy configuration of trimethylene oxide is seen to be a slightly puckered ring, the non-planarity being caused by hydrogen–hydrogen repulsion, but zero-point energy is sufficient to flip the ring freely between the two conformations.

TABLE 5.1

Far infrared spectrum of trimethylene oxide (Chan *et al.*, 1966)

Transition	$\tilde{\nu}/cm^{-1}$	Transition	$\tilde{\nu}/cm^{-1}$
$0 \rightarrow 1$	53.4	$6 \rightarrow 7$	147.3
$1 \rightarrow 2$	89.9	$7 \rightarrow 8$	155.0
$2 \rightarrow 3$	104.7	$8 \rightarrow 9$	161.6
$3 \rightarrow 4$	118.1	$9 \rightarrow 10$	168.2
$4 \rightarrow 5$	128.8	$10 \rightarrow 11$	174.9
$5 \rightarrow 6$	138.9	$11 \rightarrow 12$	180.4

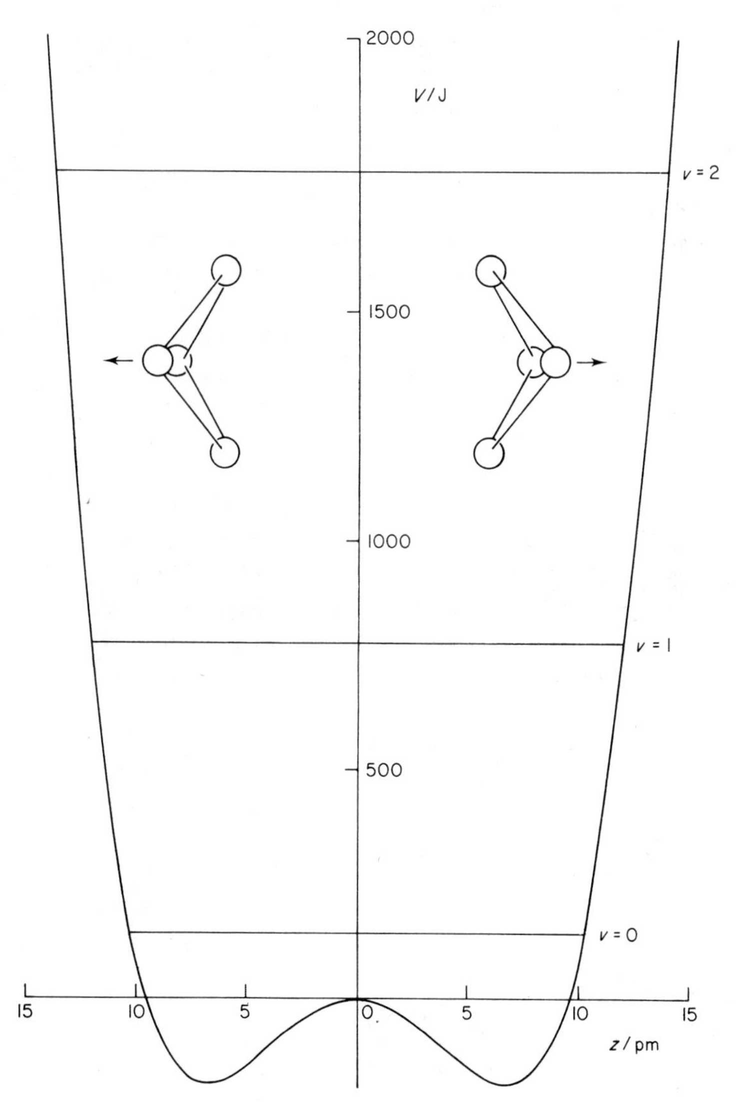

FIG. 5.4 The ring-puckering of trimethylene oxide.

5.C.2 Molecular vibrational modes

By virtue of the relationship

$$\omega = (1/2\pi)(k/\mu)^{\frac{1}{2}} \tag{2.18}$$

low-frequency vibrations causing absorption in the far infrared are to be expected for molecules containing heavy atoms and/or weak bonds. The mid infrared region is of interest predominantly to organic chemists, since hydrogen, carbon, oxygen, and nitrogen are all comparatively light atoms. By contrast, the far infrared region is of more interest to inorganic chemists, since it is in this region that vibrations of bonds between transition-metal ions and their ligands are found. For example, the antisymmetric Zn–I stretch in $[ZnI_4]^{2-}$ is at 165 cm^{-1}, and the antisymmetric U–Cl stretch in $[UCl_6]^{2-}$ is at 260 cm^{-1}. Unfortunately, water, the most common solvent for transition-metal complexes, absorbs heavily in the far infrared, and the need to find a suitable solvent restricts the applicability of this technique. Solid-state samples can be used, but intermolecular interactions in the crystal lattice perturb the intramolecular vibration frequencies. A powerful technique for the study of metal–ligand vibrations is Raman spectroscopy, which is discussed in detail in Chapter 8. Raman spectroscopy is a complementary technique to infrared spectroscopy and the results are discussed in more detail in Chapter 8. An interesting molecule which has been studied by far infrared spectroscopy is XeF_2, whose existence was for many years thought to be a theoretical impossibility. A set of P and R bands is found centred at 557 cm^{-1}, attributed to the antisymmetric stretch, and a PQR set of bands is found centred at 213 cm^{-1}, attributed to the degenerate bending mode. XeF_2 is thus a linear molecule, like CO_2. The Xe–F bond length is calculated to be 170 pm (Agron *et al.*, 1963).

Low-frequency vibrations are also found in far bonds of low force constant. The hydrogen bond is much weaker than the typical covalent bond, having an energy of the order of 20 kJ mol^{-1}. The effect of hydrogen-bonding can be seen on the frequency of the O–H stretching vibration, which, as shown in Table 4.3, is reduced from around 3500 cm^{-1} to around 3000 cm^{-1}. The hydrogen bond itself gives rise to absorption in the far infrared. For example, in the gaseous state at comparatively low temperatures, formic acid exists as the dimer, bound by a pair of hydrogen bonds:

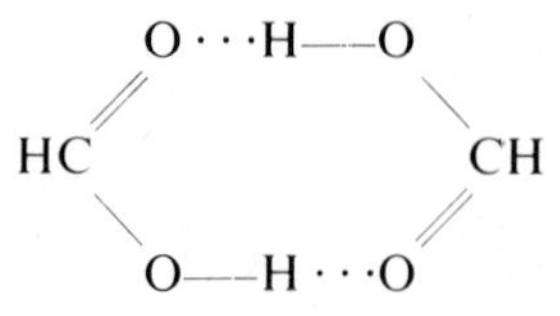

```
      O···H—O
     //      \
   HC         CH
     \       //
      O—H···O
```

The far infrared spectrum shows absorption at 248, 173, and 68 cm^{-1}, attributed to stretching, bending, and twisting respectively of the hydrogen bonds (Jakobsen *et al.*, 1967).

5.C.3 Crystal lattice vibrations

Ions in the lattice of an ionic crystal, e.g. NaCl, vibrate about their equilibrium positions. The vibration of one ion may be related to that of its neighbours in many ways, two of which are illustrated in Fig. 5.5. The motion in one of these produces oscillating dipoles, which may interact with electromagnetic radiation. Because of this interaction, this type of mode is termed an *optical mode*, by contrast with modes of the other type in Fig. 5.5, which interact with sound waves, and so are termed *acoustic modes*. Electromagnetic radiation of a frequency similar to an optical mode is heavily absorbed, and its velocity is much reduced during its passage through the crystal. As discussed in Section 1.A, a low velocity of transmission manifests itself as a high refractive index. Since the fraction of radiation reflected from a surface is related to the difference in refractive index between the two substances on either side of the surface, the crystal reflects nearly all the incident radiation in this frequency region. This phenomenon is termed *Reststrahlen* behaviour, and the frequency at which the reflection is at a maximum is termed the Reststrahlen frequency. The Reststrahlen frequency varies with the mass and charge of the ions in a manner predictable from Eqn. 2.18. For example, the maximum reflectance for NaCl is around 180 cm^{-1}, whereas the greater mass of the ions in KI causes the maximum reflectance to be around 110 cm^{-1}. The Reststrahlen frequency is higher for multi-charged ions; for example, it

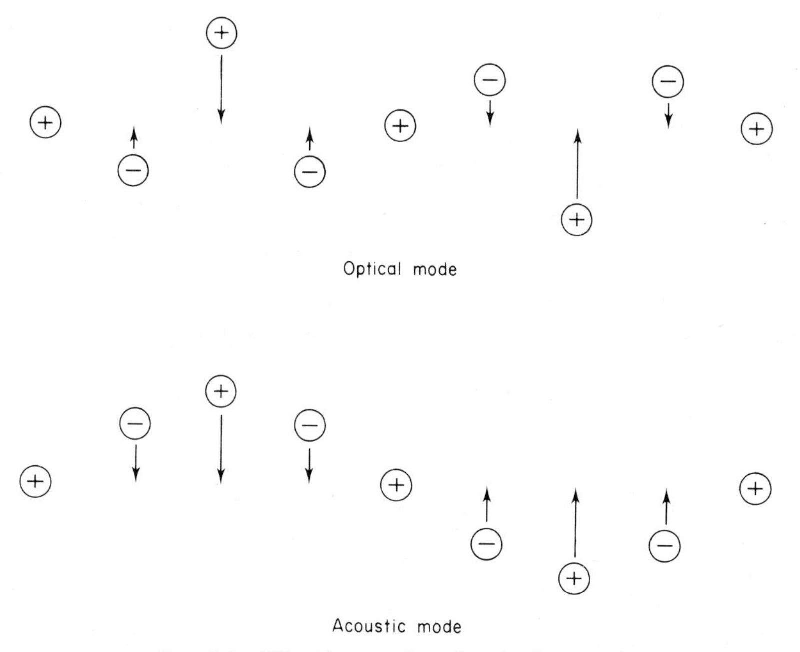

FIG. 5.5 Vibration modes of an ionic crystal.

is around 500 cm^{-1} for MgO. The high-frequency edge of the absorption by the optical modes of vibration is the effective low-frequency limit for the use of ionic crystals for windows and prisms in the mid infrared. The decrease in the frequencies of the optical modes with increasing ionic mass explains why such materials as the thallium halides are used rather than NaCl for the optics of spectrometers with range extended towards the far infrared.

It is possible to use Reststrahlen curves to calculate the effective charge on the ions, since the greater this is the greater the dipole charge for a given displacement, and the more intense the absorption. As expected, the effective ionic charge is greater for small ions than it is for larger, more polarisable, ions. For example, the effective ionic charge in the NaF lattice is 0.93 electron per atom, whereas that in the KI lattice is 0.69 electron per atom.

Molecular crystals, such as those of non-ionic organic compounds, may store energy in several ways. It is possible to distinguish between vibrations within each molecule and vibrations of the molecule as a whole in the crystal lattice. The latter are of lower frequency, since the displaced masses are greater and the restoring forces weaker. Some molecules and complex ions of roughly spherical shape can rotate freely in their sites on the lattice, and these rotatory modes may also give rise to absorption in the far infrared. Far infrared spectroscopy thus provides valuable data for the study of intermolecular forces in the solid state.

REFERENCES

Agron, P. A., Begun, G. M., Levy, H. A., Mason, A. A., Jones, G. G. and Smith, D. F. (1963) *Science* **139**, 842.

Chan, S. I., Borgers, T. R., Russel, J. W., Strauss, H. L. and Gwinn, W. D. (1966) *J. Chem. Phys.* **44**, 1103.

Gordy, W. (1971) *Phys. Rev.* **A3**, 1849.

Herschbach, D. R. (1956) *J. Chem. Phys.* **25**, 358.

Jakobsen, R. W., Mikawa, Y. and Brasch, J. W. (1967) *Spectrochim. Acta* **23A**, 2199.

Richards, P. C. (1964) *J. Opt. Soc. Amer.* **54**, 1474.

Sage, G. and Klemperer, W. (1963) *J. Chem. Phys.* **39**, 371.

BIBLIOGRAPHY

Bell, R. J. (1972) "Introductory Fourier Transform Spectroscopy", Academic Press, New York and London.

Chantry, G. W. (1971) "Submillimetre Spectroscopy", Academic Press, New York and London.

Ferraro, J. R. (1971) "Low-frequency Vibrations of Inorganic and Coordinate Compounds", Plenum, New York.

Finch, A., Gates, P. N., Radcliffe, K., Dickson, F. N. and Bentley, F. F. (1970) "Chemical Applications of Far Infrared Spectroscopy", Academic Press, New York and London.

Griffiths, P. R. (1975) "Chemical Infrared Fourier Transform Spectroscopy", Wiley, Chichester.

Kimmit, M. F. (1970) "Far Infrared Techniques", Pion, London.

Müller, K. D. and Rothschild, W. G. (1970) "Far Infrared Spectroscopy", Wiley-Interscience, New York and London.

6 The Visible and Ultraviolet Region

6.A Transitions Between Electronic States of Molecules

6.A.1 The magnitude of electronic transitions

For a typical molecule the energy gap between the highest filled and the lowest empty molecular orbital is of such a magnitude that excitation of an electron into an empty molecular orbital corresponds to absorption of electromagnetic radiation in the visible and ultraviolet region of the spectrum. Radiation of the longest wavelength perceptible to the human eye, around 750 nm, has a quantum size of 160 kJ mol^{-1}, whereas radiation of the shortest perceptible wavelength, around 400 nm, has a quantum size of 300 kJ mol^{-1}.

If the possibility of energy storage by electronic excitation is to be considered, the total energy of a molecule is given by

$$E = E_{el} + E_{vib} + E_{rot} \tag{6.1}$$

where E_{vib} and E_{rot} are given, to a first approximation, by Eqns. 2.20 and 2.3 respectively, and E_{el} is the electronic excitation energy. According to the Born–Oppenheimer approximation, these three contributions to the total energy may be considered separately; in particular E_{el}, which is concerned with the motion of the electrons, may be considered independent of E_{vib} and E_{rot}, which are concerned with the motion of the nuclei. This is valid because the nuclei are much more massive than the electrons. In general, the quantum size for electronic excitation is much greater than that for vibrational excitation, which in turn is much greater than that for rotational excitation. Typical values are 300 kJ mol^{-1}, 30 kJ mol^{-1}, and 3 J mol^{-1} respectively. The absorption of a quantum of radiation of sufficient energy to produce an electronic excitation is thus very likely to produce excitation of vibrational and rotational energies also. The absorption spectrum of a molecule in the gas phase in the visible and ultraviolet regions thus does not consist of a few lines corresponding to pure electronic transitions, but of vast numbers of lines, grouped into bands. Each electronic transition has a coarse structure, consisting of a set of bands, each band corresponding to a vibrational transition, and each band has a fine structure, consisting of many lines corresponding to rotational transitions. Only in favourable circumstances, with high-resolution instruments, is it possible to resolve the fine structure.

6.A.2 The electronic energy levels of an atom or ion

Individual atoms and ions cannot store energy in vibrational or rotational modes, but their electronic energy levels are spaced at such intervals that transitions between them give rise to absorption and emission in the visible and ultraviolet regions of the spectrum. It is necessary to have a notation to describe the various possible distributions of electrons among the orbitals of an atom. The simplest way to describe the electronic configuration of an atom or ion is by the number of electrons in each orbital. For example, the Ti^{2+} ion may be written as $1s^2\ 2s^2\ 2p^6\ 3s^2\ 3p^6\ 3d^2$. However, this is not only a cumbersome description but also an incomplete one. Each of the two electrons in the 3d orbitals can be in any one of the five d orbitals, and can have spin quantum number $+\frac{1}{2}$ or $-\frac{1}{2}$. If the electrons could be distinguished, the first could be allocated to any one of 5×2 possibilities, and the second to any one of $(5 \times 2 - 1)$ possibilities (the *Pauli exclusion principle* forbidding any two electrons on the same nucleus to have all four quantum numbers the same). There would thus be 90 ways of distributing the electrons, but the fact that the two electrons are indistinguishable reduces the number of ways to 45. These 45 ways do not all give states of the same energy, so a further classification is needed, as described below.

Electrons in inner filled orbitals, e.g. $3s^2$ and $3p^6$, are ignored for the purposes of notation, since their effects cancel out. The orbital angular momenta of all the electrons in the partly filled orbitals couple together to give a total orbital angular momentum, characterised by the quantum number L, and the spins of all the electrons in the partly filled orbitals couple together to give a total spin, characterised by the quantum number S. According to the *Russell–Saunders coupling scheme*, L and S are independent. However, spin and orbital angular momenta may couple through the magnetic dipoles they produce. This *spin–orbit coupling* is small for lighter atoms and ions, e.g. transition-metal ions of the first row, but is so large for heavier atoms and ions that L and S lose their identity. For these heavier atoms, only J, which is the quantum number for the vector sum of the orbital and spin momenta, has any real meaning. If Russell–Saunders coupling is assumed to be valid, a particular arrangement of electrons within a partly filled orbital or set of orbitals is characterised by the *spectroscopic term* ^{2S+1}L. The terms for which $L = 0, 1, 2, 3, 4, \ldots$ are referred to as S, P, D, F, G, . . . respectively. (The student should not confuse these two uses of the letter S.) The terms for which $(2S+1) = 1, 2, 3, \ldots$ are referred to as singlet, doublet, triplet, . . . respectively. Spin–orbit coupling removes the degeneracy of a term, so each value of J, which may range from $L+S$ to $|L-S|$, has a different energy. The full term description is then $^{2S+1}L_J$. It is implicit in the use of this notation that, since spin–orbit coupling is small, the energy of a state depends mostly on the values of S and L, the different values of J merely providing a "fine structure".

The simplest spectroscopic term is that for an s^1 configuration. The spin quantum number of the electron can be either $+\frac{1}{2}$ or $-\frac{1}{2}$, so $S = \frac{1}{2}$, and the orbital quantum number of the electron is 0, so L is 0, and the spectroscopic term is ^2S. Similarly for the d^1 configuration, $S = \frac{1}{2}$ and $L = 2$, so the spectroscopic term is ^2D. The d^2 configuration is more complex. The various ways of allocating the two electrons to the five d orbitals are shown in Table 6.1, the symbol ↿ representing an electron of spin $+\frac{1}{2}$ and the symbol ⇂ representing an electron of spin $-\frac{1}{2}$. The degeneracy of the spectroscopic term is given by $(2S+1)(2L+1)$. It can be seen that there are five spectroscopic terms for the d^2 configuration, namely ^1G, ^3F, ^1D, ^3P, and ^1S, which differ in energy. According to *Hund's rules*, the term with the highest spin multiplicity is the lowest in energy; if there are several terms of the same multiplicity, that of highest degeneracy is of lowest energy. The ground state of the d^2 configuration is thus ^3F. It can also be seen in Fig. 6.1 that the total number of possible ways of distributing the two electrons is $9+21+5+9+1 = 45$, as previously found in the first paragraph of this section. Similar considerations make it possible to tabulate the various possible electronic energy levels for atoms with other numbers of d electrons. A useful simplifica-

TABLE 6.1

Spectroscopic terms for the d^2 configuration

Spectroscopic term	S	L	Degeneracy $(2S+1)(2L+1)$	Magnetic quantum number 0	± 1	± 2
^1G	0	4	×9			⇅
^3F	3	3	×21		↿	↿
^1D	0	2	×5		⇅	
^3P	3	1	×9	↿	↿	
^1S	0	0	×1	⇅		

tion is that an "electron hole" behaves like an electron for the purpose of constructing spectroscopic terms, so d^9 has the same term as d^1, d^8 the same as d^2, and so on. A discussion of ways in which the Russell–Saunders term symbols can be obtained from electronic configurations is given by Hyde (1975).

Spin–orbit coupling causes the loss of degeneracy of spectroscopic terms of more than singlet multiplicity. For example, for the ^3F term, for which $S = 1$ and $L = 3$, J may have the value 4, 3, or 2. The energy gap between level J and level $(J+1)$ is $(J+1)\lambda$, where λ is the *spin–orbit coupling coefficient*. For first-row transition-metal ions, λ is of the order of 50–800 cm^{-1}, much less than the spacing between the main spectroscopic terms, which is of the order of 10 000 cm^{-1}. Each $^{2S+1}L_J$ term is $(2J+1)$-degenerate, so the $^3F_4$

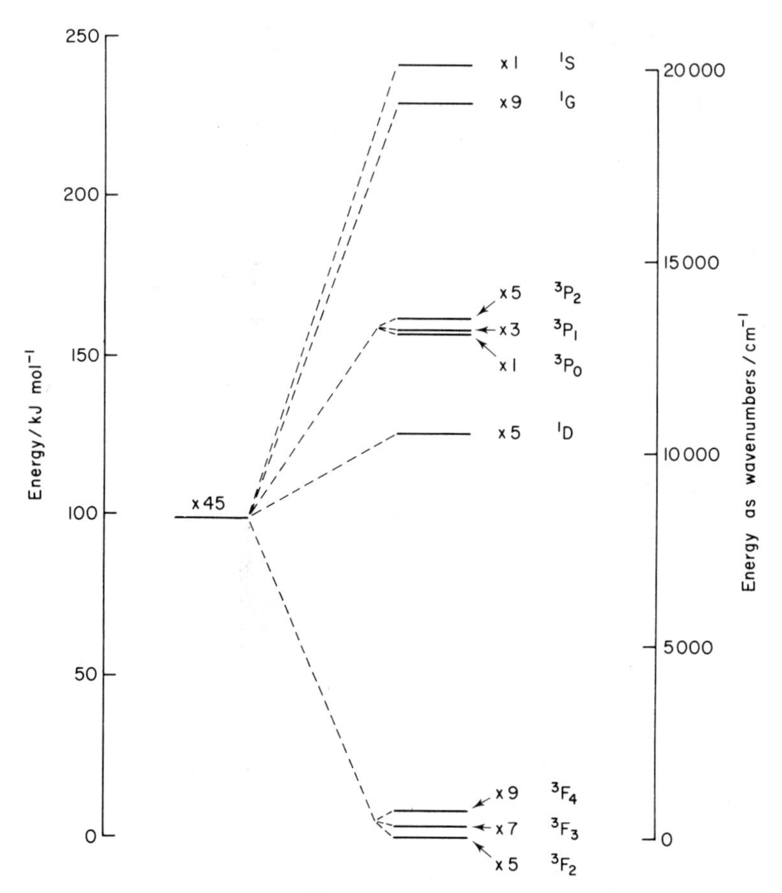

FIG. 6.1 Energy levels for the titanium ion, Ti^{2+}.

term is $\times 9$ degenerate, the $^3F_3$ term is $\times 7$ degenerate, and the $^3F_2$ term is $\times 5$ degenerate. It can be seen that $9+7+5 = 21$, the total degeneracy of the ^3F term. This final degeneracy is removed in the presence of a magnetic field, the *Zeeman effect*, which gives 45 different energy levels for the d^2 configuration. The energy levels for the various terms of Ti^{2+}, a typical d^2 species, are illustrated in Fig. 6.1.

6.A.3 The electronic energy levels of a diatomic molecule

One can visualise a diatomic molecule as being made out of a hypothetical *united atom* which has been pulled apart to a certain extent. The united atom, like a real atom, has orbitals, each characterised by a value of the total angular momentum of quantum number $L = 0, 1, 2, 3, \ldots$ corresponding to s, p, d, f, . . . sub-orbitals. When the united atom is pulled apart, an electrostatic field develops along the direction of the internuclear axis. The component of the total angular momentum along this axis must also be quantised, with quantum number Λ, where Λ is a positive integer of any value between and including 0 and L. The values $\Lambda = 0, 1, 2, \ldots$ correspond to Σ, Π, Δ states of the molecule. The total spin of all the electrons present, which in an atom is denoted by S, is denoted by Σ. (The student should not confuse these two uses of the letter Σ.) The motion of the electrons can be such that the total spin is aligned in $(2\Sigma+1)$ ways with respect to the internuclear axis. If a molecule is in a Σ state, i.e. $\Lambda = 0$, this may be not only because σ molecular orbitals are present in the molecule, but because the angular momentum vectors of two opposing π or δ molecular orbitals cancel each other out. There are two ways in which this cancellation may occur, corresponding respectively to molecules which are either symmetric or antisymmetric with respect to a plane through the molecular axis. Σ states of this type are labelled Σ^+ and Σ^- respectively. Σ states arising exclusively from σ molecular orbitals must necessarily be Σ^+.

The electronic state of a diatomic molecule is given as $^{2\Sigma+1}\Lambda$. States for which $2\Sigma+1 = 1$ are called singlet states; for which $2\Sigma+1 = 2$, doublet states; for which $2\Sigma+1 = 3$, triplet states. For example, LiH is bound by a single σ molecular orbital, formed by overlap of two atomic orbitals of s type, and there is one pair of electrons of opposing spins in the valency orbitals. The ground state of LiH is thus $^1\Sigma^+$. Electronic excitation of LiH raises one electron to a hitherto empty σ orbital formed by overlap of atomic orbitals of p type. This electron can have the same spin as the other, giving the triplet $^3\Sigma^+$ state, or the opposing spin, giving the singlet $^1\Sigma^+$ state. A more complex example is provided by the free radical CH, which in the ground state has an sσ and pσ molecular orbital, both occupied by a pair of electrons of opposing spin, and a pπ molecular orbital containing only one electron. This state is thus denoted by $^2\Pi$. Excitation of one of the electrons in the pσ orbital to the pπ orbital leads to one of the four excited states $^4\Sigma^-$, $^2\Delta$, $^2\Sigma^+$, $^2\Sigma^-$.

The notation is slightly different for homonuclear diatomic molecules, owing to their overall symmetry. If the molecular orbitals are such that interchanging the nuclei does not change the sign of the wave function, the molecule is *gerade* or g; if the sign is changed the molecule is *ungerade*, or u. Thus the ground state of the H_2 molecule is denoted by $^1\Sigma_g{}^+$. Excitation of one electron from the σ bonding orbital to an empty σ orbital formed by overlap of 2p atomic orbitals gives the excited states $^1\Sigma^+{}_u$ (spins opposed, singlet) and $^3\Sigma^+{}_u$ (spins parallel, triplet).

6.A.4 Selection rules

The selection rules for electronic transitions within atoms are that $\Delta S = 0$; $\Delta L = 0, \pm 1$. The $\Delta S = 0$ rule only holds for perfect Russell–Saunders coupling, in which S is not coupled at all to the orbital momenta of the electrons. The $\Delta S = 0$ rule means that transitions between singlet and triplet states, for example, are forbidden. In the notation for a transition the final state is by convention written first, with an arrow to indicate whether energy is being absorbed or emitted by the atom. The generalised transition, including both absorption and emission, is denoted by a dash, —.

The selection rules for electronic transitions within molecules are that $\Delta\Sigma = 0$; $\Delta\Lambda = 0, \pm 1$. The transition Σ^+—Σ^- is forbidden, but Σ^+—Σ^+ and Σ^-—Σ^- are allowed. For homonuclear diatomic molecules, u—g is allowed, but u—u and g—g are forbidden. These selection rules can be explained in terms of the symmetry of the wave-functions. They are true only in the first approximation; forbidden transitions are often observed, but give rise to much less intense absorption and emission than allowed transitions.

Whereas for purely vibrational transitions the selection rule $\Delta v = \pm 1$ holds, vibrational transitions which accompany electronic transitions, sometimes called *vibronic* transitions, may have many possible values of Δv. These transitions are subject to the *Franck–Condon principle*, which states that the internuclear distance is not altered during the course of an electronic transition. In other words, if the Morse curves of the ground and excited states are plotted on the same axes, as has been done for I_2 in Fig. 6.2, the electronic transition is represented by a vertical line. It must be noted that even for the harmonic oscillator approximation the probability of finding the nuclei at a given separation is not equal to that calculated by classical mechanics. If the diatomic molecule could be exactly described by a model comprising two masses linked by a spring, the internuclear separation for the molecule in the lowest energy level would be exactly r_e. As can be seen in Fig. 6.3, the solution of the wave equation for the harmonic oscillator shows that, although the probability of finding the nuclei at this separation is greater than for another, there is a finite probability of finding the nuclei at some other separation. Similarly, two masses linked by a spring, moving in simple harmonic motion,

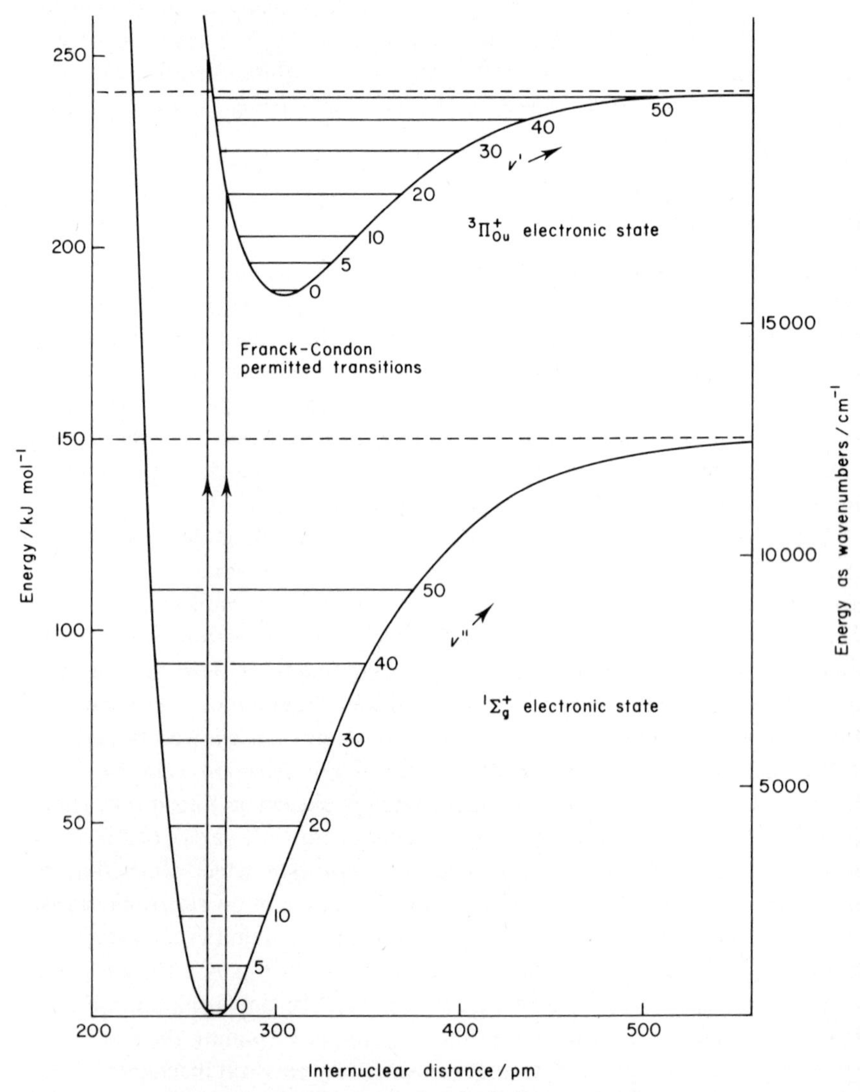

FIG. 6.2 Energy levels for the iodine molecule, I_2.

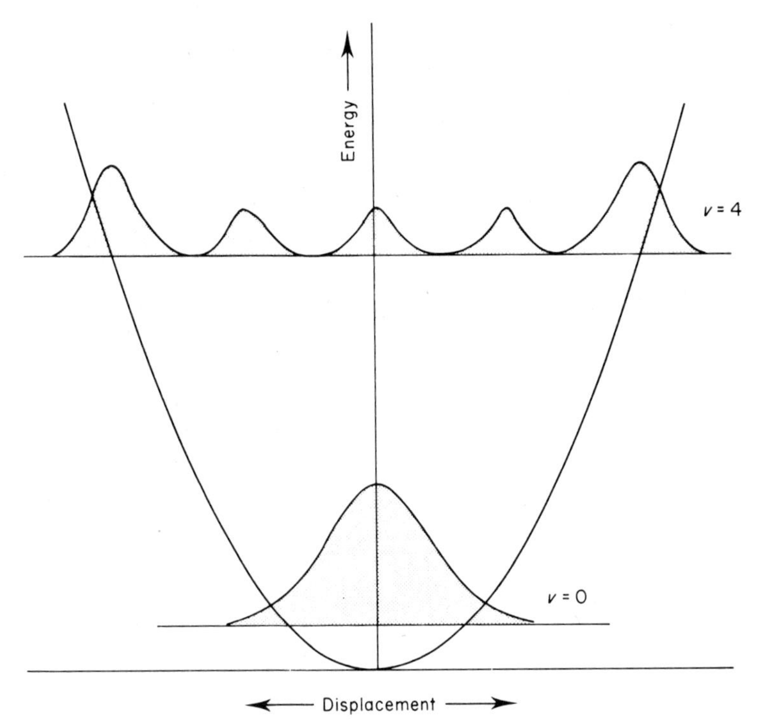

FIG. 6.3 Density distributions for the atoms in a diatomic molecule.

move most rapidly at the equilibrium separation, and come momentarily to rest at the extremes. The probability distribution obtained by solution of the wave equation, as shown in Fig. 6.3, has a series of modes between the limits, the number of modes being equal to $(v+1)$. Because of these probability distributions, excitations, even from the ground vibrational state, can occur over a range of internuclear separations, and lead to production of the electronically excited state with a range of vibrational energies as can be seen in Fig. 6.2 for excitations from the ${}^1\Sigma^+{}_g$ state of I_2 to the ${}^3\Pi_{0u}$ state. Transitions to vibrational levels of the ${}^3\Pi_{0u}$ state of quantum number less than about 15 are not seen, since the probability is very low for the internuclear distance in the ${}^1\Sigma^+{}_g$ state to be sufficiently large. If the internuclear distance in the ${}^1\Sigma^+{}_g$ state is near its minimum value, excitation is accompanied by the acquisition of sufficient vibrational energy to dissociate the molecule.

6.A.5 Intensity of absorption due to electronic transitions

The intensity of absorption of electromagnetic radiation at frequency ν is measured by the extinction coefficient, ε, as discussed in Section 4.E.2. The extinction coefficient is a function of frequency, so the total amount of radiation absorbed is given by $\int \varepsilon \mathrm{d}\nu$. It can be shown by the methods of quantum mechanics, using a simple model for electronic excitation in which it is assumed that the electron oscillates harmonically within the atom, that

$$\int \varepsilon \mathrm{d}\nu = \frac{10 f N_0 e^2}{2.303 \times 4\varepsilon_0 mc} \ \mathrm{s^{-1}\ l\ mol^{-1}\ cm^{-1}} \qquad (6.2)$$

where N_0 is the Avogadro number, e is the charge on the electron in coulombs, m is the mass of the electron in kg, c is the velocity of light in m s^{-1}, and ε_0 is the permittivity of free space in F m^{-1}. f is a dimensionless constant, called the *oscillator strength*, of maximum value unity, which is a measure of the efficiency of the coupling between the electronic transition and the electromagnetic radiation. Equation 6.2 gives the value of $\int \varepsilon \mathrm{d}\nu$ if ε is in the conventional units of l mol cm^{-1}, referring to a substance whose concentration is in units mol l^{-1} in a cell of path length 1 cm, and ν is in units Hz. It is more convenient for ν to be in wavenumber units, so that

$$\int \varepsilon \mathrm{d}\tilde{\nu} = \frac{f N_0 e^2}{2.303 \times 400\varepsilon_0 mc^2} \ \mathit{cm}^{-1}\ \mathrm{l\ mol^{-1}\ cm^{-1}} \qquad (6.3)$$

The wavenumber unit, *cm*$^{-1}$, has been italicised to avoid confusion with the reciprocal length unit, cm^{-1}. Substitution of numerical values into Eqn. 6.3, with f at its maximum value of unity, gives a value for $\int \varepsilon \mathrm{d}\tilde{\nu}$ of 2.31×10^8 *cm*$^{-1}$ l mol^{-1} cm^{-1}. If the half-height width of the absorption peak is taken to be of the order of 2000 cm^{-1}, a typical value for a substance in solution, then the maximum value of the extinction coefficient is of the order of 10^5 l mol^{-1} cm^{-1}. This value is approached for some dyestuffs. Comparisons between theoretical and experimental values of $\int \varepsilon \mathrm{d}\tilde{\nu}$ enable values of f to be calculated. Low values of f, of the order of 10^{-4}, are found for forbidden transitions.

6.B Apparatus

6.B.1 Optical components

The light source for absorption spectroscopy in the visible and near ultraviolet region, i.e. in the wavelength range 750–350 nm, is almost universally the tungsten filament lamp, which radiates approximately as a black body at about 2800 K. Below 350 nm the hydrogen or deuterium gas discharge lamp is a preferred source. Pyrex glass is a satisfactory material for windows, lenses, and prisms down to about 350 nm, below which quartz, i.e. a glass of pure silicon dioxide, must be used. Quartz prisms were used in early studies, but have now been almost completely replaced by diffraction gratings as the dispersive element, for the reasons discussed in Section 4.B.3. A 50-mm wide grating ruled at 600 grooves mm^{-1} has a resolution of 30 000, as discussed in Section 4.B.3, so its resolving power at 300 nm is 0.01 nm, or, in wavenumber terms, its resolving power at 21 000 cm^{-1} is 0.7 cm^{-1}. High resolution is obtained in apparatus in which there is a large distance (typically over 1 m) between dispersive element and detector, so that a small angular displacement in the dispersive element produces a large linear displacement at the detector. The *reciprocal linear dispersion* is defined as the difference in wavelength between two lines divided by the observed separation. This is constant for a grating, being typically around 1 nm mm^{-1}, but for a prism is much less in the ultraviolet than in the visible. Typical values range from 0.5 nm mm^{-1} at 200 nm to 5 nm mm^{-1} at 450 nm.

Air absorbs ultraviolet light of wavelength shorter than about 180 nm, so studies at shorter wavelengths than 180 nm require the use of an evacuated spectrometer. This region is thus termed the *vacuum ultraviolet*.

6.B.2 Detectors

The most common detectors for visible and ultraviolet radiation are based on the *photoelectric effect*. The impact of a photon of sufficient energy on a metal surface causes the ejection of an electron. If the metal surface is enclosed in an evacuated vessel containing an anode, i.e. a wire maintained at an electric potential which is positive with respect to the metal surface, the ejected electron travels to the anode. The consequence is that an electric current flows from the anode to the photosensitive cathode. The greater the number of photons incident on the cathode in unit time, the greater the current, and the proportionality is exact over a very wide range of radiation intensities. The photocell is only sensitive to radiation of frequency greater than ν where $h\nu$ is equal to the *work function* of the photosensitive cathode, i.e. the energy required to liberate an electron from the cathode surface. Metals of low ionisation potential are thus most suitable, and early photocells used a caesium-coated cathode. Modern photosensitive cathodes are coated with mixtures of alkali metals, antimony, and their oxides, and some are sensitive to radiation of as long a wavelength as 1 μm. The lower wavelength limit is set by the window material rather than the cathode. For special applications in the far ultraviolet, a windowless photocell may be mounted inside an evacuated enclosure containing the other components of the spectrophotometer. The current from the simple *vacuum photocell* is very small, usually less than a nanoampere when the photocell is used as a detector in a spectrophotometer. The extremely high output impedance of the photocell means that this tiny current can be amplified without excessively elaborate electronic circuitry. However, this circuitry contains very high-resistance components which are easily short-circuited by a film of moisture, and must be kept in a dry atmosphere. The amplifier generates noise in its own account, which is added to the shot noise from the photocell.

The vacuum photocell has been developed into a much more satisfactory device, the *photomultiplier*. The photomultiplier resembles a vacuum photocell, except that the electrons emitted from the cathode do not go to the anode, but instead hit an intermediate electrode, called a *dynode*, which is at a potential of 100 V or so above the cathode. The impact of the electron on the dynode surface, which has a low work function, causes the emission of three or four electrons of low energy. These then travel to another dynode, at a potential 100 V or so above the first dynode, acquiring considerable kinetic energy as they do so. Each secondary electron causes the emission of three or four more from the second dynode, and so on. A typical photomultiplier has about a dozen dynodes in sequence, so that the emission of a single electron from the cathode causes the emission of a million or so electrons from the final dynode, which are collected by the anode. The photomultiplier thus acts as a vacuum photocell with a built-in high-gain amplifier. The amplification process adds very little noise to the signal, and is completely reliable as long as constant voltages are applied to the dynodes. The gain is exponentially proportional to the dynode voltage, so a highly stabilised power supply is essential.

Much early work on visible and ultraviolet spectroscopy was performed using a photographic plate as detector, the apparatus with quartz prism dispersive element and photographic plate detector being termed a *quartz spectrograph*. The light which has passed through the sample is dispersed, and the whole spectrum is allowed to irradiate a photographic plate, which thus acts as both detector and display element. The plate is blackened except where absorption at some particular wavelength has reduced the intensity of the light incident on the plate. The spectrum thus appears as a set of white lines on a black background. The wavelength calibration is obtained by taking the spectrum of a calibrated source, for example a mercury discharge lamp, and the wavelength of a line is formed by interpolation, which is linear if a grating is used as the dispersive element. The degree of blackening of the plate is a measure of the intensity of the radiation, so an approximate estimation of the line intensity can be made. A *microdensitometer* may be used to measure the blackening of the plate as a function of distance, so as to obtain a more conventional display of the spectrum. Despite its inconvenience in use, the photographic plate still has its uses as a detector. The linear displacement of the lines can be measured with a travelling microscope to a limit set only by the graininess of the photographic emulsion, so a much greater precision in wavelength measurement is obtained than if a photoelectric detector is used. The whole spectrum is observed simultaneously, so the spectrum of a short-lived species can be observed. This is useful in the technique of flash photolysis (Porter, 1968), in which transient chemical species, of lifetimes as short as a few nanoseconds, are produced by irradiation by an intense flash of light of duration less than the lifetime of the transient species. The photographic plate has a multiplex advantage, as discussed in Section 5.C, because the whole spectrum is observed simultaneously, and so may be used to obtain the emission spectra of very faint sources, which would otherwise be swamped by detector noise. The photographic plate is thus still useful for astrophysical studies.

In general the light sources are so intense and the photomultiplier is so sensitive that the noise problem is not severe for spectrophotometry in the visible and ultraviolet. Signal modulation is thus not used in the apparatus, which saves trouble and expense. Double-beam operation is customary for scanning spectrophotometers. Low-cost single-beam spectrophotometers, in which the absorbance of the sample is measured at a predetermined wavelength, are in common use for quantitative analysis.

6.C Molecular Structure Determination

6.C.1 Information from the fine structure of spectra for molecules in the gas phase

The absorption spectrum of I_2 in the vapour phase in the visible region of the spectrum is shown in Fig. 6.4. The spectrum consists of a series of bands, representing transitions to different vibrational levels in the upper electronic state, as illustrated in Fig. 6.2. The standard nomenclature for such transitions is that the vibrational quantum number in the upper electronic state is v', and in the lower, v''. A set of bands with constant v'' is termed a *progression*. For most of the bands in Fig. 6.4, $v'' = 0$, but at wavelengths longer than 540 nm another progression may be seen, the intensity of the bands increasing at longer wavelengths. These are hot bands, for which $v'' = 1$. The vibrational frequency of I_2 in the ground electronic state is low, since the nuclei are heavy and the bond is weak, having the value 215 cm^{-1}, so at room temperature the population of molecules in the $v'' = 1$ level is, by Eqn. 2.12, about 37% of that for $v'' = 0$. The band spacing decreases with increasing v' as the vibrational levels pack more closely together towards the top of the potential-energy curve. The rotational fine structure of each band cannot be seen in

Fig. 6.4, since the resolving power of the spectrometer used is insufficient, but the shape of each band shows that the lines are well spaced out on the long wavelength side, and packed closely together on the short wavelength side. This is due to the effect discussed in Section 4.D.1, the maximum of each band being a band head. The formation of a band head may be seen on a Fortrat diagram, on which the wavenumber of the transition from a rotational state of quantum number J'' in the ground electronic state to a rotational state of quantum number J' in the excited electronic state is plotted against J''.

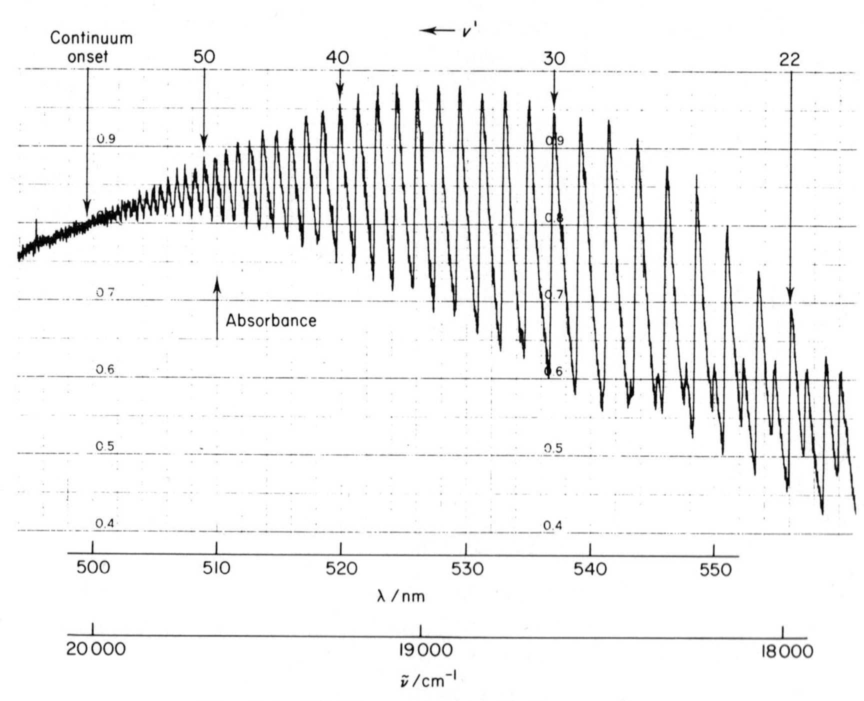

FIG. 6.4 Visible spectrum of iodine vapour.

The selection rule is that $(J'-J'') = +1$ for the P branch and $(J'-J'') = -1$ for the R branch, as discussed in Section 4.D.1. The presence of J^2 terms in Eqns. 4.12 and 4.15 means that the plot is parabolic in shape. The P and R branches form parts of the same parabola, which is called a *Fortrat parabola*. In the construction of a Fortrat parabola from an experimentally observed spectrum, lines are given arbitrary numbers in a running sequence, and the wavenumber of each line is plotted against its sequence number. The absence of the P(0) line enables the arbitrary sequence number to be assigned to values of R(J'') and P(J''). The Fortrat parabola for I_2 (using calculated values of wavenumbers for $v' = 30$) is shown in Fig. 6.5 for the first few lines, and in Fig. 6.6 for the whole width of the band for $v' = 30$. The corresponding spectrum is shown schematically below the Fortrat parabola (only lines for J' a multiple of 10 are shown in Fig. 6.6 for simplicity); the formation of the characteristic shape of the band can be seen. For some diatomic molecules, containing an odd number of electrons, coupling between the orbital momentum of the electrons and the angular momentum of the molecule as a whole permits ΔJ to be zero, so a Q branch also appears.

For pure vibration–rotation spectra, B for the lower level is always greater than B for the upper level, as the molecule elongates with increasing vibrational

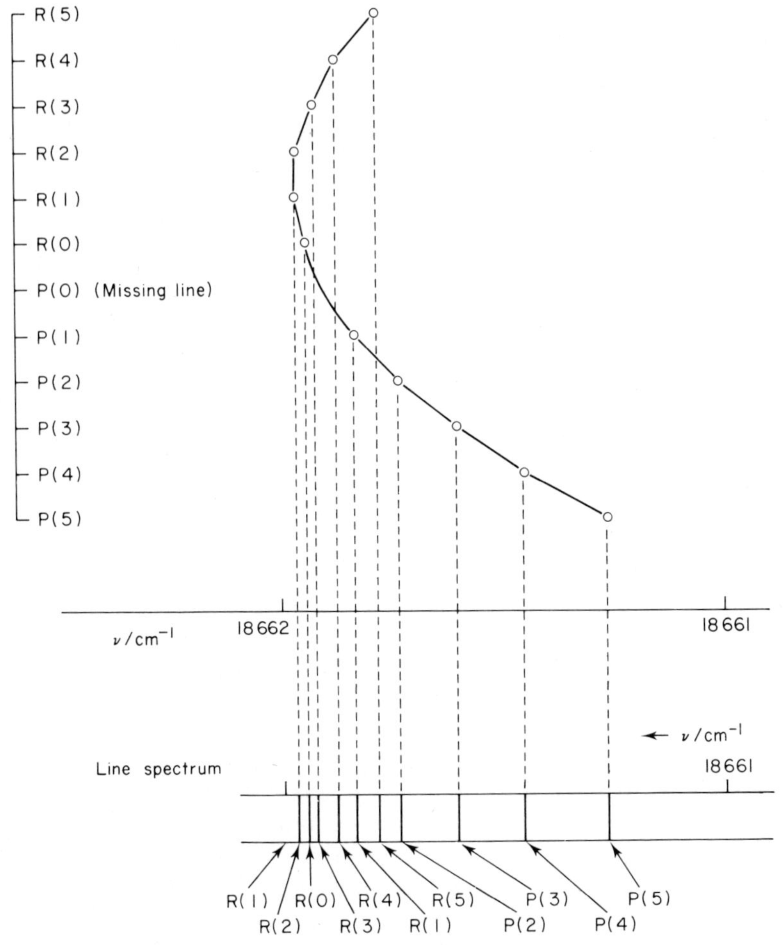

FIG. 6.5 The Fortrat parabola for I_2 (details of band head).

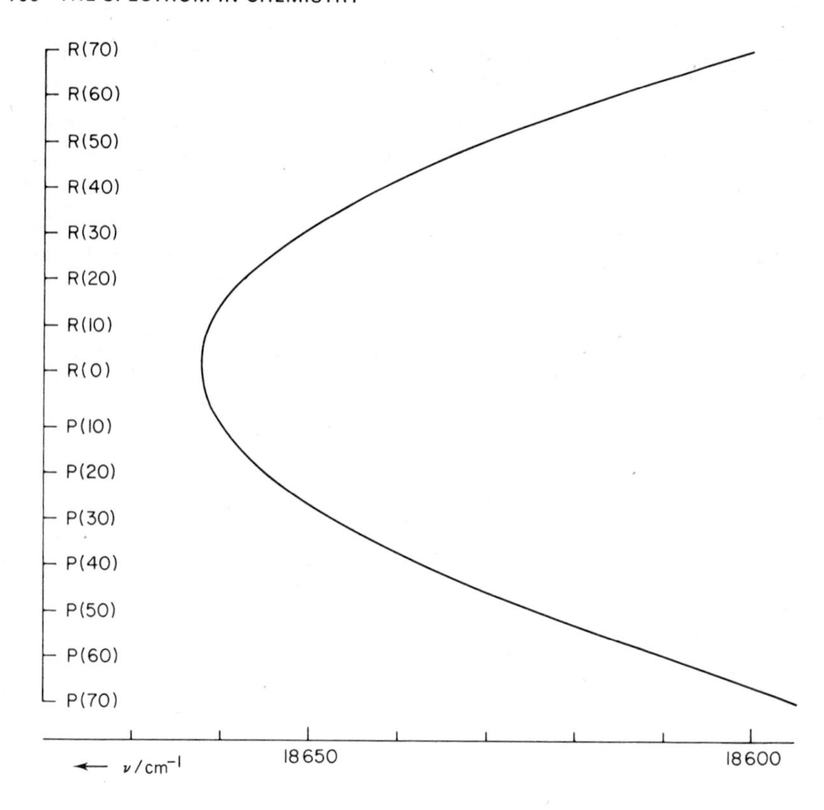

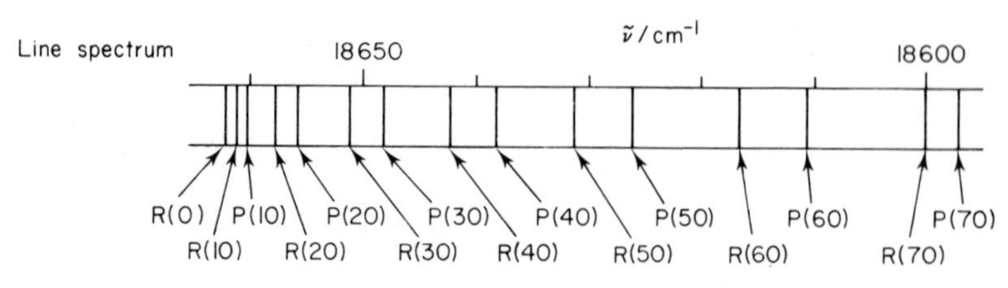

FIG. 6.6 The Fortrat parabola for I_2 (whole band for $v' = 30$).

excitation. This is not necessarily true for vibronic spectra. Electronic excitation causes the strength of the internuclear forces to alter, which may cause the mean internuclear distance to be greater or less than in the ground state. The $^3\Pi_{0u}{}^+$ state of I_2 is much less strongly bound than the $^1\Sigma_g{}^+$ state, so the value of r_e increases from 266.7 to 301.6 pm on excitation, and the value of $\tilde{B}_e$ correspondingly drops from 0.03736 to 0.02920 cm^{-1}. By contrast, the value of r_e for the $^2\Sigma^+$ ground state of CN is 117.8 pm, whereas that for an excited state, also of $^2\Sigma^+$ type, is 115.1 pm. The band head is on the low wavenumber side of the band at 25 751 cm^{-1}, the band extending to 25 830 cm^{-1}. Even if the rotational fine structure of the bands cannot be resolved, the shape of the band gives useful information. If the band head is to the long wavelength (low wavenumber) side, $B_v' < B_0''$, so it can be deduced that $r_e' > r_e''$, as for I_2, whereas if the band head is to the short wavelength (high wavenumber) side, $r_e' < r_e''$, as for CN. The value of J at the band head is found by differentiating Eqn. 4.12.

$$J_{bh} = \frac{3B_v' - B_0''}{2(B_0'' - B_v')} \tag{6.4}$$

If there is a large change in equilibrium internuclear distance on excitation, as for I_2, the band head occurs at low values of J', whereas if the change is small, the band head is at high values of J', as illustrated by CN for which the band head is around $J' = 30$.

For any one molecule, several electronic transitions may be observed, each giving a progression of bands in some region of the visible and ultraviolet spectrum. The spacing of the bands gives a set of values for ω for a range of values of v, and the spacing of the lines in each band gives a set of values of B_v for a range of values of J. A vast amount of information is therefore available. It is possible to study the effect of electronic excitation on bond length and bond stiffness, and the effect of centrifugal stretching and vibrational anharmonicity on bond length. Since homonuclear diatomic molecules do not absorb in the microwave or infrared region, the visible and ultraviolet regions are the only regions of the spectrum from which values of their molecular parameters may be obtained. This technique has had great historical importance for the investigation of other molecules also, since the techniques of microwave and infrared spectroscopy were not developed until after the quartz spectrograph had reached a high degree of perfection. These studies have been extended to small polyatomic molecules, but the wealth of data obtained is such that interpretation is a formidable task. Absorptions due to a vast number of possible transitions overlap to such an extent that the observed spectrum is very complex.

More information is available from emission spectra. These are produced when molecules in the ground state are excited to upper electronic levels by an electrical discharge, and emit radiation as they fall back to the ground state. The bands have a range of values of v'' as well as v', so emission spectra are more complex than absorption spectra. The assignment of vibrational quantum numbers is generelly done by arranging the frequencies in at *Deslandres table*, in which the differences between the frequencies in adjacent columns are approximately constant, and the differences between the frequencies in adjacent rows are approximately constant. An extract from a Deslandres table for I_2 is given in Table 6.2. Each row corresponds to a transition of constant v', and each column to a transition of constant v''. A summary of the work on I_2 is given by Mulliken (1971), and a general review of results obtained from the vacuum ultraviolet is given by Wilkinson (1961).

TABLE 6.2

Deslandres table for I_2 (extract) (Haranath and Tiruvenganna Rao, 1958)

$v'' =$	6	$\delta\tilde{\nu}$	7	$\delta\tilde{\nu}$	8	$\delta\tilde{\nu}$	9
v'							
0			43 709	206	43 503		
$\delta\tilde{\nu}$			91		91		
1			43 800	206	43 594	192	43 402
$\delta\tilde{\nu}$					93		90
2	44 104				43 687	195	43 492
$\delta\tilde{\nu}$	91				91		84
3	44 195	213	43 982	204	43 778	202	43 576
$\delta\tilde{\nu}$	84		88				88
4	44 279	209	44 070				43 664
$\delta\tilde{\nu}$	80		82				83
5	44 359	207	44 152	205	43 947	200	43 747

Wavenumbers of lines in the emission spectrum of I_2. The electronic transition is $^3\Sigma^+_n \leftarrow {}^1\Sigma^+_g$.

6.C.2 Dissociation energies

At high values of v the spacing between the vibrational levels decreases, and the levels converge towards the limit of dissociation. A diatomic molecule dissociates into two free atoms or ions, the vibrational energy of the molecule appearing as translational energy of the atoms or ions. The quantum size for translational energy is inversely proportional to the size of the space in which the translation is confined, and is infinitesimal for vessels of macroscopic size. The translational energy of the free atoms or ions is thus effectively not quantised, and the spectrum shows continuous absorption, with no fine structure. The convergence limit for the dissociation of I_2 can be seen in Fig. 6.5. A linear plot of ΔE_{vib} as a function of v, the *Birge–Sponer extrapolation*, is sometimes used to evaluate the convergence limit if it cannot be seen directly. This is only valid if the potential energy curve has a constant value of $\omega_e'x_e'$, which is so for the Morse potential, as in Eqn. 4.4. As can be seen from Fig. 6.3, the convergence limit $E_{\infty,0}$ gives a value for the dissociation energy of the upper electronic state relative to the $v'' = 0$ level of the ground electronic state. The energy $E_{0,0}$ of the $v' = 0$ level relative to the $v'' = 0$ level is given by the frequency of the 0—0 transition, which may be observed directly if $\tilde{B}'$ is not too dissimilar to $\tilde{B}''$, or obtained by extrapolation. The dissociation energy D_e' of the upper electronic state is thus given by

$$D_e' = E_{\infty,0} - E_{0,0} + \tfrac{1}{2}h\omega' \tag{6.5}$$

The energy $E_{\infty,0}$ is that of the separated I atoms relative to that of the I_2 molecule in the ground vibrational and electronic state. Experiment shows that this is considerably greater than the value of the dissociation energy obtained by thermodynamic techniques. One of the atoms produced by the dissociation must therefore be itself in an excited electronic state. An examination of the energy levels of atomic iodine, obtained from the spectrum at high temperature, shows that this excited state is the $^2P_{\frac{1}{2}}$ state, the ground state being $^2P_{3/2}$. Subtraction of the energy for the $^2P_{\frac{1}{2}}$—$^2P_{3/2}$ transition from $E_{\infty,0}$ gives the dissociation energy of the ground state, to a much higher degree of accuracy than that from thermodynamic methods.

Some electronic transitions lead to excited states which are non-bonding, so dissociation occurs for all excitations, as illustrated in Fig. 6.7(a). The absorption spectrum is then a structureless continuum. Conversely, a continuous emission spectrum is given if the excited state, which corresponds to a bonded molecule, can undergo a transition to a non-bonded ground state, as in Fig. 6.7(b). A familiar example is provided by the H_2 molecule, which is often used as a source of ultraviolet radiation in the 160–350 nm range. The

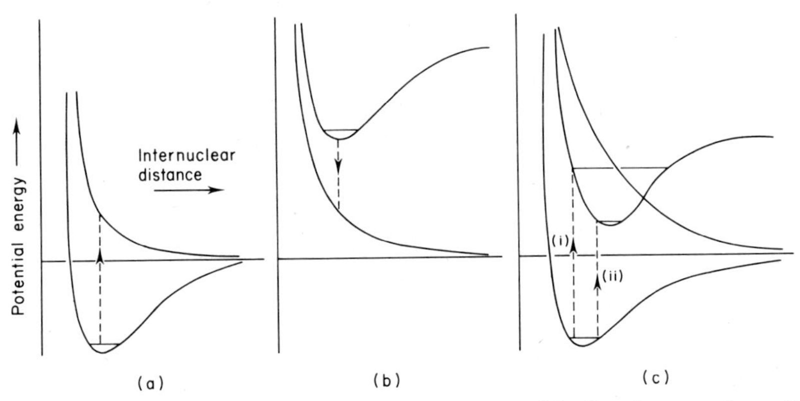

FIG. 6.7 Transitions causing continuous spectra. (a) Continuous absorption spectrum; (b) continuous emission spectrum; (c) predissociation.

upper $^3\Sigma_g^+$ state is a triplet bonded state, which can only undergo a transition to the lower $^3\Sigma_u^+$ state, a non-bonded triplet state corresponding to two separated atoms. If sufficient energy is supplied to He, for example by an electric discharge, some He atoms are excited to a state in which one electron has been raised from the 1s orbital to the 2s orbital. These excited atoms combine with He atoms in the ground state to make stable He_2 molecules in the $^1\Sigma_u^+$ state, which are bound by one 1σ bond and one 2σ bond. The transition from this state to the unstable ground state gives an intense continuous emission in the far ultraviolet, in the 60–100 nm region.

If two excited states exist, one bonding and one non-bonding, such that their potential energy curves cross, as illustrated in Fig. 6.7(c), it is possible for a molecule in the bonding state to cross over to the non-bonding state at the intersection point. Since no energy change occurs, this is a *radiationless transfer*. A molecule can thus dissociate even though it has not sufficient energy to do so in the electronic state to which it has been excited. This is termed *predissociation*. Although the vibrational structure can still be seen, the rotational fine structure is lost. This is because the time for a rotation is two orders of magnitude greater than the time for a vibration, so the molecule does not have time to rotate before dissociation occurs. The selection rules for radiationless transfer are the same as those for an electronic transition leading to absorption or emission of radiation, as discussed in Section 6.A.4. Although many molecules have excited states with potential energy curves of the types illustrated in Fig. 6.7(c), predissociation is comparatively rarely observed because of the stringency of the selection rules. The Franck–Condon principle applies in that the internuclear distance must be unaltered by the radiationless transfer. The transition (i) in Fig. 6.7(c) leads to predissociation at the cross-over point, where the potential energy and internuclear distances are equal for the two states. The kinetic energy of the vibration, represented in Fig. 6.7(c) by the vertical distance between the cross-over point and the upper vibrational level, serves to separate the atoms. The transition (ii) in Fig. 6.7(c) does not lead to predissociation, since the correct internuclear distance is not found at any time during the vibration.

6.C.3 Information from the spectra of organic molecules in solution

As has been discussed in Section 4.D.1, collisions between solvent and solute molecules blur the structure of the spectra of molecules in solution. The rotational fine structure is completely lost, and only some features of the vibrational structure can be seen, to an extent depending on the magnitude of the interaction between solute and solvent. It is usually, but not invariably, found that the more polar the solvent the less clearly the vibrational structure can be seen. For example, the vibrational structure in the spectrum of phenol can be clearly seen in hexane solution, but not in ethanol solution. The information obtained from the ultraviolet and visible spectra of species in solution is nearly always limited to that concerning electronic energy levels.

Molecular orbitals commonly encountered are of two types, either σ, i.e. cylindrically symmetrical about the internuclear axis, or π, i.e. with twin regions of high electron density on either side of the axis and zero electron density on the axis. Molecular orbitals are either of lower energy than the atomic orbitals from which they are constructed, in which case they are bonding, or of higher energy, in which case they are antibonding. Antibonding orbitals are denoted by σ^*, π^*. Two electrons, one of which is in a bonding orbital and one of which is in the corresponding antibonding orbital, constitute a lone pair denoted by n. σ orbitals are generally of lower energy than π orbitals, so excitation of an electron from a σ into a σ^* orbital requires a photon of wavelength less than 200 nm, the practical working limit for a standard spectrophotometer. Alkanes thus do not have absorption spectra in the 200–750 nm region. Absorption does occur in this region by molecules in which various groups termed *chromophores* are substituted into the alkane skeleton. Chromophores contain empty π^* orbitals into which electrons from σ or π orbitals can be excited. Typical chromophores include $C{=}O$, $C{=}C$, $N{=}N$, and NO_2. Other substituents, which are not themselves chromophores, modify the absorption of molecules containing chromophores. These are termed *auxochromes*. Typical examples are CH_3, Cl, NH_2, and OH. Their effect is due to the ability to push electrons in or out of the molecular orbitals in the chromophore, thereby lowering or raising the energy of the empty π^* orbital. Auxochromic effects are sometimes described by the terms:

hyperchromic, i.e. an increase in intensity of absorption;
hypochromic, i.e. a decrease in intensity of absorption;
bathochromic, i.e. a shift to longer wavelength;
hypsochromic, i.e. a shift to shorter wavelength.

It is often found that a hyperchromic effect is associated with a bathochromic shift, being caused by an electron-donating auxochrome such as CH_3, and a hypochromic effect is associated with a hypsochromic shift, being caused by an electron-withdrawing auxochrome such as Cl.

The isolated $C{=}C$ chromophore causes absorption at around 190 nm, due to a $\pi \rightarrow \pi^*$ transition. (Unfortunately, in contrast to the convention for spectra in the gas phase, the excited state is placed second in the description of a transition for a molecule in solution.) This absorption is intense, as $\pi \rightarrow \pi^*$ transitions are allowed, and has values of the extinction coefficient ε, as defined by Eqn. 4.27, at around 10 000 l mol^{-1} cm^{-1}. Conjugation of two or more C=C bonds has a hypochromic and bathochromic effect. This can be explained in terms of the "free electron in a box" model. Solution of the Schrödinger wave equation for an electron constrained to exist in a box of length a shows that the electron can only have energy E_i, where

$$E_i = i^2h^2/8ma^2 \tag{6.6}$$

where m is the mass of the electron and i is a quantum number. A system of conjugated π orbitals may be thought of as a box extending the whole length of the system, along which the π electrons move freely. If there are n electrons in the π orbitals, the energy gap between the lowest unfilled and the highest filled orbital corresponds to absorption of light of frequency ν, where

$$h\nu = [(n+1)^2 - n^2]h^2/8ma^2$$

i.e.

$$\nu = (2n+1)h/8ma^2 \tag{6.7}$$

Since n is proportional to a for a conjugated chain, the longer the chain the lower the value of ν. An extreme example is provided by β-carotene, responsible for the orange colour of carrots.

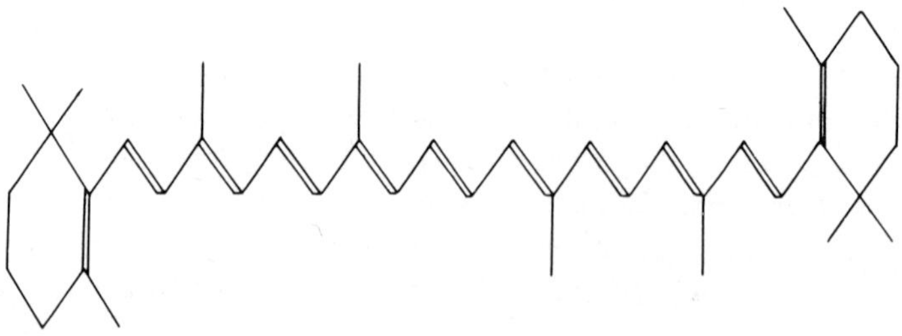

For β-carotene, n has the value 11. An estimate can be made of the value of a using standard values for the lengths of C—C and C=C bonds, and substitution into Eqn. 6.7 gives a value of $\tilde{\nu}$ of 15 000 cm^{-1} corresponding to a wavelength of 667 nm. The observed spectrum of β-carotene shows an intense absorption in the 400–500 nm region, with a maximum at 453 nm in hexane solution, corresponding to an extinction coefficient of 134 000 l mol^{-1} cm^{-1} (Isler *et al.*, 1956). The rough calculation based on Eqn. 6.7 thus gives a reasonably good value, although it does predict that carrots are blue in colour! This discrepancy perhaps reflects more on the sensitivity of the human eye to wavelength in the visible region than on the failure of the electron-in-a-box model. The agreement between observed and calculated wavelengths for maximum absorption is especially good for a range of cyanine dyes, in which two nitrogen atoms are linked by a conjugated carbon chain (Brooker, 1947).

Steric repulsion between substituents on a conjugated system may twist the π orbitals out of alignment, with the result that overlap is not so efficient and conjugation is not so effective. This is probably the cause of the discrepancy between the observed and calculated values for the absorption peak of β-carotene. *cis*-Dienes absorb at shorter wavelengths than their *trans* isomers since steric repulsion is more pronounced, so the ultraviolet spectrum gives information about the stereochemistry of a molecule.

Aromatic compounds form an important class of conjugated molecules. The conjugation of two or more benzenoid rings by N=N or C=C links, with bathochromic auxochromes substituent on the rings, gives molecules for which ν is low enough for absorption to occur in the visible part of the spectrum. Dyes and indicators are important examples of such compounds. For example, phenolphthalein in the acidic form does not absorb in the visible region, but proton loss enables a conjugated system to be set up, and in alkaline solution phenolphthalein gives an intense red colour.

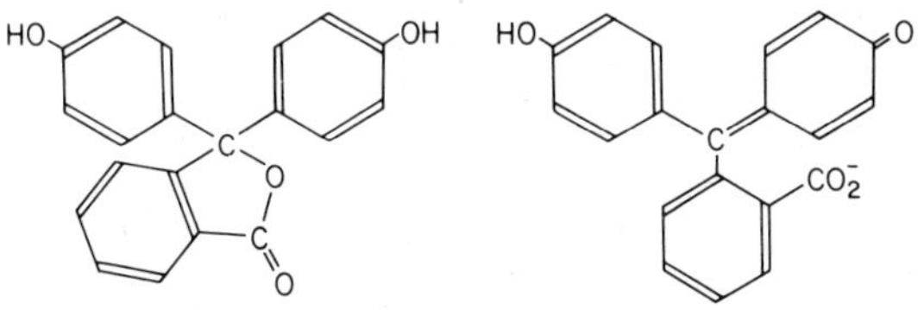

In acidic solution

In alkaline solution

Two types of excitation of an electron in the lone pair on oxygen in carbonyl compounds are observed. The n $\rightarrow \sigma^*$ transition, which is permitted, occurs in the 170–190 nm region, at the limits of the vacuum ultraviolet, and so is not easily seen, despite the high values of ε, of the order of 10 000 l mol^{-1} cm^{-1}. The n $\rightarrow \pi^*$ transition, which gives absorption around 270 nm, is a more familiar feature of the spectrum of carbonyl compounds, but since it is forbidden, values of ε are low, of the order of 20 l mol^{-1} cm^{-1}.

Solvent effects are quite different for $\pi \rightarrow \pi^*$ and n $\rightarrow \pi^*$ transitions. Increasing polarity of the solvent produces a hyperchromic bathochromic effect on $\pi \rightarrow \pi^*$ transitions, but a hypochromic hypsochromic effect on n $\rightarrow \pi^*$ transitions. The effect on n $\rightarrow \pi^*$ transitions can be readily explained. In polar solvents the lone pair is attracted to the positive ends of the solvent molecule dipoles, so more energy is required to remove one of the electrons.

The effect of the solvent is more noticeable for polar solutes than for non-polar solutes, or for polar solutes in which the chromophore is isolated from the polar group. For example, λ_{max}, the wavelength of maximum absorption, for the non-polar solute flavoxanthin is only changed from 448.5 to 446.5 nm when the solvent is changed from hexane to methanol. By contrast, λ_{max} for the $n \rightarrow \pi^*$ transition in acetone is changed from 279 to 267 nm for the same change in solvent. The spectrum of 1-ethyl-4-carbomethoxypyridinium iodide is especially sensitive to solvent, λ_{max} ranging from 342 nm for methanol solvent to 438 nm for methylene chloride solvent. The magnitude of the shift has been proposed as a measure of solvent polarity, the *Z*-value (Kosower, 1958).

6.C.4 Information from the spectra of transition-metal complexes

Absorption by transition-metal ions in the visible and ultraviolet region is due to excitation of an electron in one of the d orbitals. The distribution of electron density for the various d orbitals is sketched in Fig. 6.8. In the free ion the five d orbitals are degenerate, but in complex ions the presence of the ligands tends to remove the degeneracy. For instance, the ligands in octahedral complexes, as shown in Fig. 6.8, are oriented so as to be closer to the $d_{x^2-y^2}$ and d_{z^2} orbitals, which are classified as the e_g group, than to the d_{xy}, d_{yz}, and

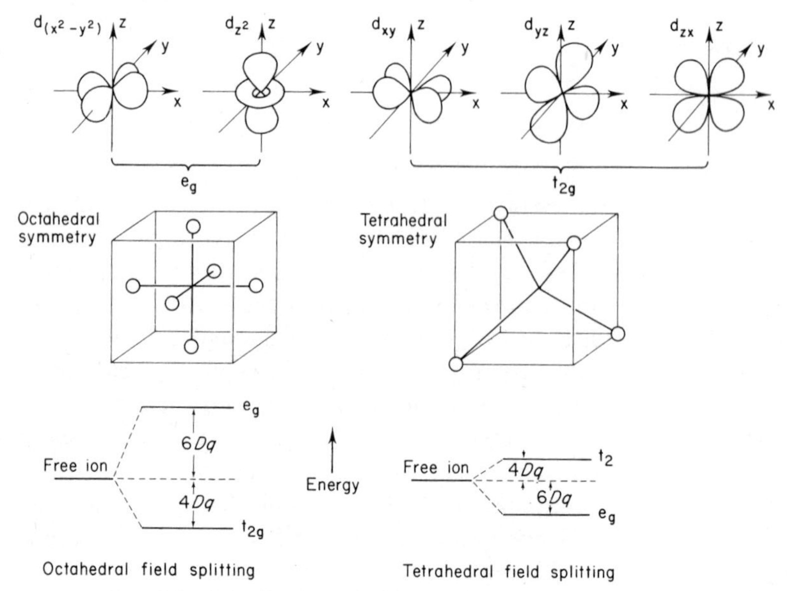

FIG. 6.8 Distribution of electron density in d orbitals.

d_{zx} orbitals, which are classified as the t_{2g} group. Thus the energy of the doubly degenerate e_g group is raised and the energy of the triply degenerate t_{2g} group is lowered. The energy gap so produced is denoted by Δ or $10Dq$. Because of the principle of conservation of energy, the e_g group is $(3/5)\Delta$, or $6Dq$ units, above, and the t_{zy} group is $(2/5)\Delta$, or $4Dq$ units, below the original 5-degenerate level. In tetrahedral complexes neither the e nor the t_2 group of orbitals (the g subscript is dropped because a tetrahedron does not have a centre of symmetry) points directly at a ligand, but the t_2 orbitals point at a narrower angle to the ligands than do the e orbitals. The t_2 orbitals are thus raised in energy, and the e orbitals lowered, but the magnitude of the splitting is less than in octahedral complexes. A calculation based on a simple electrostatic model predicts that Δ for tetrahedral complexes is 4/9 that for octahedral complexes observed.

The *Jahn–Teller effect* is a further source of complexity in the electronic energy levels of a complex ion. If the ground state is degenerate, the complex may distort until it has a configuration of lower symmetry, and hence lower degeneracy and lower energy. Consider, for example, a d^4 ion in an octahedral complex. Three electrons are in the triply degenerate t_{2g} level and one is in the doubly degenerate e_g level. If two ligands opposite (*trans*) to each other move outwards and the other four ligands which are coplanar move in towards the metal ion, the degeneracies of t_{2g} and e_g levels are lifted. Two t_{2g} levels are further lowered, and one is raised in such a way that the total energy of the three electrons in the t_{2g} levels is unaltered. However, one e_g level is raised and one lowered, so the energy of the electron in the e_g levels is lowered, and the complex ion as a whole is stabilised by the loss of symmetry.

According to Hund's rules, as discussed in Section 6.A.2, the state of maximum multiplicity is the state of lowest energy. There is thus a tendency for electrons to distribute themselves as widely as possible in the d orbitals of a transition-metal ion. Pairing of electrons is an unfavoured process, and energy must be expended to make it happen. This energy may be supplied by the stabilisation of t_{2g} orbitals with respect to e_g orbitals in octahedral complexes, or vice versa for tetrahedral complexes. For example, a d^4 ion can have configuration either $(t_{2g})^3(e_g)^1$ or $(t_{2g})^4(e_g)^0$ depending on whether the energy gap Δ is greater or less than the pairing energy. The magnitude of Δ is affected by the nature of the ligands, so the same d^4 ion may have either configuration depending on its ligands. Complex ions in which the pairing energy outweighs Δ, so that the d electrons have maximum multiplicity, are termed *high-spin*, and those in which the d electrons are forced to pair up by a high value of Δ are termed *low-spin*. In general, the more strongly bound the ligands, the more probable is a complex to be low-spin. For example, $Fe(H_2O)_6{}^{2+}$ is high-spin and $Fe(CN)_6{}^{4-}$ is low-spin. The distinction between high-spin and low-spin complexes is only valid for d^4, d^5, d^6, and d^7 octahedral complexes, and for d^3, d^4, d^5, and d^6 tetrahedral complexes.

A transition-metal ion with only one d electron and no ligands to remove the degeneracy of the d orbitals has an electronic configuration denoted by ^2D, since $L = 2$ and $S = \frac{1}{2}$. In an octahedral complex of this ion the d orbitals split, so the electron may be in a t_{2g} or an e_g orbital, giving configurations $(t_{2g})^1$ or $(e_g)^1$. These correspond to energy levels denoted by $^2T_{2g}$ and ^2E_g respectively, so an electronic transition $^2E_g \leftarrow {}^2T_g$ corresponds to energy absorption by the ^2T_g ground state. This transition is spectroscopically forbidden, since the selection rule $\Delta l = \pm 1$, sometimes known as the *Laporte selection rule*, means that d—d transitions are forbidden. "Uncomplexed" transition metal ions in aqueous solution, in which they occur as the aquo-complexes, thus absorb electromagnetic radiation extremely weakly, having extinction coefficients of the order of $10\ l\ mol^{-1}\ cm^{-1}$ or less. The magnitude of Δ is such that absorption occurs in the visible or near ultraviolet region of the spectrum. If the ion is complexed in such a way that the resulting complex lacks a centre of symmetry, the selection rule is relaxed since the d orbitals themselves lose their symmetry. Even if the complex ion does have a centre of symmetry, metal–ligand vibrations distort the complex so that it temporarily lacks symmetry, and again the selection rule is relaxed. The d orbitals may also lose their symmetry as the result of covalent bonding of π type between the ligand and the ion. The greater the covalent bonding, the greater the intensity of absorption. A familiar example is provided by Fe^{3+} which as the aquo-complex $Fe(H_2O)^{3+}$ is feebly coloured ($\varepsilon_{max} = 9 \times 10^{-2}\ l\ mol^{-1}\ cm^{-1}$, $\lambda_{max} = 550$ nm) but is an intense blood-red in the form of $Fe(SCN)_3$ ($\varepsilon_{max} = 9 \times 10^3\ l\ mol^{-1}\ cm^{-1}$, $\lambda_{max} = 500$ nm). The other important selection rule is that $\Delta S = 0$, i.e. singlet–triplet transitions, which involve a change in multiplicity, are forbidden. The least intense absorptions, e.g. that of $Mn(H_2O)_6{}^{2+}$, for which $\varepsilon_{max} = 10^{-2}\ l\ mol^{-1}\ cm^{-1}$, are those forbidden by both the spin rule and the Laporte rule.

The electronic absorption spectrum of a complex ion can be used to evaluate Δ. It is found that Δ increases with the oxidation number of the ion, being, for the first transition series, in the range 7500–12 500 cm^{-1} for divalent ions and in the range 14 000–21 000 cm^{-1} for trivalent ions. Values of Δ for corresponding complexes of ions in the second transition series are some 30% greater than for those in the first series, and the corresponding values for ions in the third transition series are some 30% greater still. The effect of the ligand on the value of Δ follows the *spectrochemical series*:

$$I^- < Br^- < Cl^- < F^- < OH^- < H_2O < NCS^- < \text{pyridine} < NH_3 < \text{ethylenediamine} < o\text{-phenanthroline} < NO_2{}^- < CN^-$$

The higher the position of the ligand in the series, the greater the value of Δ. Some values of Δ illustrative of the spectrochemical series are given in Table 6.3.

A different type of transition is sometimes observed, which causes intense absorption in the visible or ultraviolet region. This is the excitation of an electron from an orbital on the ligand to an orbital localised on the metal ion (or vice versa). Since charge has been transferred from one atom to another, these spectra are termed *charge-transfer* spectra. The selection rules do not forbid this type of transition, so absorption is intense, and values of the extinction coefficient in the range 10^2–10^4 l mol^{-1} cm^{-1} are found. The absorption frequency depends on the relative energies of the electrons on the two sites, and this is comparatively low if the ligand is easily oxidisable and the metal easily reducible, or vice versa. The intense colorations of MnO_4^-, CrO_4^{2-}, and MnO_4^{2-}, for example, are due to charge-transfer processes. Ions with a full d shell, which might be expected to have colourless complexes since d—d transitions are impossible, sometimes are found to have coloured complexes as a consequence of charge-transfer processes. For example, Hg^{2+} is easily reduced, and I^- is easily oxidised, so HgI_2 has an intense yellow colour.

TABLE 6.3

Values of Δ illustrative of the spectrochemical series (Lever, 1968)

Octahedral V^{3+} complexes (d^2)	Δ/cm^{-1}
VCl_6^{3-}	12 000
VF_6^{3-}	16 100
$V(H_2O)_6^{3-}$	19 000
$V(CN)_6^{3-}$	23 850
Octahedral Co^{2+} complexes (d^2)	**Δ/cm^{-1}**
$CoBr_2$	6490
$CoCl_2$	7640
$Co(H_2O)_6^{2+}$	9200
$Co(NH_3)_6^{2+}$	10 200
Octahedral Ni^{2+} complexes (d^8)	**Δ/cm^{-1}**
$NiBr_2$	6800
$Ni(H_2O)_6^{2+}$	8500
$Ni(NH_3)_6^{2+}$	10 800
$Ni(en)_3^{2+}$	11 500
$Ni(bipy)_3^{2+}$	12 650

en = ethylenediamine; bipy = bipyridyl.

6.D Analytical Applications

6.D.1 Atomic emission spectroscopy

Every student has at some time tested for the presence of sodium ions by dipping a platinum wire into the solution to be analysed, then holding the wire in a Bunsen flame. The characteristic yellow coloration of the flame, due to the $^2P_{3/2,1/2} \rightarrow {}^2S_{1/2}$ transition shows the presence of sodium ions in the solution. This experiment illustrates the principle of an old-established technique of qualitative and quantitative analysis, atomic emission spectroscopy. The substance to be analysed is supplied with sufficient energy to cause the constituent atoms to be raised to upper electronic states, and the radiation emitted as they fall to the ground state is monitored. The energy can be supplied in two ways. If the substance to be analysed is in solution, the solution can be sprayed as a fine mist into a flame burning under carefully controlled conditions. If the substance to be analysed is an insoluble solid, it can be placed in a little cup forming part of a carbon electrode. An electric discharge is then passed between this electrode and a pointed carbon electrode above it, and the arc spectrum contains lines due to emission from atoms in the sample. If metallic samples are to be analysed, an electrode made of the sample material itself can be used.

Emission spectroscopy has been used in the analysis for nearly all the elements in the Periodic Table, the exceptions being the rare gases, the halogens, oxygen, and sulphur. The high resolution obtainable in the visible and ultraviolet, and the sharpness of the lines due to atomic transitions, mean that many elements may be identified in a single emission spectrum. Both photographic and photoelectric detection have been widely used. The technique has been very popular for the analysis of metallurgical and mineralogical specimens. The sensitivity depends on the particular element in question. The alkali metals give the strongest signals, and can be detected at levels down to 0.5 part per million (p.p.m.); other metals can be detected in the 1–100 p.p.m. range.

Atomic emission spectroscopy is also used for quantitative analysis, the intensity of the lines giving a measure of the concentration of the element. The apparatus can be standardised by measurement of line intensities for samples of known concentrations of the same order of magnitude as the unknown. A more accurate technique is to add a known concentration of another element to the sample, an *internal standard*, and compare the intensities of standard and sample lines on the same spectrum. Under the most favourable conditions, using standards of the same type and having the same history as the sample, it is possible to achieve an accuracy of a few per cent.

However, for arc excitation, the line intensity is very dependent on the nature of the sample as a whole, termed the *matrix*. Elements in different matrices give line intensities varying by an order of magnitude even though excited in a standard manner.

6.D.2 Atomic absorption spectroscopy

Atomic emission spectroscopy suffers from two disadvantages as a technique for qualitative analysis. The actual quantity measured is the number of atoms in the excited state, which, because of the large energy gap corresponding to transitions in the visible and ultraviolet, is only a small fraction of the number in the ground state, as shown by Eqn. 2.12. Furthermore, since this number depends in an exponential manner on the temperature, a small change in the operating conditions of the arc or the flame produces a large change in the number of excited atoms. For example, the fraction of excited sodium atoms at 2750 K is 1.4×10^{-4}, whereas the fraction at 3000 K is 1.8×10^{-4}. Even under steady conditions, the emission intensity from an arc is not proportional to the concentration of the element over a wide concentration range. At higher concentrations the radiation emitted by atoms in the central parts of the arc is absorbed by atoms in their ground state in the cooler outer regions. In extreme cases this *self-absorption* causes the centre of an emission peak to be less intense than the shoulders.

These problems are solved by the measurement of the intensity of radiation absorbed by the atoms in the ground state. The concentration of atoms in the ground state is, to a very good approximation, independent of temperature. The free atoms are produced by spraying a fine mist of a solution of the sample into a flame. The only limitation on the flame conditions is that the flame should be hot enough to decompose the molecules of the sample into free atoms. Although a conventional oxygen–methane flame is adequate for most samples, unconventional flames, e.g. nitrous oxide–acetylene, are necessary for the detection of some elements, such as aluminium and silicon. These elements present a special problem owing to the high thermal stability of their oxides.

The original obstacle to the use of atomic absorption spectroscopy was based on the limitation of Beer's Law to monochromatic radiation, as discussed in Section 4.E.2. For substances in solution, the absorption peaks are so wide that the extinction coefficient is essentially constant over the bandwidth of a simple monochromator, say 0.5 nm. However, the linewidths of atomic absorption spectra are extremely narrow. There are several causes for the finite width of spectral lines. The most fundamental is the Heisenberg Uncertainty Principle, by which $\Delta E \times \Delta t$ is of the order of $\hbar$, Δt being the lifetime of the excited state and ΔE being the uncertainty in the energy of the excited state. This uncertainty gives atomic spectral lines a *natural width* of about 10^{-5} nm. Two factors that much increase the width are pressure broadening, as discussed in Section 2.A.4, and the *Doppler effect*. The Doppler effect is the change in frequency of a radiation from a source caused by the relative motion of source and observer. An atom moving towards the detector seems to be emitting radiation of a higher frequency than a stationary atom, and vice versa, so the translational motion of the atoms in the flame, due to thermal agitation, causes a broadening of the spectral line. The Doppler width $\Delta\lambda$ of a line of wavelength λ is given by

$$\Delta\lambda = 1.67(\lambda/c)(RT/M)^{\frac{1}{2}} \tag{6.8}$$

where R is the gas constant, T the absolute temperature, M the atomic weight, and c the velocity of light. Under the conditions typically used for atomic absorption spectroscopy, $\Delta\lambda$ is of the order of 10^{-3} nm. At atmospheric pressure, pressure broadening causes a linewidth of up to 10^{-2} nm. Thus the use of atomic absorption for quantitative analysis requires resolution of the order of 10^{-3} nm, which is well beyond the power of a general-purpose monochromator.

The problem is solved by the use of a light source that emits radiation at exactly the right wavelength over an extremely narrow spectral range. This light source is the *hollow-cathode lamp*. The hollow-cathode lamp consists of a sealed tube containing argon or neon and a pair of electrodes, the cathode being in the form of a cup, the inside of which is lined with the element to be determined. When a voltage is applied across the electrodes, the gas is ionised, and the positive ions so formed bombard the cathode, causing the emission of atoms of the element of interest. These atoms are in an excited state, and emit radiation of the required wavelength as they fall to the ground state. The gas pressure is in the range 200–800 Pa, i.e. of the order of one-thousandth of atmospheric pressure, so the emission lines are not subject to appreciable pressure broadening, and have a width of the order of 10^{-3} nm, narrower than the corresponding absorption lines for the atoms of sample in the flame. Beer's Law is thus applicable. The only drawback to the use of the hollow-cathode lamp is that a different lamp is required for the analysis for each element. The total number of lamps required may be reduced by the use of lamps with two or three elements on the cathode.

Atomic absorption spectroscopy can be used to detect elements and measure concentrations down to very low levels. The *relative detection limit* is defined as the concentration in $\mu g\ cm^{-3}$ in aqueous solution that will yield a signal equal to twice the noise in a signal from a blank solution. This limit depends on the element. A typical modern atomic absorption spectrometer has relative detection limits ranging from 7×10^{-4} $\mu g\ cm^{-3}$ for sodium to 0.1 $\mu g\ cm^{-3}$ for titanium, most metals having a limit of the order of 10^{-2} $\mu g\ cm^{-3}$.

6.D.3 Atomic fluorescence spectroscopy

The radiation absorbed by an atom in an atomic absorption spectroscopy experiment is re-emitted in all directions as the atom falls to its ground state. Usually the transition resulting in emission is between the same pair of levels as that resulting in absorption. The atom is then said to exhibit *resonance fluorescence*. The concentration of atoms in the flame can thus be measured by measuring the intensity of radiation emitted perpendicular to the incident radiation. This technique is not so widely used as atomic absorption spectroscopy, to which it is very closely allied. It does however have the advantage in principle that zero concentration corresponds to total darkness, whereas zero concentration in atomic absorption spectroscopy corresponds to 100% transmission of the incident light. The noise level at low concentrations is thus much less for atomic fluorescence spectroscopy. A simple numerical example may clarify this point. Consider an optical system in which 100% transmission causes a signal of 100 μA from the detector, with an r.m.s. noise current of 0.1 μA. A certain concentration of sample which causes 1% absorption can be measured to $\pm 10\%$ by measurement of the fraction of radiation absorbed. Suppose the same concentration of sample caused the emission of 1% of the incident light, giving a signal of 1 μA in the detector set to measure the emission. The r.m.s. noise current is proportional to the square-root of the total current, by Eqn. 1.3, so the noise is now 0.01 μA, and the sample concentration can be measured to $\pm 1\%$ by emission spectroscopy.

6.D.4 Qualitative analysis of organic compounds

By contrast with infrared spectroscopy, spectroscopy in the visible and ultraviolet is of much more use for quantitative than for qualitative analysis. Absorption bands for liquid samples are too broad and featureless to serve as "fingerprints" for the identification of unknown materials. The spectrum tends to be that of the molecule as a whole, rather than the sum of that of the individual functional groups, so the method of group assignment commonly used in infrared spectroscopy cannot usually be applied. However, the visible or ultraviolet spectrum can have occasional use in confirming a structure already deduced by other means, or in deciding between two possible structures if one has a chromophore or conjugated system and the other has not. For example, the lack of the characteristic carbonyl absorption at 270 nm for trichloroacetaldehyde in aqueous solution shows that the molecule is hydrated to form $CCl_3CH(OH)_2$. Some empirical rules have been devised to predict λ_{max} for various classes of unsaturated compounds. For example, the Woodward–Fieser rules predict λ_{max} for conjugated dienes in ethanol solution. The parent diene has λ_{max} of 217 nm, on condition that the two double bonds are not both in the same ring. Alkyl groups and double bonds not in the ring each add 5 nm to λ_{max}, and chlorine and bromine substituents each add a further 17 nm. A value of λ_{max} can thus be calculated for a suggested structure and compared with the experimental value to see if this structure is correct.

6.D.5 Quantitative analysis

Spectrophotometry in the visible and ultraviolet is a very popular method of quantitative analysis, widely used in both organic and inorganic chemistry.

A notable early example in organic chemistry occurred in the characterisation of vitamin A. A correlation was found between the biological activity of crude extracts from fish oils and their absorption at 325 nm. The replacement of bio-assay by spectrophotometric analysis was of great help in the isolation of the pure vitamin.

If the species to be analysed has little or no absorption in the accessible parts of the visible and ultraviolet spectrum, it is often possible to convert it into a species of high extinction coefficient by a simple chemical reaction. Techniques for the analysis of a wide range of materials in this way have been worked out. An important class of these methods is the analysis of transition-metal ions as their highly coloured complex ions. For example, the addition of 1,10-phenanthroline to a solution of Fe^{2+} causes the formation of a complex ion having maximum extinction coefficient 10 000 l mol^{-1} cm^{-1} for the charge-transfer band at 510 nm.

The stability and linear response of the photomultiplier detector enable results of high accuracy to be obtained, $\pm 0.2\%$ in favourable cases. The sensitivity depends on the extinction coefficient. If a value of 10 000 l mol^{-1} cm^{-1} is taken as typical, the relative detection limit is of the order of 10^{-6} mol l^{-1}.

REFERENCES

Brooker, L. G. S. (1947) *Chem. Rev.* **41**, 325.

Haranath, P. B. V. and Tiruvenganna Rao, P. (1958) *J. Mol. Spectroscopy* **2**, 428.

Hyde, K. E. (1975) *J. Chem. Educ.* **52**, 87.

Isler, O., Lindlar, H., Montavon, M., Rüegg, R. and Zeller, P. (1956) *Helv. Chim. Acta* **39**, 249.

Kosower, E. M. (1958) *J. Amer. Chem. Soc.* **80**, 3253.

Mulliken, R. S. (1971) *J. Chem. Phys.* **55**, 288.
Porter, G. (1968) *Science* **160**, 1299.
Wilkinson, P. G. (1961) *J. Mol. Spectroscopy* **6**, 1.

BIBLIOGRAPHY

Ahrens, L. H. and Taylor, S. R. (1961) "Spectrochemical Analysis", Pergamon, Oxford.

Grun, F. (1972) in "Techniques of Chemistry, Vol. I, Physical Methods of Chemistry, Part IIIB, Optical, Spectroscopic and Radioactivity Methods" (A. Weissberger and B. W. Rossiter, eds.), pp. 207–428, Wiley-Interscience, New York and London.

Herzberg, G. (1950) "Molecular Spectra and Molecular Structure, Vol. I, Spectra of Diatomic Molecules", Van Nostrand, Reinhold, New York.

Herzberg, G. (1966) "Molecular Spectra and Molecular Structure, Vol. III, Electronic Spectra and Electronic Structure of Polyatomic Molecules", Van Nostrand, Reinhold, New York.

Kirkbright, G. F. and Sargent, M. (1974) "Atomic Absorption and Fluorescence Spectroscopy", Academic Press, London and New York.

Lever, A. B. P. (1968) "Inorganic Electronic Spectroscopy", Elsevier, Amsterdam.

Maass, D. H. (1974) in "An Introduction to Spectroscopic Methods for the Identification of Organic Compounds", Vol. 2 (F. Scheinmann, ed.), pp. 93–140, Pergamon, Oxford.

Rao, C. N. R. (1974) "Ultraviolet and Visible Spectroscopy; Chemical Applications", Butterworths, London.

Stern, E. S. and Timmons, C. J. (1970) "Electronic Absorption Spectroscopy in Organic Chemistry", Edward Arnold, London.

7 Fluorescence and Phosphorescence

7.A The Fate of Electronic Excitation Energy

A coloured solution is capable of continuously absorbing energy from electromagnetic radiation in the visible, so there must be one or more processes that return the electronically excited molecules to the ground state so as to be available for re-excitation (Jaffé and Miller, 1966). Some of these processes are shown in schematic form in Fig. 7.1. Excitation from the ground singlet state is to an upper singlet state, since the selection rule forbids change in spin multiplicity. As discussed in Section 6.A.4, excitation to many different vibrational levels of the upper singlet state is possible, subject only to the constraints of the Franck–Condon principle. The excited molecules lose vibrational energy by collisions with the solvent molecules, or, if the molecules are in a solid matrix, by exchange with the vibrational modes of the crystal. This *vibrational relaxation* is a rapid process, being complete in about 10^{-12} s. A molecule in the ground vibrational level of the upper singlet state can lose energy by several mechanisms. It can collide with another molecule, to which its energy is transferred. This process is termed *external quenching*. It can undergo a radiationless transfer, that is a process involving little or no energy change, so energy is neither gained nor lost by exchange with electromagnetic radiation, to the ground singlet state in a highly excited vibrational level. This can then lose vibrational energy to reach the ground state, the process being termed *internal quenching*. The molecules in the upper singlet state can also return to the ground singlet state by emission of radiation. This process is termed fluorescence. The transition may be to any one of a number of vibrational levels of the ground singlet state, subject of course to the constraints of the Franck–Condon principle. The emitted radiation thus covers a wide range of frequencies. The transition of lowest energy in the absorption spectrum is the same as the transition of highest energy in the emission spectrum, being the transition between the ground vibrational levels of the two electronic states. This transition is not always observed, since it may be forbidden by the Franck–Condon principle. Collisional broadening means that the transitions do not give sharp lines, but broad bands with irregular contours, the peaks showing the location of the vibrational levels. If the vibrational levels are similarly spaced in the lower and upper singlet states, it can be seen from Fig. 7.1 that the emission spectrum will be the mirror image of the absorption spectrum, the 0—0

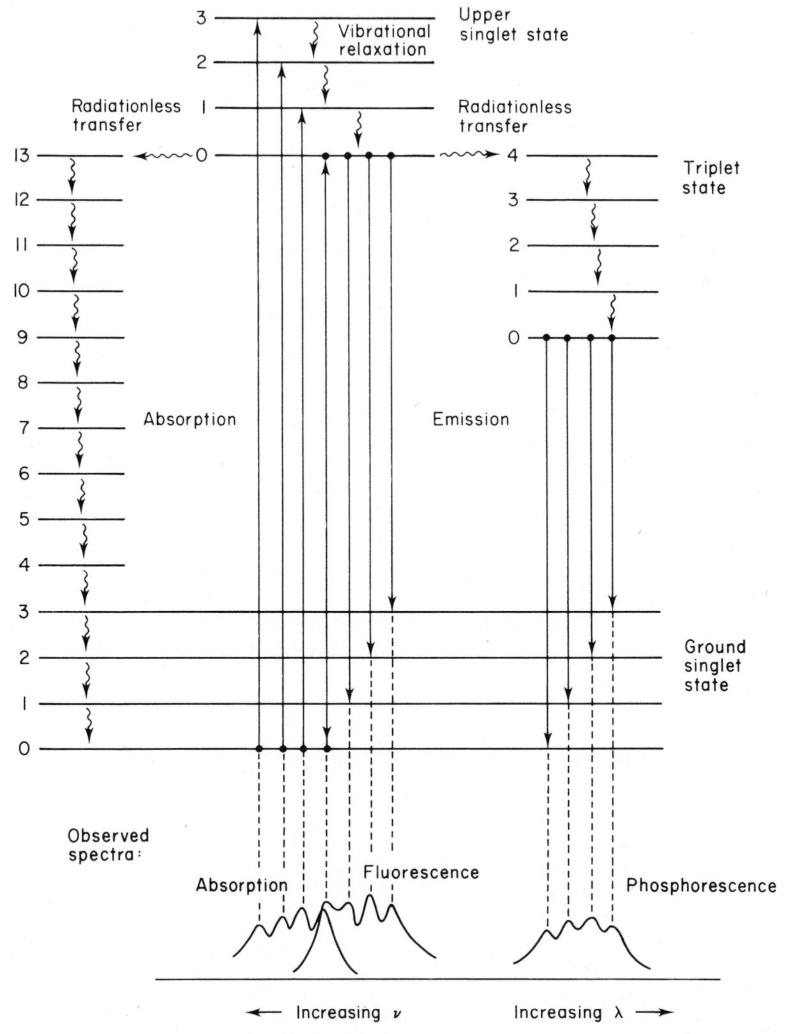

FIG. 7.1 Electronic and vibrational energy levels. Transitions and corresponding spectra for absorption, fluorescence, and phosphorescence.

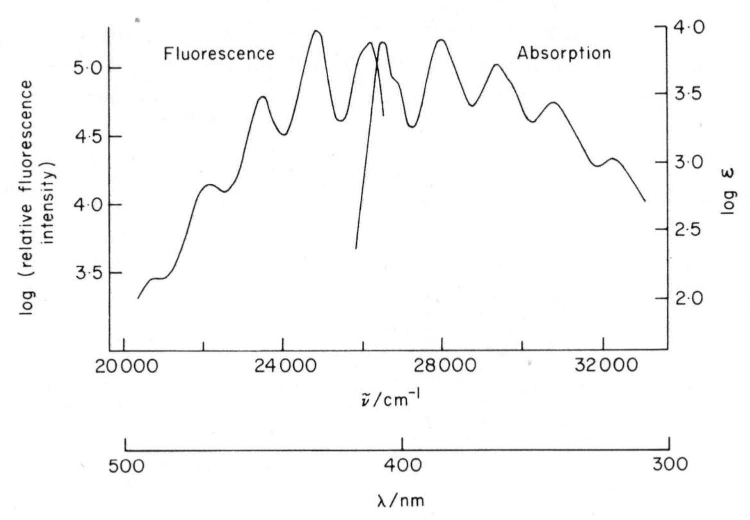

FIG. 7.2 Absorption and fluorescence spectra of anthracene (Kortum and Finckh, 1942).

transition giving the plane of symmetry. An example, the absorption and emission spectra of anthracene, is given in Fig. 7.2.

Only a limited number of species exhibit fluorescence, since, for most species, internal quenching is a more rapid process. The lifetime of the upper singlet state of most fluorescent molecules is in the range 1–100 ns. Fluorescent molecules are typically large and flat with extensive π conjugation, e.g. derivatives of polycyclic aromatic hydrocarbons. Internal rotation seems to facilitate internal quenching. Phenolphthalein, for example, is not fluorescent, whereas fluorescein, which closely resembles phenolphthalein except in that two benzene rings are prevented from internal rotation by a single bond, is fluorescent, as its name implies.

HO OH HO O OH

C O C O C O C O

Phenolphthalein Fluorescein

Another possible fate of the upper singlet state is a radiationless transfer to a triplet state in a high vibrational level. Since this *intersystem crossing* is a forbidden transition, owing to the change in spin multiplicity, usually only a small fraction of the molecules arrive in the triplet state. These undergo rapid vibrational relaxation down to the ground vibrational level of the triplet state. Since the transition down to the ground singlet state is forbidden, owing to the change in spin multiplicity, the triplet state has a potentially much greater lifetime than the excited singlet state. In solution, the triplet state is likely to be quenched by collision with solvent molecules or dissolved oxygen, but in a rigid matrix it can survive for comparatively long periods, up to several seconds, losing energy only by the emission of radiation. This process is termed phosphorescence. As can be seen from Fig. 7.1, phosphorescence emission occurs at lower frequencies (longer wavelengths) than fluorescence, which in turn occurs at lower frequencies than the absorption of radiation by that species. Phosphorescence is favoured if the selection rule is relaxed by the presence of a paramagnetic centre in the molecule or if the upper singlet is of comparatively long lifetime. Fluorescence generally occurs if the lowest excited singlet state is of π,π^* type, i.e. the excitation transition is $\pi^* \leftarrow \pi$. For species, e.g. carbonyl compounds, in which the lowest excited singlet is of n,π^* type, intersystem crossing to the triplet state is favoured, and phosphorescence is observed rather than fluorescence.

7.B Experimental Techniques

7.B.1 Spectrofluorimetry

Since fluorescence occurs in the visible and ultraviolet, the components of the apparatus required for its observation are those described in Section 6.B, and the difference in instrumentation lies in the way in which those components are arranged.

Since only a minute proportion of the radiant energy emitted by the source finds its way to the detector in fluorescence spectroscopy, the most intense sources and most sensitive detectors available are used. The excitation source is, except in the least expensive instruments, the xenon arc, since this is the most intense source of radiation in the 200–800 nm region. In the more elaborate instruments, called *spectrofluorimeters*, radiation in a narrow wavelength range is selected by an excitation monochromator to impinge on the sample. Cheaper instruments have a simple broad-band optical filter for this purpose. If the sample is a transparent liquid it is contained in a quartz cuvette with windows at an angle of 90° to each other, so that the radiation emitted at 90° to the incident can be observed. If the sample is an opaque solid, it is mounted so that its front face is at such an angle to the incident radiation that the radiation emitted at 90° to the incident can be observed.

The emitted radiation is focused on the entrance slit of a second monochromator, termed the emission monochromator, at the exit slit of which is the detector, almost universally a photomultiplier.

A spectrofluorimeter can be used in either of two modes. If the excitation wavelength is fixed, scanning the emission wavelengths gives the emission spectrum. If, on the other hand, the emission wavelength is fixed, scanning the excitation wavelengths gives the absorption spectrum, since the intensity of emitted radiation at a fixed wavelength is proportional to the amount of incident light absorbed by the sample. The absorption spectrum thus obtained however will not be exactly comparable to that obtained from a standard double-beam spectrophotometer. Most spectrofluorimeters are essentially single-beam devices. The signal amplitude at some given wavelength is a function of detector sensitivity at that wavelength and of source intensity at the excitation wavelength. Thus both emission and excitation spectra are distorted, the distortion being greatest towards 200 nm where both source intensity and detector sensitivity decrease rapidly with decreasing wavelength. Spectra can be corrected by multiplying signal intensity at each wavelength by some factor determined for the particular instrument used. Automatically corrected spectra can be obtained by use of a *quantum converter*. A fraction of the radiation incident on the sample is used to excite fluorescence in a reference substance whose quantum yield is known to be unity over the whole wavelength range of interest. The intensity of the reference fluorescence is monitored by a second photomultiplier, whose spectral response and sensitivity is matched with that of the photomultiplier monitoring the sample fluorescence. The ratio of the signals from the two photomultipliers gives a corrected excitation spectrum. A concentrated solution of Rhodamine B in ethylene glycol provides a suitable quantum converter (Melhuish, 1962). An alternative technique for obtaining corrected spectra involves the use of a thermocouple as detector since its response is independent of the wavelength and directly proportional to the energy with which it is irradiated. A fraction of the output from the excitation monochromator is sent directly to the detector to give the reference signal. The detector may be irradiated alternately with sample and reference beam, and the out-of-balance signal used to drive the recorder pen. If the signal is displayed as a function of wavenumber rather than wavelength, the spectrum is presented in the form required to provide data for Eqn. 7.6.

7.B.2 Phosphorimetry

The apparatus required for the measurement of phosphorescence is similar to that required for fluorescence, with the addition of a cylindrical can, with an aperture in the curved side, rotating round the sample cell. At one instant the aperture is in such a position that the sample cell is irradiated, but the fluorescence radiation is blocked from reaching the emission monochromator. A short time later, the can has rotated through 90°, and the phosphorescence radiation is allowed to reach the emission monochromator, but the excitation beam is blocked off so that no fluorescence can occur. The phosphorescence lifetime can be measured by monitoring the intensity of phosphorescence as a function of the angular velocity of rotation of the can. The sample cell is usually of Dewar vacuum-jacketed construction, since the sample is often frozen into a glass of organic solvents at liquid-nitrogen temperature, 77 K.

7.B.3 Determination of quantum yield

Corrected spectra give values of relative quantum yields, but an absolute measurement on at least one substance is required for the general evaluation of absolute quantum yields. Of the various methods used, the *integrating-sphere* technique is probably the most soundly based (Förster and Livingston, 1952). The fluorescent sample is at the centre of a large hollow sphere, some 50 cm in diameter, whose inside surface is coated with a highly reflective matt white substance, e.g. $BaSO_4$. Excitation radiation is admitted into the sphere through a small aperture, say 3 cm in diameter, and the radiation filling the sphere is sampled through another small aperture. A thermocouple detector, whose response is independent of wavelength, is used to monitor the radiation. The amount of radiation emitted by the sample is proportional to the detector signal if the scattered excitation radiation is prevented from reaching the detector by a suitable filter. The amount of radiation absorbed by the sample is measured by the change in detector signal when the sample solution is replaced by pure solvent, a suitable filter preventing all radiation except that at the excitation wavelength from reaching the detector. The use of the integrating sphere eliminates errors due to uncertainty as to what fraction of the emitted radiation reaches the detector.

7.B.4 Measurement of lifetimes of excited states

The most direct way to measure the lifetime of an excited state is to create it by a short pulse, or flash, of radiation, whose duration is much shorter than the lifetime, and observe the consequent disappearance of the excited species. The concentration of the excited species can be monitored either by the absorption or by the emission of radiation. Triplet states in solution typically have lifetimes in the region of 1–100 μs, and have been much studied by this technique. They are usually destroyed by external quenching rather than by phosphorescence emission, so kinetic absorption spectroscopy is used to monitor the rate of disappearance. Upper singlet states have lifetimes in the 1–100 ns region; at such short times, problems of instrumentation become quite formidable. The de-excitation to the ground state is monitored by fluorescence intensity. Nanosecond pulses are necessarily of low energy, so the detector signal is weak, but the bandwidth must be high to follow such a rapid process, so noise levels are high, as shown by Eqns. 1.3 and 1.4. Repetition techniques are used to improve the signal-to-noise ratio. The sample is irradiated by a steady stream of flashes, at a repetition rate of, say, 10 kHz, and signal-averaging techniques are used to build up a comparatively low-noise plot of fluorescence intensity as a function of time after each flash. If the repetition rate is 10 kHz, a run of duration of 1 second produces an improvement in the signal-to-noise ratio of a hundred-fold over that for a single flash. A sophisticated version of the signal-averaging technique borrows technology from nuclear physics, and is known as the *photon-counting technique*. Each photon emitted by the sample causes a current pulse in the detector, a high-gain photomultiplier, and these pulses are sorted, counted, and stored by a *multi-channel analyser*, which measures the time interval between the excitation flash and the arrival of each photon. A typical multi-channel analyser has 1024 channels, so the numbers of photons arriving after any one of 1024 separate times from each flash can be stored. The total numbers stored in each channel after a large number, up to many millions, of flashes give a set of 1024 points on the fluorescence decay curve. Fluorescence decay is usually a first-order process, so a plot of the logarithm of the number of pulses in each channel as a function of the time interval from the excitation pulse is a straight line whose gradient gives a value of τ.

7.C Factors Affecting Fluorescence Spectra

7.C.1 Kinetic aspects of fluorescence

Fluorescence is a kinetic phenomenon, only observable if the quenching processes require a longer time than τ_0, the intrinsic lifetime of the upper singlet state. If the first-order rate constants for the disappearance of the upper singlet state are k_e and k_i for fluorescence emission and internal quenching respectively, and if the rate of disappearance due to external quenching by species Q is $k_c[Q]$, then the total rate of disappearance is $k_e + k_i + k_c[Q]$. At equilibrium this is equal to the rate of absorption of quanta of radiation. The rate of emission of quanta as fluorescence is proportional to k_e, so the *quantum yield* ϕ, defined as (number of photons emitted)/(number of photons absorbed), is given by

$$\phi = k_e/(k_e + k_i + k_c[Q]) \tag{7.1}$$

Quantum yields approaching the limit of unity are found for some species in the absence of quenching agents. Efficient quenching agents are usually polarisable species, such as I^-, or paramagnetic species, such as O_2 or some transition-metal ions. Increasing viscosity of solvent increases quantum yield, since both internal rotation, a powerful factor favouring internal quenching, and the rate at which quenching molecules can diffuse towards the excited singlet states are reduced in viscous solution. Theory predicts that rates of molecular motion in liquids are proportional to $T\eta^{-1}$, where η is the viscosity, and it is often observed experimentally that ϕ^{-1} is directly proportional to $T\eta^{-1}$.

The observed lifetime of the upper singlet in the absence of external quenching agent is τ, where

$$\tau = (k_e + k_i)^{-1} \tag{7.2}$$

If the quantum yield in the absence of external quenching agent is ϕ_0, and that in the presence of quenching agent at concentration [Q] is ϕ, then Eqns. 7.1 and 7.2 can be combined to give the *Stern–Volmer equation*:

$$\phi_0/\phi = 1 + \tau k_c[\mathrm{Q}] \tag{7.3}$$

If the intensity of the exciting radiation is kept constant, the relative quantum yield is equal to the relative fluorescence intensity, so values of k_c can be obtained experimentally for a range of quenching agents if the value of τ is known. Because values of τ are so small, very large values of k_c are accessible to experimental measurement. Values of k_c as large as 10^{10} l mol^{-1} s^{-1} can be measured in this way, and observed values agree with those calculated by the Smoluchowski equation for the rate of a diffusion-controlled bimolecular reaction (Bowen, 1954):

$$k_c = 8000RT/3\eta \tag{7.4}$$

where R is the gas constant in units J K^{-1}, and η is the viscosity in units kg m^{-1} s^{-1}.

The value of τ can be determined by direct observation, as described in Section 7.B.4, or by calculation. The intrinsic lifetime τ_0, which is that in the absence of both internal and external quenching processes, is related to the observed lifetime τ, in the absence of external quenching agents, by

$$\tau_0 \phi_0 = \tau \tag{7.5}$$

It can be shown by the methods of quantum mechanics that τ_0 is related to the shapes of the absorption and emission spectra by the equation (Strickler and Berg, 1962):

$$\tau_0^{-1} = (2.880 \times 10^{-9}) n^2 \langle \tilde{\nu}_f^{-3} \rangle (g_l/g_u) \int \varepsilon \,\mathrm{d} \ln \tilde{\nu} \tag{7.6}$$

where n is the refractive index of the medium, g_l and g_u are the degeneracies of the lower and upper states, $\langle \tilde{\nu}_f^{-3} \rangle$ is the mean value of $\tilde{\nu}^3$ in the emission spectrum, and $\int \varepsilon \mathrm{d} \ln \tilde{\nu}$ is the integral of the extinction coefficient with respect to the logarithm of the wavenumber for the absorption spectrum. Values calculated from Eqn. 7.6 agree well with those found by direct observation.

7.C.2 The effect of pH

An acid–base indicator is a substance that can exist in either protonated (acidic) or deprotonated (basic) form, the two forms having widely different electronic absorption spectra. The $\mathrm{p}K_a$ of such an indicator, as measured by absorption spectrophotometry, is the pH at which the concentration of the acid form is equal to that of the basic form. If either form is fluorescent, it is also possible to define a $\mathrm{p}K_a$ that is the pH at which the fluorescent intensity is half the maximum value. It is found that the absorption $\mathrm{p}K_a$ and the emission $\mathrm{p}K_a$ are not identical; the latter is usually much less than the former for phenols, and vice versa for carboxylic acids. For example, for 2-naphthol the absorption $\mathrm{p}K_a$ is 9.5 and the emission $\mathrm{p}K_a$ is 3.1 (Jackson and Porter, 1961). This is because the emission $\mathrm{p}K_a$ is that for the excited singlet state, and should therefore be written $\mathrm{p}K_a^*$, whereas the absorption $\mathrm{p}K_a$ is that for the ground singlet state. The rate of proton loss from the acidic form is greater than the rate of de-excitation of the excited singlet state, so there is sufficient time for acid–base equilibrium to be set up before fluorescence emission. For some fluorescent acids and bases, the concentrations can be adjusted until the rate of ionisation is similar to the rate of de-excitation, so that measurement of fluorescence intensity gives a measure of the rate of ionisation (Weller, 1961). The $\mathrm{p}K_a$ of an acid in a triplet state is usually intermediate between those for the ground and excited singlet states; e.g. for 2-naphthol the $\mathrm{p}K_a$ of the triplet state has the value 8.1.

The relationship between the change in $\mathrm{p}K_a$ on excitation and the fluorescence spectra is shown in Fig. 7.3, which is approximately to scale for 2-naphthol. It can be seen that

$$\Delta E_{\mathrm{HA}} - \Delta E_{\mathrm{A}} = \Delta H^0 - \Delta H^{0*} \tag{7.7}$$

where ΔE_{HA} is the energy gap between the ground vibrational levels of the ground and excited singlet states of the acid, ΔE_{A} is the corresponding value for the anion, ΔH^0 is the enthalpy of ionisation of the acid in the ground singlet state, and ΔH^{0*} is the enthalpy of ionisation of the acid in the upper singlet state. If entropy effects are ignored,

$$\Delta H^0 - \Delta H^{0*} \simeq \Delta G^0 - \Delta G^{0*} = RT(\mathrm{p}K_a - \mathrm{p}K_a^*) \tag{7.8}$$

ΔE_{HA} and ΔE_A can be evaluated from either the absorption or the emission spectrum of the acid and anion respectively, but it is more satisfactory to use the latter. The Franck–Condon principle means that the values of ΔE_{HA} and ΔE_A from the absorption spectra refer to non-equilibrium solvation of the excited state whereas the solvent molecules have had time to organise themselves around the excited state before emission occurs, as discussed in Section 7.C.3 (Wehry and Rogers, 1965).

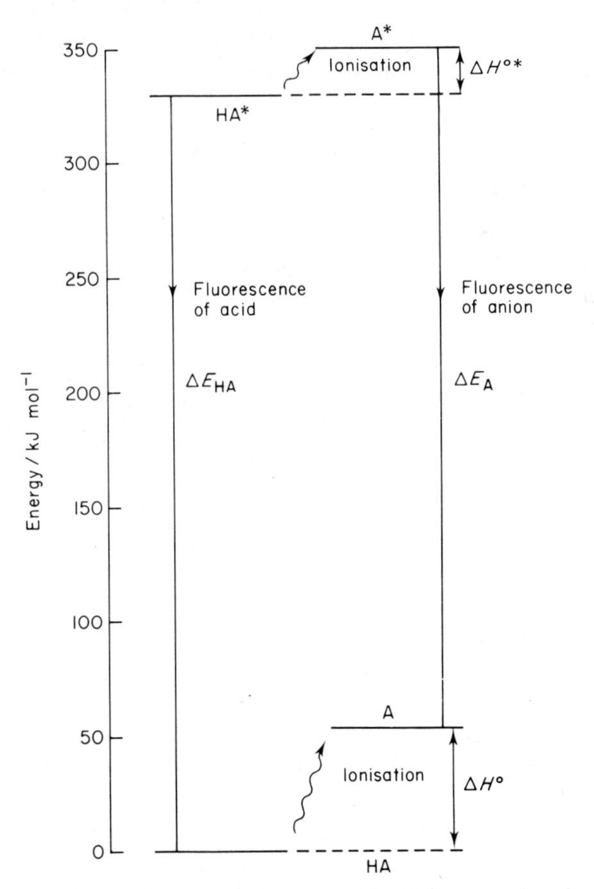

FIG. 7.3 Energy diagram for dissociation of acid HA in ground and excited states, approximately to scale for 2-naphthol.

7.C.3 Solvent effects

There is an important difference between the effects of the solvent on absorption and emission spectra (Van Duuren, 1963). The excited state is generally more polar than the ground state, and so has a different pattern of solvent molecule orientation about it. By virtue of the Franck–Condon principle, solvent reorientation does not have time to occur during excitation, so the energy gap measured by absorption spectroscopy corresponds to that for a non-equilibrium solvation of the excited state. However, solvent reorientation has had time to occur before emission, since it requires a time of only about 10^{-12} s in non-viscous solvents, so the energy gap measured by fluorescence corresponds to that for a properly solvated excited state. It is generally found that the more polar the solvent, the greater the shift to longer wavelengths of the fluorescence emission spectrum. This is attributable to the increased solvation of the excited state, which lowers its energy. As mentioned in Section 6.C.3, solvent effects for absorption spectra due to $n \rightarrow \pi^*$ transitions show a shift to shorter wavelengths, i.e. the transition energy is increased in more polar solvents. This is because only the ground state solvation is of importance in absorption spectroscopy. Both ground and excited states are stabilised by polar solvents, but the excited state, being more polar, is stabilised to a greater extent.

The solvent also has a marked effect on the intensity of fluorescence, i.e. the quantum yield. There are several ways in which this can occur. Many species containing nitrogen or oxygen, e.g. acridine and 2-naphthol, show much more intense fluorescence in hydroxylic solvents than in hydrocarbon solvents. This is because in hydrocarbon solvents the lowest energy transition is of $n \rightarrow \pi^*$ type, the excited electron coming from the lone pair on the oxygen or the nitrogen. As discussed in Section 7.A, this excitation is more likely to lead to triplet state formation than to fluorescence. In hydroxylic solvents the lone pair electrons are involved in hydrogen bonding to the solvent, so the lowest energy transition is of $\pi \rightarrow \pi^*$ type, and fluorescence is favoured instead. Increasing solvent viscosity increases fluorescence intensity, as discussed in Section 7.C.1. Solvent quenching is also found, especially for solvents whose molecules contain large polarisable atoms such as iodine.

The sensitivity of fluorescence to environment may be regarded as a nuisance, but it affords a powerful technique for the investigation of macromolecular structure. A fluorescent molecule acts as a *probe*, and its emission spectrum can be utilised to obtain information about the local structure of its surroundings, in particular whether the probe is in a water-like (hydrophilic) or hydrocarbon-like (hydrophobic) environment. For example, the fluorescence spectrum of anilinonaphthalenesulphonate bound to the enzyme liver alcohol dehydrogenase shows that the binding site is hydrophobic, i.e. the probe is in a hydrocarbon-like environment (Brand *et al.*, 1967).

7.C.4 Polarisation effects in fluorescence

If plane-polarised light, i.e. that in which the oscillation of the electric field is oriented in one direction only, is used to excite fluorescence, it might be expected that the emitted radiation would be similarly polarised, but this is not observed. The *degree of depolarisation*, P, is defined as

$$P = (I_{\parallel} - I_{\perp})/(I_{\parallel} + I_{\perp}) \quad (7.9)$$

where $I_{\parallel}$ and $I_{\perp}$ are intensities of emitted radiation polarised respectively parallel and perpendicular to the plane of polarisation of the excitation radiation. Firstly, the axes in a molecule for absorption and emission are not necessarily coincident, since the excited state has a different electronic structure from that for the ground state. This causes an *intrinsic degree of depolarisation*, P_0. Secondly, the molecule usually has time to rotate between absorption and emission, and measurement of the degree of depolarisation gives a measure of the ability of the molecule to rotate during the lifetime of the upper singlet state. The ability of a molecule to rotate is expressed by the *rotational relaxation time*. Suppose that it were possible to arrange all the molecules in a system with their axes pointing in the same direction. When the restraint is removed they rotate so as to take up the equilibrium, random, distribution of orientations. This is a first-order process, and the reciprocal of the first-order rate constant is the rotational relaxation time. The rotational relaxation time, τ_R, is related to the degree of depolarisation by

$$P/P_0 = 1 + (3 - P_0)(\tau/\tau_R) \quad (7.10)$$

If the fluorescent molecule is firmly fixed to a macromolecule such as a protein, measurement of P leads to measurement of τ_R for the protein. The larger the molecule, the greater is τ_R. It has been found empirically that τ_R in nanoseconds is 1.2–2 times the molecular weight in thousands for globular proteins (Weber and Young, 1964). For example, the partial hydrolysis of bovine serum albumin, for which τ_R had the value 127 ns, gave three main products for which τ_R had the values 20.5, 42, and 58 ns respectively.

7.C.5 Excimer formation

The fluorescence spectra of some hydrocarbons show an interesting new feature at higher concentrations. For example, whereas pyrene at 10^{-4} mol l^{-1} has a fluorescence emission spectrum in the region 360–420 nm, with a structure attributable to vibrational levels, pyrene at 10^{-2} mol l^{-1} has a structureless fluorescence emission in the region 440–540 nm (Parker and Hatchard, 1963). The intensity of this new emission is proportional to the square of the pyrene concentration. The new emission is attributed to excited dimers, or excimers, which are complexes consisting of one molecule in the ground state and one molecule in the excited singlet state. Since these complexes have no ground state, emission causes dissociation, and does not correspond to a transition to a well-defined lower level, so the spectrum lacks vibrational structure. Excimer formation is possible only if the lifetime of the upper singlet state is comparatively long; e.g. it is 290 ns for pyrene.

7.D Analytical Applications

7.D.1 Qualitative analysis

Fluorescence emission spectra are little used in qualitative analysis, since they are too broad and featureless, and too solvent-dependent. However, the mere fact that a molecule is fluorescent implies something about its structure, namely that it contains a conjugated π system of electrons and lacks internal rotation. Corrected excitation spectra are potentially useful in that they are equivalent to electronic absorption spectra, but may be obtained from much more dilute solutions, as discussed in Section 7.D.2.

7.D.2 Quantitative analysis

Fluorimetry, the measurement of the intensity of fluorescence emission, is a powerful technique for quantitative analysis. It is inherently more sensitive than electronic absorption spectroscopy for the reason discussed in Section 6.D.3; zero concentration of fluorescent material corresponds to zero signal, whereas zero concentration of absorbent material corresponds to maximum signal. In favourable instances concentrations down to 10^{-12} mol l^{-1} can be measured. Turbid samples can be used, since the scattered radiation is of different wavelength from the fluorescence radiation. Many substances for which the analysis of trace quantities is important, such as drugs and pesticides, and many substances of biological interest, such as proteins and nucleic acids, are fluorescent. Several amino acids, e.g. phenylalanine, tyrosine, and tryptophan, are fluorescent, and this is exploited in the analysis of the products of hydrolysis of proteins.

Many species that are not fluorescent in themselves can be estimated fluorimetrically by the formation of fluorescent compounds. Metal ions may be complexed with non-fluorescent ligands to make fluorescent complexes. The organic ligands used, often derivatives of quinolin-8-ol are non-fluorescent because of oxygen or nitrogen atoms in the molecule which cause the upper singlet state to be of n,π^* type, as discussed in Section 7.A. The formation of the complex with the metal ion involves the donation of the lone pairs of electrons on the oxygen or nitrogen, so these are less available for excitation, and the lowest available upper singlet state is of the fluorescent π,π^* type. Sensitivities of the order of parts per 10^9 may be achieved in this way.

It is important that the absorbance of the solution to be analysed should not be too high. If the excitation radiation is all absorbed in the first millimetre or so of a 1 cm square cross-section cuvette, a smaller proportion of the emitted radiation is incident on the emission monochromator slit than if the whole cell is luminescent. Thus an experimental plot of signal against concentration will be linear at low concentrations, but curve down, and even show a maximum, at high concentrations. A simple solution to this problem is to use a different excitation wavelength, at which the extinction coefficient of the fluorescent species is less. The apparatus must be calibrated by measuring the signal for known concentrations of the species of interest in the analysis, at the same excitation and emission wavelengths, and preferably at a similar concentration to that of the sample.

REFERENCES

Bowen, E. J. (1954) *Trans. Faraday Soc.* **50**, 97.
Brand, L., Golke, J. R. and Rao, D. S. (1967) *Biochemistry* **6**, 3510.
Förster, L. S. and Livingston, R. (1952) *J. Chem. Phys.* **20**, 1315.
Jackson, G. and Porter, G. (1961) *Proc. Roy. Soc. A* **260**, 13.
Jaffé, H. H. and Miller, A. L. (1966) *J. Chem. Educ.* **43**, 469.
Kortum, G. and Finckh, B. (1942) *Z. Phys. Chem. B* **52**, 263.
Melhuish, W. H. (1962) *J. Opt. Soc. Amer.* **52**, 1256.
Parker, C. and Hatchard, C. (1963) *Trans. Faraday. Soc.* **59**, 284.
Strickler, S. and Berg, R. (1962) *J. Chem. Phys.* **37**, 814.
Van Duuren, B. L. (1963) *Chem. Rev.* **63**, 325.
Weber, G. and Young, L. (1964) *J. Biol. Chem.* **239**, 1424.
Wehry, E. and Rogers, L. (1965) *Spectrochim. Acta* **21**, 1976.
Weller, A. (1961) *Progr. Reaction Kinetics* **1**, 189.

BIBLIOGRAPHY

Becker, R. S. (1969) "Theory and Interpretation of Fluorescence and Phosphorescence", Wiley-Interscience, New York and London.
Guilbaut, G. G. (1973) in "MTP International Review of Science, Physical Chemistry Series One, Vol. 12, Analytical Chemistry, Part 1" (T. S. West, ed.), pp. 127–250, Butterworths, London.

Pesce, A. J., Rosén, C.-G. and Pasby, T. L. (1971) "Fluorescence Spectroscopy", Marcel Dekker, New York.

Wotherspoon, N., Oster, G. K. and Oster, G. (1972) in "Techniques of Chemistry, Vol. I, Physical Methods of Chemistry, Part III B, Optical, Spectroscopic and Radioactivity Methods" (A. Weissberger and B. W. Rossiter, eds.), pp. 429–484, Wiley-Interscience, New York and London.

8
Raman Spectroscopy

8.A The Theory of Raman Spectroscopy

8.A.1 The Raman effect

When electromagnetic radiation of any frequency impinges on an atom or molecule, its electrons are forced to oscillate at the frequency of the incident radiation, and so emit radiation of that frequency in all directions. This phenomenon of the scattering of radiation by molecules is known as *Rayleigh scattering*. The fraction scattered is proportional to the fourth power of the frequency. Both the blue colour of the sky and the red colour of the setting sun are due to Rayleigh scattering of sunlight by air molecules. The blue components of sunlight are scattered more intensely than the red, so only the red components of sunlight are directly transmitted through the atmosphere at grazing incidence, where the distance to be traversed through the atmosphere is greatest.

The electric field of the radiation, of strength E, induces a dipole moment M in the molecule. The more easily deformed the electron shells in the molecule, the greater the dipole moment induced by a given field, and this is measured by the *polarisability* α, where

$$M = \alpha E \tag{8.1}$$

Not all the scattered radiation is of the same frequency as the incident radiation, since energy may be gained or lost by exchange with energy stored by the molecule. Although the energy thus available for exchange may be stored in the molecule in electronic, vibrational, or rotational modes, it is exchange with the vibrational modes that is of the greatest interest to chemists. A small proportion of the scattered radiation is of frequency $(\nu \pm \omega)$, where ν is the frequency of the incident radiation and ω is the frequency of a vibrational mode. If the molecule was originally in the ground vibrational level, it acquires energy from the incident radiation, so that the frequency of the scattered radiation is $(\nu - \omega)$. On the other hand, if the molecule was originally in the first vibrational level, it can give energy to the scattered radiation, which thus has frequency $(\nu + \omega)$. The scattering of radiation at a frequency different from the incident radiation is termed the Raman effect, after its discoverer. The Raman spectrum, i.e. the plot of intensity of scattered radiation as a function of the frequency difference between the incident and scattered radiation, thus falls into two sections. Peaks in the spectrum of lower frequency than the Rayleigh scattered peak are termed *Stokes peaks*, whereas those of higher frequency are termed *anti-Stokes peaks*. Since most molecules in a sample are in the ground vibrational level, Stokes emission is much less intense than anti-Stokes emission. The intensity of Raman scattering in general is very low. Only about 10^{-5} of the incident radiation is Rayleigh-scattered, and the total amount of Raman-scattered radiation is only about 10^{-2} that of the Rayleigh-scattered radiation. It is customary to observe only Stokes emission, since anti-Stokes emission for vibrational energy transfer is usually too weak to be detected.

It can be seen that there are certain resemblances between Raman and fluorescence spectroscopy. In both, the radiation emitted from an irradiated sample is observed, and in both the emitted radiation is of lower frequency than the excitation radiation as a consequence of energy exchange with vibrational modes. The nomenclature of Stokes and anti-Stokes emission is based on this similarity, Stokes being the original investigator of fluorescence who observed that the emitted radiation was of lower frequency than the excitation radiation. However, the two phenomena are different in principle. Raman emission takes place extremely rapidly, in a time similar to that at which absorption occurs, whereas fluorescence emission occurs at a comparatively long time after absorption. In the Raman effect, a molecule does not absorb a quantum of energy from the incident radiation, so radiation of any frequency can be used. By contrast, the excitation of fluorescence requires the use of radiation whose frequency exactly corresponds to the energy gap between the ground and upper singlet states. Despite the difference in theoretical principles, the apparatus required for Raman spectroscopy resembles that for fluorescence spectroscopy in that both comprise an intense source of radiation and a monochromator and detector to monitor the radiation emitted perpendicular to the incident radiation. It is thus possible inadvertently to run a Raman spectrum under the impression that one is running a fluorescence spectrum, and vice versa. Every fluorescence spectrum contains a peak due to Raman emission by solvent or solid matrix. This is usually much less intense than the true fluorescence spectrum, but may be noticeable if the sample is only weakly fluorescent so that a high signal amplification is employed. It is generally distinguishable from the fluorescence spectrum by its closeness to the Rayleigh peak. The frequency difference between excitation and emitted radiation for fluorescence usually corresponds to several vibrational quanta, whereas that for Raman emission corresponds to only one. Conversely, it is possible to obtain a fluorescence spectrum superimposed on a Raman spectrum if the excitation radiation happens to be at a frequency at which the sample absorbs. This possibility is avoided by careful selection of the correct Raman excitation frequency.

The neat distinction between Raman and fluorescence emission becomes somewhat more subtle if the Raman excitation radiation is absorbed by the sample, i.e. if it is of frequency corresponding to the energy gap between the ground and upper singlet states. The Raman emission is then much more intense than if the excitation radiation were at a very different frequency. The amplification is inversely proportional to the fourth power of the frequency difference. This is an example of the general principle that the coupling, i.e. ease of energy exchange, between two energy levels of different types is large if the two levels are of similar energy, and at a maximum if the energies are equal. Energy exchange thus facilitated is termed *resonance*; Fermi resonance, discussed in Section 4.C.4, is another example. Hence, Raman emission whose intensity is enhanced by the similarity between the energy of a photon of excitation radiation and the energy gap between ground and upper singlet state is termed *resonance Raman emission*. Enhancement due to an approximate similarity of energies is termed *pre-resonance Raman emission*, and enhancement due to an exact equality of energies is known as *rigorous resonance Raman emission*. Resonance Raman emission may be accompanied by fluorescence emission, or the latter may be quenched. The effect of an external quenching agent provides a simple experimental distinction between resonance Raman and fluorescence emission. The most fundamental theoretical distinction is that resonance Raman emission occurs simultaneously with excitation, whereas fluorescence emission occurs after a definite time interval, even though it be of short duration (Behringer, 1974). Students of chemical kinetics will notice the analogy with the distinction between transition state and reaction intermediate. Resonance Raman emission has a practical utility in that it provides a means of amplifying the intensity of an observed spectrum by several orders of magnitude, so spectra of substances in dilute solution can be run.

Beginning students sometimes form the impression that Raman spectroscopy is "a kind of infrared spectroscopy" because Raman shifts are quoted as wavenumbers in cm^{-1}. It is important to remember that a "Raman frequency" of 2000 cm^{-1} is not measured in the infrared region, but corresponds to, say, an emission at a wavelength of 540.8 nm for excitation by an Ar laser source at 488.0 nm, or an emission at 723.4 nm for excitation by a He–Ne laser source at 632.8 nm.

8.A.2 The selection rules for rotational Raman spectroscopy

Energy exchange of the Raman type is possible between the incident radiation and rotational levels of the molecule. Raman interaction occurs only if there is a change of polarisability of the molecule in the direction of the incident radiation. Rotation of a non-spherical molecule provides such a change. The electrons in, say, the bond of a hydrogen molecule are more easily polarised by the incident radiation when the axis of the molecule is oriented along the electric field vector than when it is oriented perpendicular to it. Thus the end-over-end rotation of the molecule produces a sinusoidally varying polarisability. Rotation about the axis does not produce a change in polarisability in any direction, and so does not give a Raman effect.

The selection rule for permitted transitions is that $\Delta J = 0, \pm 2$. It should be noted that this is different from that for the absorption of radiation by excitation of rotational energy, for which, as discussed in Section 2.A.2, $\Delta J = +1$. The doubling of ΔJ occurs because the polarisability change goes through two cycles during one rotation of the molecule. The positions of peaks in the rotational Raman spectrum are given by Eqn. 8.2, which is derived from substitution of $\Delta J = \pm 2$ into Eqn. 2.7.

$$\Delta \nu = \pm B(4J+6) \tag{8.2}$$

where $\Delta \nu$ is the Raman shift, i.e. the frequency difference between emitted and incident radiation. The negative and positive signs refer to Stokes and anti-Stokes emission respectively. Because of the appreciable population of upper rotational levels, as discussed in Section 2.A.4, anti-Stokes peaks for rotation are not much less intense than Stokes peaks. The observed spectrum thus consists of two sets of peaks spaced at intervals of $4B$, symmetrically arranged about the Rayleigh peak, with a gap of $6B$ between the Rayleigh peak and the first Raman peak on each side. By contrast with the situation for microwave spectroscopy, the molecule does not need to have a permanent dipole moment to give a Raman spectrum, so values of B for homonuclear diatomic molecules may be obtained. An example, the rotational Raman spectrum of N_2, is given in Fig. 8.1. The equilibrium bond length was calculated from this spectrum to be 109.7700 ± 0.0007 pm (Bendtsen, 1974).

8.A.3 The selection rules for vibrational Raman spectroscopy

The selection rules for Raman interaction with vibrational modes are that the quantum number v changes by ± 1, and that the vibration should produce a change in polarisability of the molecule. The single stretching vibration of a diatomic molecule produces the required change. As the bond stretches, the electrons are further away from the nuclei, and so are less strongly attracted

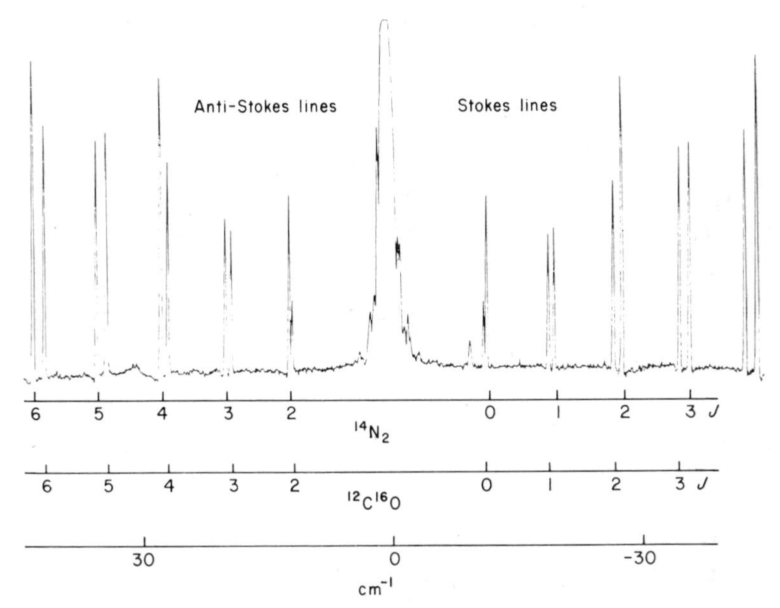

FIG. 8.1 Rotational Raman spectrum of $^{14}N_2$, with excitation wavelength 514.5 nm (Bendtsen, 1974). The sample is a mixture of $^{14}N_2$ and $^{12}C^{16}O$, the latter providing a frequency calibration since its spectrum has been well-characterised by microwave spectroscopy.

to the nuclei and are more polarisable. Conversely, the electrons become less polarisable as the bond shortens. Whereas homonuclear diatomic molecules do not have a vibrational absorption spectrum, since the selection rule for absorption requires a change in dipole moment with vibration, these molecules do exhibit a vibrational Raman spectrum. Only one of the three vibrational modes of linear triatomic molecules of XYZ type, namely the symmetric stretch as illustrated in Fig. 4.6, is *Raman active*, i.e. is observable in the Raman spectrum. The other two modes, namely the antisymmetric stretch and the bend, do not cause a change in the polarisability of the molecule as a whole. In the bending mode the bond lengths stay constant, and in the antisymmetric mode the increasing polarisability of one Y–X bond is offset by the decreasing polarisability of the other. A more detailed and rigorous analysis shows that the preceding remarks are somewhat oversimplified. What determines whether a vibrational mode is Raman active is whether $d\alpha/dx$ is not zero at the equilibrium bond lengths and bond angles, where x is some parameter measuring displacement from the equilibrium position. The bending and antisymmetric stretching of a linear XYZ molecule does produce appreciable changes in polarisability at large displacements, but $d\alpha/dx$ is zero at equilibrium.

Linear molecules of XYZ type provide a simple example of a general rule, the *rule of mutual exclusion*, by which a vibrational mode of a centrosymmetric molecule cannot be both infrared and Raman active. Another illustration is provided by oxalyl chloride, the Raman and infrared spectra of which are shown in Fig. 8.2. As mentioned in Section 4.C.1, a deeper understanding of the effects of symmetry on infrared and Raman vibrational spectra requires a knowledge of group theory, which is outside the scope of this book. It is unfortunate that CO_2, which one might hope would provide a neat example of a single Raman peak due to the symmetric stretch, shows in fact two Raman

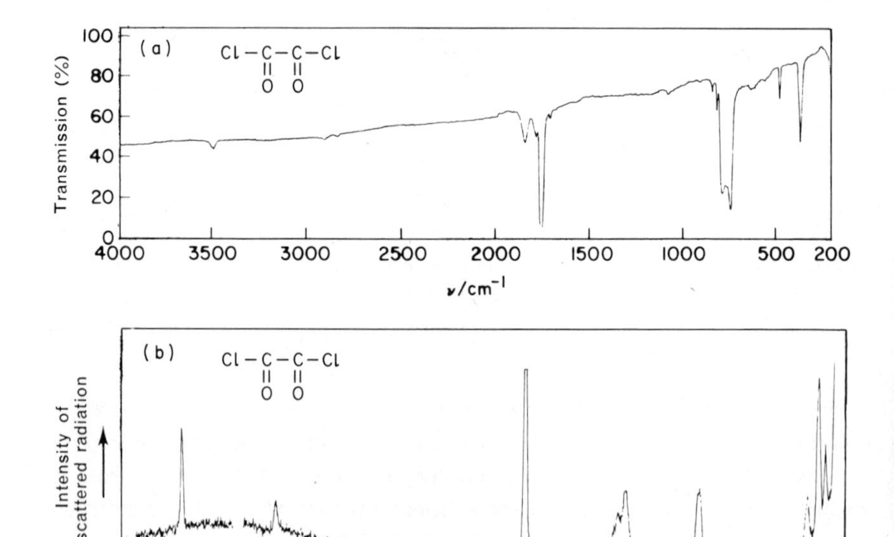

FIG. 8.2 (a) Infrared and (b) Raman spectra of solid oxalyl chloride. Peak assignments (Durig and Hannum, 1970)

Infrared active	cm^{-1}	Raman active	cm^{-1}
C=O antisym. stretch	1769	C=O sym. stretch	1762
C—Cl antisym. stretch	756	C—C stretch	1093
Cl—C=O antisym. bend	490	C—Cl sym. stretch	620
Cl—C=O sym. out-of-plane bend	391	Cl—C=O antisym. out-of-plane bend	442
Cl—C=O antisym. rock	220	Cl—C=O sym. bend	427
torsion about C—C	55	Cl—C=O sym. rock	287

peaks of similar intensity at 1285 and 1388 cm^{-1}. This is because the wave-number of the bending vibration, 667.3 cm^{-1}, is almost exactly half that of the symmetric stretch, so Fermi resonance can occur between the overtone of the bending vibration and the symmetric stretching vibration as discussed in Section 4.C.4.

Vibrational Raman spectra of gases show fine structure due to rotational transitions, as do infrared spectra, as discussed in Section 4.D. The selection rule $\Delta J = \pm 2$ holds, so three types of branch, labelled O, Q, and S, are observed, corresponding to $\Delta J = +2, 0$, and -2 respectively. Whereas the Q branch is not observed in the infrared spectra of diatomic molecules, it is observed in the Raman spectra.

8.A.4 The polarisation of Raman-scattered radiation

Consider radiation, travelling in a direction designated as the x axis, which is polarised in the xz plane, i.e. the oscillating electric field vector always lies in the xz plane. When this radiation is incident on a spherically polarisable molecule, i.e. a molecule that has the same polarisability in the z direction whatever orientation it assumes as the result of molecular motion, an oscillating dipole is induced in the z direction. This dipole emits radiation, polarised in the yz plane, in the y direction, as shown in Fig. 8.3. A proportion of this radiation exchanges energy with the molecule, and so appears as Raman-scattered radiation. Now consider the effect of incident radiation polarised in the xy plane. This induces an oscillating dipole in the y direction, which cannot emit radiation in the y direction. Radiation from most sources is unpolarised, i.e. the amplitudes of the electric vector are equal in the y and z directions as the radiation travels in the x direction. The Raman-scattered radiation, as

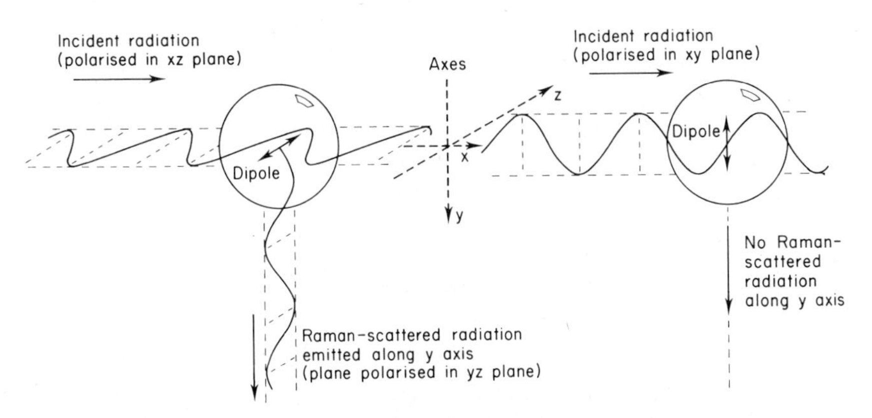

FIG. 8.3 Polarisation effects on Raman scattering.

observed in the y direction, produced by unpolarised radiation will thus be polarised in the yz plane. If plane-polarised radiation is used, Raman-scattered radiation will be observed only in a direction perpendicular to the plane of polarisation of the incident radiation. This situation occurs only if the molecule is spherically polarisable. If the molecule lacks this symmetry, the induced dipole moment is greatest along the molecular axis of maximum polarisability. For molecules in the gaseous or liquid phase this axis is oriented at random to the incident radiation owing to molecular motion, so the induced dipole is randomly oriented, and unpolarised radiation is emitted in all directions. Even a molecule that at equilibrium is symmetrical may become temporarily unsymmetrical as the consequence of bond bending and stretching in some vibrational modes, and these modes will give unpolarised Raman scattering. It is thus possible to distinguish between Raman peaks for totally symmetrical and other vibrational modes (Tobias, 1967). For example, methane has a totally symmetrical vibrational mode in which the C–H bonds lengthen and contract all in phase together, the bond angles remaining constant. Such a mode is called a *breathing mode*, and gives polarised Raman scattering. In the antisymmetric stretching mode of methane, one bond lengthens while the other three contract, and this mode gives unpolarised Raman scattering.

Detailed mathematical analysis shows that the Raman-scattered radiation may be polarised to a certain extent when due to the symmetric vibration even of a molecule lacking perfect symmetry. A quantitative measure of the extent of polarisation is given by the *depolarisation ratio* ρ, defined by

$$\rho = I_{\parallel}/I_{\perp} \tag{8.3}$$

where $I_{\perp}$ is the intensity of Raman scattering in the y direction for incident radiation polarised in the xz plane, and $I_{\parallel}$ is the intensity in the y direction for incident radiation polarised in the xy plane. For a symmetric vibration of a spherically symmetrical molecule, ρ is zero. It can be shown that the maximum value of ρ for unpolarised incident radiation is 6/7, not, as might be expected, unity. For a laser source, which gives plane polarised radiation, the maximum value of ρ is 3/4. Intermediate values of ρ, as are often observed, give an estimate of the degree of symmetry of the vibrational mode responsible.

8.B Apparatus for Raman Spectroscopy

8.B.1 Radiation sources

It can be seen from the discussion in Section 8.A.1 that the ideal source for Raman spectroscopy is both very intense and highly monochromatic.

Although there is no limitation on the frequency of the radiation, it is experimentally convenient to use radiation in the visible and near ultraviolet, since monochromators and detectors for this region have been developed to a high degree. The variation of scattered intensity with the fourth power of the frequency indicates that as high a frequency as possible should be used, with the proviso that the frequency should not be so high as to excite fluorescence. Early work on Raman spectroscopy was performed using a mercury arc lamp as a source, the line at 435.8 nm being selected by optical filters. A tremendous impetus to Raman spectroscopy was provided by the development of the laser which is an ideal source, giving radiation that is intense, monochromatic, and polarised. A further advantage is that the radiation is highly directional, which simplifies instrument design, and may be focused down to a very small area, so minute samples can be studied. Whereas sample volumes of 20 cm^3 of solution were common in the days when the mercury arc was the source, liquid samples of a few mm^3 can be studied using a laser source, and spectra have been obtained from single crystals of microscopic size ($\sim 5 \times 10^{-3}$ mm^3). Rare-gas lasers are used, a range of wavelengths being available depending on the nature of the gas. The He–Ne laser at 632.8 nm and the Ar laser at 488.0 nm are especially popular.

8.B.2 Sample handling

Gaseous, liquid, and solid samples can be studied. The use of excitation radiation in the visible means that glass sample cells may be employed, a major advantage over infrared spectroscopy. Ordinary melting-point capillaries, which can be used for liquids or powders, provide cheap sample cells requiring only a few mm^3 of sample. Single crystals may be mounted on the end of a pin, whose orientation can be adjusted by a goniometer head for studies of the angular dependence of Raman emission. It is even possible to run a Raman spectrum of a substance in an unopened sample bottle, or sealed in an ampoule.

8.B.3 Monochromators

The observation of a Raman spectrum makes considerable demands on the monochromator. The range of Raman shifts from 0 to 3500 cm^{-1} corresponds to the spectral region from 488 to 590 nm for the Ar laser as source, so a resolution of ± 1 cm^{-1} in the Raman spectrum corresponds to a resolution of the order of $\pm 3 \times 10^{-2}$ nm in the observed radiation. In order to improve the resolution, it is customary to employ a double monochromator, the output from one monochromator serving as the input to the second. The use of a high-resolution monochromator is of benefit not only in giving a sharper resolution of peaks, but also in reducing the observed width of the Rayleigh peak, so that small Raman shifts can be observed. As can be seen in Fig. 8.1, shifts as small as 3 cm^{-1} can be observed by modern instruments, so there is an overlap between the frequencies observable by Raman and by microwave spectroscopy.

8.B.4 Detectors

Earlier workers used the photographic plate as detector; this has the advantage that a long exposure can be used to detect weak Raman emission. Features in the Raman spectrum are sometimes referred to as lines because of this. The photomultiplier is now the standard detector in instruments for routine use. If the signal is extremely weak, photon-counting techniques, as discussed in Section 7.B.4, are used. The apparatus required is expensive, but has the advantage that the information is stored in digital form, so the output can be linked directly to a computer for further processing.

8.C Determination of Molecular Structure

8.C.1 Determination of molecular symmetry and shape

Analysis of vibrational modes may be used to determine molecular structure. It is often profitable to run both the Raman and the infrared absorption spectrum, since bands can be assigned to vibration modes according to whether they are Raman active or infrared active, or both. Measurement of the depolarisation ratio is also an aid to mode assignment, since it enables a distinction to be made between totally symmetric modes and others. This technique may be illustrated by one example taken from the large number of such studies. The structure of $Si(NCO)_4$ has been largely elucidated by this technique. The features observed in the spectra are summarised in Table 8.1. The first point to be settled is whether the molecule is $Si(NCO)_4$ or $Si(CNO)_4$, i.e. a tetraisocyanate or a tetracyanate. This is soon decided; cyanates do not show any absorption in the region 1200–2000 cm^{-1}, whereas isocyanates absorb strongly around 1400 cm^{-1}. The absorption at 1481 cm^{-1} shows that the species is an isocyanate. The next question is whether the molecule is of overall tetrahedral symmetry, with the four NCO groups directed towards the corners of a tetrahedron, or planar. The lower symmetry of planar configurations means that far more vibrational modes are to be expected than for tetrahedral configurations, and the comparatively small number of observed frequencies strongly suggests tetrahedral symmetry. A more difficult question is that of whether the molecule has true tetrahedral symmetry, i.e. each

TABLE 8.1

The infrared and Raman spectra of $Si(NCO)_4$ (Miller and Snider, 1974)

Wavenumber/cm^{-1}	Raman	Infrared	ρ	Assignment on tetrahedral model
253	125	—	0.65	SiN_4 bending deformation
300	35	—	0.74	SiNCO bending deformation
340	2	m	dp	SiN_4 bending deformation
496	1000	—	0.004	SiN sym. stretch
542	—	vw	—	SiNCO bending deformation
608	—	s	—	SiNCO bending deformation
619	110	vs	0.64	SiNCO bending deformation
730	15	vs	?	SiN_4 degenerate stretch
815	9	—	?	Unassigned
1120	16	w	?	Combination (496 + 730)
1474	940	—	0.16	NCO sym. stretch, in phase
1481	?	s	?	NCO sym. stretch, out of phase
2277	14	vvs	?	NCO antisym. stretch, out of phase
2350	28	—	0.073	NCO antisym. stretch, in phase
2763	—	w	—	Combination (496 + 2277)
3000	—	m	—	Combination (730 + 2277)
3738	—	m	—	Combination (1474 + 2277)

Numbers in Raman column are relative intensities. — means not observed. ? means detectable but too weak for accurate measurement. dp = depolarised. Relative intensities: s = strong; m = medium; w = weak; v = very.

Si–N–C–O group of atoms is in a straight line, or whether this line is kinked at some point. Earlier studies (Miller and Carson, 1961) favoured true tetrahedral symmetry, but a more recent investigation (Miller *et al.*, 1974) showed that five of the Raman peaks were depolarised whereas theoretical analysis of the vibrations of a molecule with true tetrahedral symmetry predicts that only three polarised Raman peaks will be observed. It is therefore concluded that the nitrogen atoms are arranged in tetrahedral symmetry around the silicon, but the Si–N–C bond angle is less than 180°.

8.C.2 Determination of bond vibration frequencies

As discussed in Chapter 5, the investigation of vibrational modes of low frequencies by infrared spectroscopy is a matter of some experimental difficulty. Species in aqueous solution cannot be studied, since water is opaque in the far infrared. Raman spectroscopy, by contrast, is most convenient. It is as easy to observe a mode at 100 cm^{-1} as it is at 1000 cm^{-1}. Furthermore, water is transparent in the visible and gives only weak Raman emission. Raman spectroscopy is thus especially suitable for the study of inorganic species.

A successful early Raman study (Woodward, 1934) showed that aqueous solutions of mercurous nitrate gave an intense polarised peak at 169 cm^{-1}. This was attributed to the Hg–Hg vibration, showing that the mercurous ion was dimeric, i.e. not Hg^+ but $(Hg–Hg)^{2+}$. Since that time other metal–metal bonds have been studied by Raman spectroscopy. For example, the Raman spectra of several iron complexes, in the crystalline state, have a common feature of an intense peak at 220 cm^{-1}, which is attributed to the Fe–Fe vibration. Other metal–metal bond frequencies measured are 285 cm^{-1} for Re–Re, 345 cm^{-1} for Mo–Mo, and 337 cm^{-1} for Ru–Ru (San Filippo and Sniadoch, 1973).

The vibration frequencies for the bonds between metal ions in aqueous solution and solvating water molecules have been measured by Raman spectroscopy, and it is hard to think of any other technique suitable for this problem. For example, the symmetric stretching frequencies have been found to be 363 cm^{-1} for $Mg^{2+}(OH_2)_6$, 389 cm^{-1} for $Zn^{2+}(OH_2)_6$, 524 cm^{-1} for $Al^{3+}(OH_2)_6$, and 536 cm^{-1} for $Be^{2+}(OH_2)_4$. The comparatively high frequency for $Al^{3+}(OH_2)_6$ is due to the strength of the bond, as shown by the high enthalpy of solvation for Al^{3+} (da Silveira *et al.*, 1965).

Until the advent of the laser source, Raman spectroscopy of transition-metal complexes was limited by the intense absorption of most complexes at the 435.8 nm wavelength of the mercury arc source, but since then many Raman studies of coloured complexes have been made. For example, the use of the He–Ne laser source at 632.8 nm gave Raman spectra of the green $[CoI_4]^{2-}$ complex and the red $[NiI_4]^{2-}$ complex, both in the solid phase as alkylammonium salts. The symmetric stretching, i.e. breathing, frequencies of these tetrahedral species were found to be 118 cm^{-1} for $[CoI_4]^{2-}$ and 105 cm^{-1} for $[NiI_4]^{2-}$ (Edwards *et al.*, 1970).

8.C.3 Information from bond polarisabilities

As indicated in Section 8.A.3, the intensity of vibrational Raman emission gives a measure of the bond polarisability, or to be more exact, $d\alpha/dr$. Absolute measurement of Raman intensities is extremely difficult, for much the same reasons as those discussed for the measurement of quantum yield in Section 7.B.3. Not only do the energies of the incident and scattered radiation have to be measured absolutely, but also the proportion of the scattered radiation reaching the detector must be estimated. It is preferable to use an internal standard, and relate observed peak heights to that of the standard which is present in constant concentration. For example, cyclohexane has been used as internal standard, peaks of other species being referred to the cyclohexane peak at 806 cm^{-1} (Clark and Willis, 1971).

The polarisability of a bond may be expected to depend on its degree of covalent character. The more ionic a bond, the more the electrons are located around the nuclei, and so the less influenced by external electric fields. Clark and Willis (1971) found that for many halides of type MX_4, where M is C, Si, Ge, and Sn, and X is a halogen, there is a moderately good agreement between observed values of $d\alpha/dx$ and values calculated from the equation

$$d\alpha/dx = Cp(Z_M + Z_X) \tag{8.4}$$

where C is a constant, Z_M is the atomic number of M, and p is the fractional covalent character of the M–X bond calculated from

$$p = \exp[-(\chi_M - \chi_X)^2/4] \tag{8.5}$$

where χ_M and χ_X are the Pauling electronegativity values of M and X respectively.

The observed intensities required correction for resonance enhancement before calculation of $d\alpha/dx$. Equation 8.4 shows that the polarisability of a bond increases with the atomic number of the atoms it links, so atomic polarisability contributes to the effect. Equation 8.4 predicts that an ionic bond, for which p is zero, will be Raman inactive. Raman emission from concentrated aqueous solutions of ions or fused ionic salts may thus be attributed to ion-pair formation by transient covalent bonds, but such emission is usually too weak for satisfactory interpretation.

It is reasonable to suppose that, other things being equal, the more electrons in a bond, i.e. the higher the bond order, the higher the value of $d\alpha/dr$. This has been found for hydrocarbon molecules in the gas phase (Yoshino and Bernstein, 1959). The values of $d\alpha/dr$, in units of 10^{-16} cm^2, are 1.37, 1.54, 1.89, and 2.94 for the carbon–carbon bonds in ethane, benzene, ethylene, and acetylene respectively. (The unit for $d\alpha/dr$ of cm^2 is in the electrostatic c.g.s. system; 1 cm^2 is equal to $10^{-13}/9$ C m V^{-1} in the SI.) Values of $d\alpha/dr$ have been used to estimate bond order in oxyanions (Chantry and Plane, 1960). It has been found that in ClO_4^- the effective number of double bonds approaches the value 2, whereas the P–O bonds in PO_4^{3-} are effectively single bonds, with negligible contributions of π bonding. In such arguments it is necessary to disentangle the effects of bond order from bond polarity, since both affect $d\alpha/dr$.

8.C.4 Low-frequency studies

Raman spectroscopy is used to study many of the same problems as far infrared spectroscopy. Ring-puckering modes of cyclic molecules, as discussed in Section 5.C.1, are Raman-active. Raman spectroscopy has the advantage over far infrared spectroscopy that it is possible to observe the pure puckering modes of molecules without a permanent dipole moment. For example, cyclobutane does not absorb in the far infrared, but gives weak Raman emission at $\Delta\tilde{\nu}$ of 117, 155, 175, and 197 cm^{-1} (Stone and Mills, 1970). These energy levels fit a quartic function of the same type as Eqn. 5.1, and a barrier to inversion of 42 J mol^{-1} (503 cm^{-1}) is found. The equilibrium dihedral angle of the ring was found to be 35°. The values found agreed with those calculated from the puckering fine-structure of the vibrational bond in the infrared absorption spectrum at 1453 cm^{-1}.

Raman spectroscopy, like far infrared spectroscopy as discussed in Section 5.C.3, is also used to study vibrational modes of crystal lattices (Nakayawa, 1974) but this application belongs more perhaps to the field of physics than to that of chemistry.

8.D Analytical Applications

8.D.1 Qualitative analysis

Vibrational Raman spectra can be used for the identification of organic compounds by the method of group frequencies, as discussed in Section 4.E.1. Although the frequency of a particular vibrational mode is the same in whichever spectra it is observed, the relative peak heights are in general different. Infrared absorption is a function of bond polarity, whereas Raman emission is a function of bond polarisability. Thus, for example, –OH and –NH stretching and bending bands are strong in the infrared but weak in the Raman spectrum, whereas C═C, C≡N, and C≡C stretching bands are weak in the infrared and strong in the Raman spectrum. These intensity differences may cause apparent differences in the observed group frequencies in the two spectra. Groups of type AB_2 have two stretching modes; the symmetric stretch is stronger in the Raman, and the antisymmetric stretch is stronger in the infrared. For example, the infrared group frequencies for alkyl ethers are 1060–1150 cm^{-1}, whereas the Raman group frequency is 950 cm^{-1}, corresponding respectively to antisymmetric and symmetric stretching modes of C–O–C. The other group frequency may be seen in each spectrum, but much more faintly. Some group frequencies are difficult to observe in the infrared, but very prominent in the Raman. Examples are N═N and O–O, which give intense Raman emission at 1530–1585 and 700–900 cm^{-1} respectively. Raman spectroscopy is thus a most suitable technique for the study of diazo and peroxide compounds.

8.D.2 Quantitative analysis

Although Raman spectroscopy is not generally used for quantitative analysis, it has proved useful for the study of some ionic equilibria in aqueous solution. Although visible or ultraviolet absorption spectrophotometry is the preferred technique for such measurements in general, there are some systems for which it is not suitable; for example, neither reagent nor product may have an accessible absorption spectrum, or they may have similar spectra, so no significant spectral change is observed on reaction. The dissociation constants of several strong acids have been measured by quantitative Raman spectroscopy. The peak height for the anion in a standard aqueous solution of a salt of the acid in question is compared with the peak height for the anion in a series of solutions of the acid. A slight variation of this method was used to measure the dissociation constant of *p*-toluenesulphonic acid, in which the peak height for the C–H stretch of the methyl group was used as internal standard. With increasing acid concentration, the ratio of the height of the peak for SO_3^- to that for C–H decreased (Bonner and Torres, 1965). A value of 11.6 mol l^{-1} was found for the dissociation constant. By contrast, a Raman study of perchloric acid showed that ionisation was complete up to a concentration of 8 mol l^{-1} (Heinzinger and Weston, 1965).

The dissociation of complex ions has also been studied by Raman spectroscopy. For example, the dissociation constant of $Zn(C_2O_4)_3^{4-}$ was found to be 0.3 mol l^{-1} by measurement of the intensity of the peak at 443 cm^{-1} due to free oxalate ion (Gruen and Plane, 1967).

REFERENCES

Behringer, J. (1974) Specialist Periodical Reports of the Chemical Society, Molecular Spectroscopy, **2**, 100.
Bendtsen, J. (1974) *J. Raman Spectroscopy* **2**, 133.
Bonner, O. D. and Torres, A. L. (1965) *J. Phys. Chem.* **69**, 4109.
Chantry, G. W. and Plane, R. A. (1960) *J. Chem. Phys.* **32**, 319.
Clark, R. J. H. and Willis, C. J. (1971) *Inorg. Chem.* **10**, 1118.
da Silveira, A., Marques, M. A. and Marques, N. M. (1965) *Mol. Phys.* **9**, 271.
Durig, J. R. and Hannum, S. E. (1970) *J. Chem. Phys.* **52**, 6089.
Edwards, H. G. M., Woodward, L. A., Gall, M. J. and Ware, M. J. (1970) *Spectrochim Acta* **26A**, 287.
Gruen, E. C. and Plane, R. A. (1967) *Inorg. Chem.* **6**, 1123.
Heinzinger, K. and Weston, R. E., Jr. (1965) *J. Chem. Phys.* **42**, 272.
Miller, F. A. and Carson, G. L. (1961) *Spectrochim. Acta* **17**, 977.
Miller, F. A. and Snider, A. M., Jr. (1974) *J. Raman Spectroscopy* **2**, 377.
Nakayawa, I. (1974) *Appl. Spectroscopy Rev.*, 229.

San Filippo, J., Jr. and Sniadoch, H. J. (1973) *Inorg. Chem.* **12**, 2326.
Stone, J. M. R. and Mills, I. M. (1970) *Mol. Phys.* **18**, 631.
Tobias, R. S. (1967) *J. Chem. Educ.* **44**, 2.
Woodward, L. A. (1934) *Phil. Mag.* **18**, 823.
Yoshino, T. and Bernstein, H. J. (1959) *Spectrochim. Acta* **14**, 127.

BIBLIOGRAPHY

Anderson, A. (ed.) (1971) "The Raman Effect", Vol. I, Marcel Dekker, New York.
Anderson, A. (ed.) (1973) "The Raman Effect", Vol. II, Marcel Dekker, New York.
Dollish, F. R., Fateley, W. G. and Bentley, F. F. (1974) "Characteristic Raman Frequencies of Organic Compounds", John Wiley, Chichester, Sussex.
Durig, J. R. and Harris, W. C. (1972) in "Techniques of Chemistry, Vol. I, Physical Methods of Chemistry, Part III B, Optical, Spectroscopic and Radioactivity Methods" (A. Weissberger and B. W. Rossiter, eds.), pp. 85–206, Wiley-Interscience, New York and London.
Freeman, S. K. (1974) "Applications of Laser Raman Spectroscopy", Wiley-Interscience, New York.
Gilson, T. R. and Hendra, P. J. (1970) "Laser Raman Spectroscopy", Wiley-Interscience, New York.
Herzberg, G. (1945) "Molecular Spectra and Molecular Structure, Vol. II, Infrared and Raman Spectra of Polyatomic Molecules", Van Nostrand, New York.
Tobin, M. C. (1971) "Laser Raman Spectroscopy", Wiley-Interscience, New York.

9 Optical Rotatory Dispersion and Circular Dichroism

9.A The Theory of Optical Activity, Optical Rotatory Dispersion, and Circular Dichroism

9.A.1 Polarised radiation and optical activity

Plane polarised radiation is characterised by the electric vector being oriented along one axis only, this orientation being random in unpolarised radiation. The electric vector thus always lies in a single plane containing the direction of propagation. The passage of plane polarised radiation through some substances, which are termed *optically active*, causes this plane to be rotated about the direction of propagation. If the plane is rotated clockwise as seen by an observer looking towards the source of radiation, the substance is said to be *dextrorotatory*; if the plane is rotated anticlockwise the substance is *laevorotatory*. In order to be able to understand this phenomenon, one must consider the nature of polarised radiation.

At any point along the line of propagation the magnitude of the electric vector of plane polarised radiation oscillates in a sinusoidal fashion. This oscillation is equivalent to that of the vector sum of two vectors, of constant magnitude, whose orientation is changing at a constant angular velocity. Each of these vectors characterises *circularly polarised* radiation. If the tip of the vector rotates in a clockwise direction, as seen looking towards the source, about the axis of propagation, the radiation is said to be *right circularly polarised*; if the tip of the vector rotates anticlockwise, the radiation is *left circularly polarised*. It can be seen in Fig. 9.1(a) how the sum of right and left circularly polarised radiation gives plane polarised radiation. This is not a mere mathematical fiction; beams of right and left circularly polarised radiation can be produced and combined to give plane polarised radiation.

It was stated in Section 1.A that the velocity of light is less in any transparent medium than in a vacuum. This reduction is due to the interaction between the electric vector of the light and the polarisable electrons in the medium. The extent of the reduction is given by the refractive index. For example, the refractive index of 1-iodobutane (1.500 for light of wavelength 589 nm in air) is greater than that of 1-fluorobutane (1.342 for light of wavelength 589 nm in air) because there are many more electrons in the iodine atom than in the fluorine atom, and these electrons, being well away from the nucleus, are easily polarised.* Optically active substances have different polarisabilities for right and left circularly polarised radiation. This means that the refractive index and the velocity of propagation are different for the two components of plane polarised radiation, and one component is retarded relative to the other. The phase angle of retardation is linearly proportional to the path length of the radiation passing through the substance. The effect of the retardation can be seen in Fig. 9.1(b). The plane of polarisation is rotated by half the phase angle difference between the two components. In Fig. 9.1(b) the right circularly polarised component is retarded by 45°, causing a laevorotation of $22\frac{1}{2}°$. The relationship between the rotation ϕ, in radians, and the difference in refractive index, $(n_L - n_R)$, is given by

$$\phi = \pi d(n_L - n_R)\lambda \tag{9.1}$$

for radiation of wavelength λ in the substance passing through a distance d of the substance; $(n_L - n_R)$ is very small for most optically active substances. For example, the two refractive indices of laevorotatory α-quartz for radiation of wavelength 760 nm in air are 1.53914 for left circularly polarised light and 1.53920 for right circularly polarised light.

At any instant the electric vectors in a ray of circularly polarised light are arranged in a helical pattern, like the steps in a spiral staircase. The axis of the helix is the line of propagation of the ray. A helical pattern of atoms provides the difference in polarisability to the component vectors of plane polarised radiation necessary for optical activity. As shown in Fig. 9.2, a helix can be either *right-handed* or *left-handed*. If a point on the rim of the helix rotates in a clockwise sense as it moves away from the observer, the helix is right-handed; if it rotates in an anticlockwise sense, the helix is left-handed. The spatial distribution of the electric vectors in a ray of right circularly polarised light has a right-handed helical configuration, and the corresponding distribution for left circularly polarised light is left-handed. As can be seen in Fig. 9.2, a right-handed helix is right-handed from whichever end it is viewed; a left-handed helix is similarly always left-handed. However, the reflection of a right-handed helix in a mirror is a left-handed helix, and vice versa, as shown

* The use of the word "polarised" to describe two quite distinct phenomena, i.e. the distortion of an electron cloud by an electric field, and the orientation of the electric vector of electromagnetic radiation, is unfortunate, but should present no difficulty to the careful student.

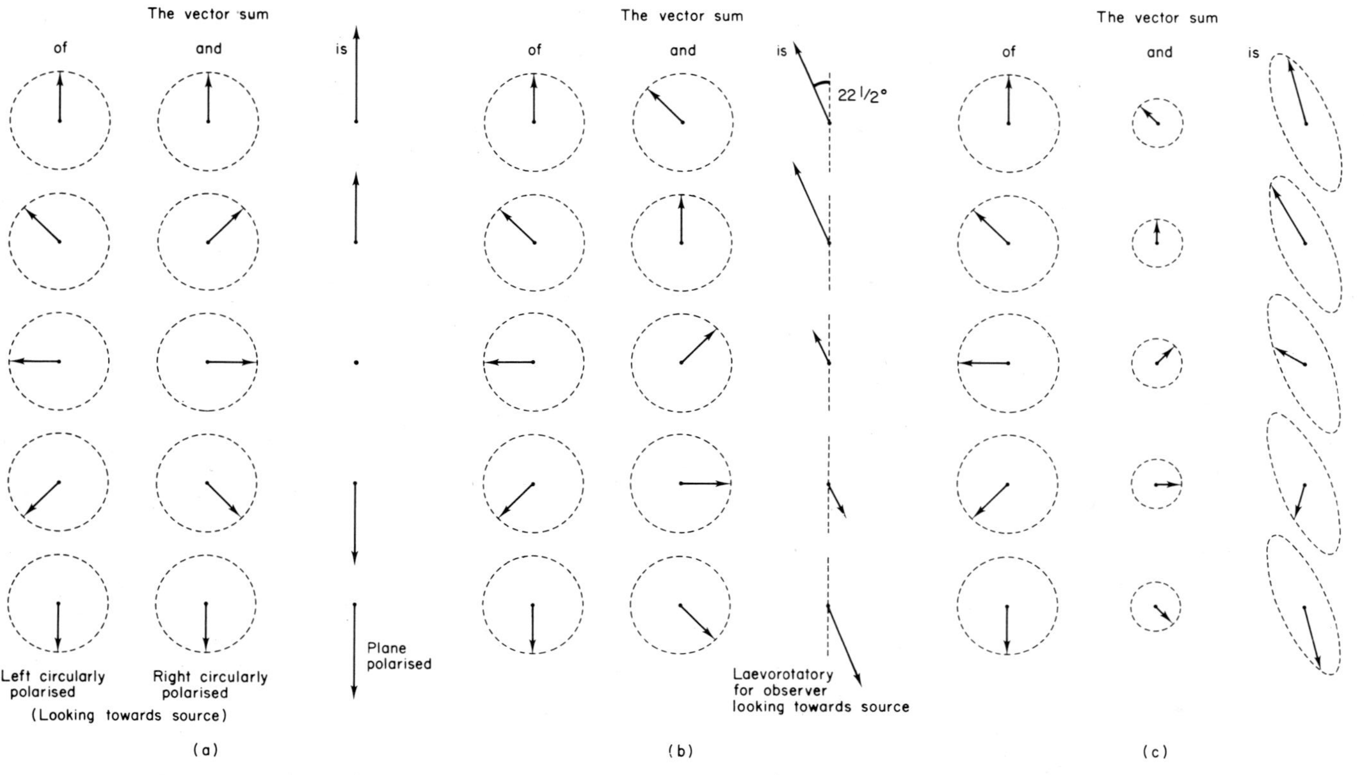

FIG. 9.1 (a) Plane polarised radiation. (b) Rotation of the plane of polarisation. (c) Elliptically polarised radiation.

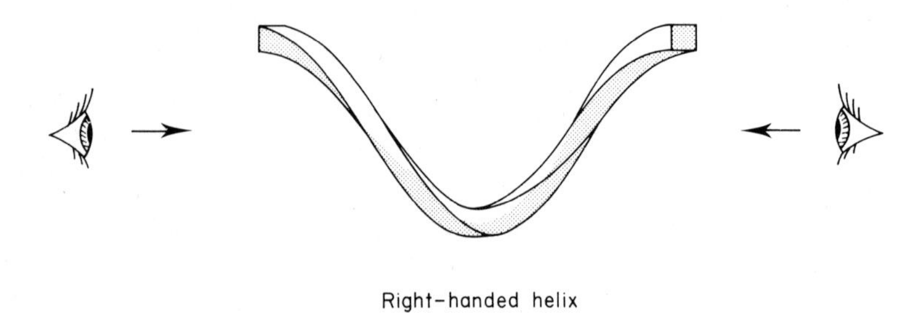

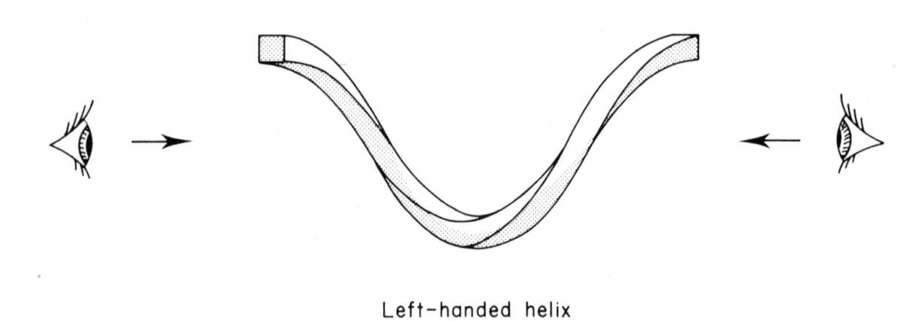

FIG. 9.2 Invariance of chirality with orientation.

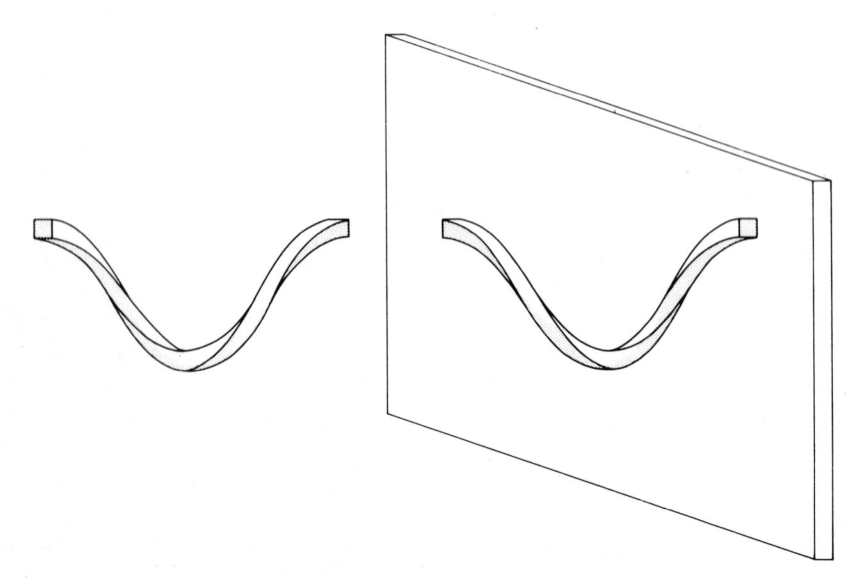

FIG. 9.3 Reversal of chirality by reflection in a mirror.

in Fig. 9.3. This phenomenon of right- and left-handedness is termed *chirality*, and, as will be seen, it is not only found in helices.

Although solutions of sodium chlorate are optically inactive, the crystals show optical activity. The structure of dextrorotatory sodium chlorate, as viewed along the y axis of the unit cell, is shown in Fig. 9.4. It can be seen that the chlorate ions are arranged in a left-handed helix in the unit cell. The effect of this helicity on the polarisability of the contents of the unit cell may be understood by consideration of a simplified model, shown in Fig. 9.5. Eight atoms are arranged in a left-handed helix and irradiated by right and left circularly polarised radiation, of wavelength equal to the pitch of the helix. Each atom is polarised by the electric field to form an electric dipole, as

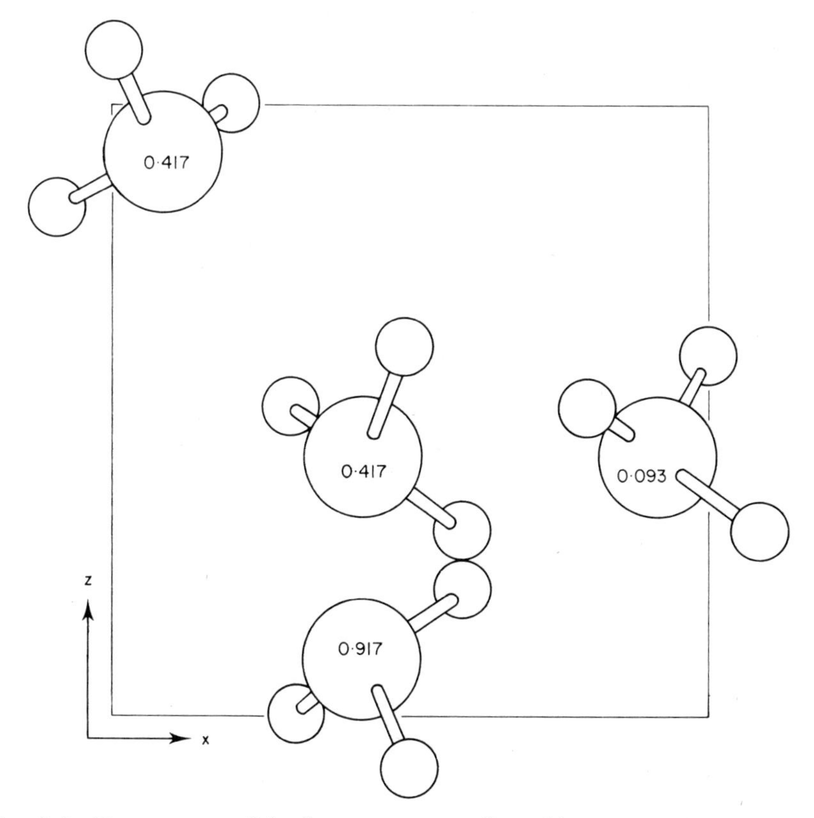

FIG. 9.4 The structure of the dextrorotatory sodium chlorate crystal. The unit cell is shown projected on the xz plane. The large circles represent Cl, the small circles represent O, and the Na^+ are omitted for clarity. The numbers represent the fractional distance of the centre of each Cl from the front face of the unit cell along the y axis.

shown in Fig. 9.5. The magnitude of a dipole is affected by neighbouring dipoles as a consequence of electrostatic induction; the closer together the negative end of one dipole and the positive end of another, the greater is the induced dipole moment. For left circularly polarised radiation the dipoles are aligned so as to give a minimum distance between the positive end of one and the negative end of the next; this component is thus preferentially retarded, and the left-handed helix is dextrorotatory. Although in practice the pitch of the helix is much less than the wavelength of the radiation, the principle is the same but the magnitude of the effect is much reduced. The *cholesteric mesophases* of liquid crystals contain molecules stacked in helical fashion in such a way that the pitch of the helices is of the same order of magnitude as the wavelengths of visible light. These exhibit extremely high rotations.

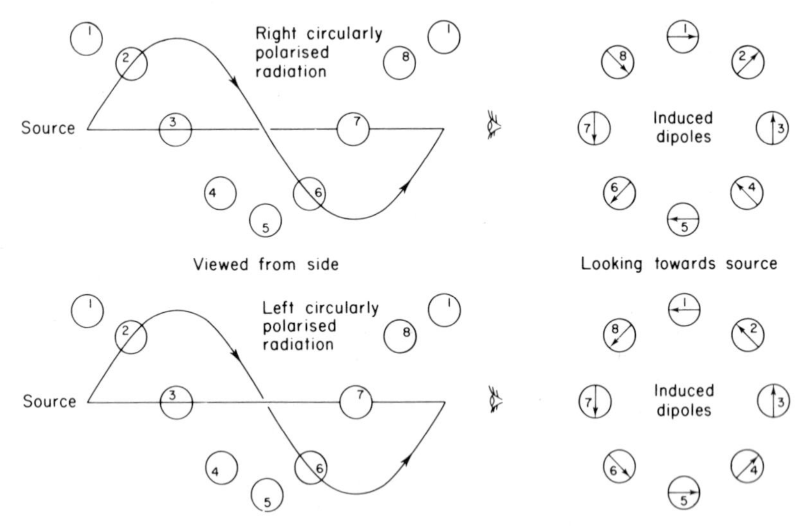

FIG. 9.5 The interaction between circularly polarised radiation and a left-handed helical arrangement of atoms.

Unfortunately, in general it is not possible to predict by inspection whether a substance is dextro- or laevo-rotatory by examining the chirality of its helical molecular structure. A detailed examination of all the polarisabilities of all the atoms and their relative positions is required.

The helical structure of sodium chlorate disappears when the crystals are dissolved in water, since the individual ClO_3^- ions do not have a helical form. However, many molecular species are known that do have a helical structure. An example, 1,3-dimethylallene, is illustrated in Fig. 9.6, in which the two chiral forms, or *optical isomers*, are shown (Jones and Walbrick,

1968; Waters and Caserio, 1968). The dextrorotatory isomer has the methyl groups arranged in a right-handed helix, viewed along the C=C=C axis, and the laevorotatory isomer has the methyl groups in a left-handed helix, viewed along the C=C=C axis. As shown in Fig. 9.2, the chirality of a helix is unchanged by changing orientation, so 1,3-dimethylallene exhibits optical activity even in solution, where the molecules are oriented at random. As was noted for crystals, it is not possible to predict which optical isomer is dextro- and which laevo-rotatory merely by inspection of the chirality. It depends on the relative polarisability and position of every atom in the molecule. The assignment of the correct structure to an experimentally prepared sample of optically active substance is known as the determination of *absolute configuration*. This can be done either by direct X-ray diffraction analysis of the crystal structure, or by following the substance through a series of chemical reactions, of types known not to affect the absolute configuration, until a substance of known absolute configuration is reached.

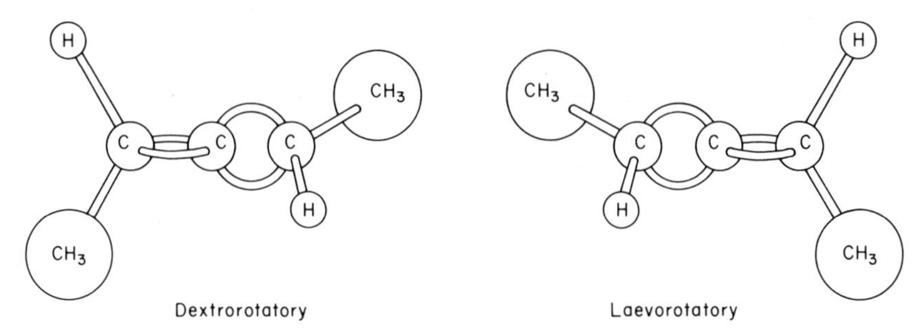

Fig. 9.6 The chirality of 1,3-dimethylallene.

The majority of optically active organic compounds do not show a spiral structure in such an obvious manner as 1,3-dimethylallene. It is usually stated in textbooks that any molecule containing an asymmetrically substituted carbon atom, i.e. a molecule of the type CWXYZ, is optically active. It is not immediately obvious how such an asymmetric carbon atom gives a helical structure to the molecule. As an aid to understanding this phenomenon, one can consider the chirality of a helix to be defined by two parameters, namely the direction in which the axis is pointing and the sense of the rotation about this axis. If one uses the C–W bond, say, to define the axis, then the sequence $X \rightarrow Y \rightarrow Z$ defines the sense of rotation as in Fig. 9.7. All four substituents on the carbon atom must be different for the molecule to be chiral; if, say, Y and Z are identical the sense of rotation is ambiguous. The general rule for a molecule to be optically active is that, like a helix, it is not superimposable on its mirror image.

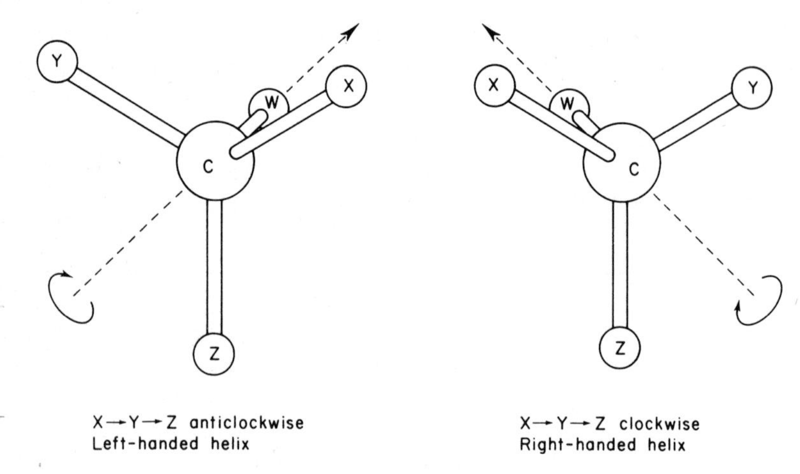

FIG. 9.7 The helicity of an asymmetrically substituted carbon atom.

The angle through which the plane of polarised light is rotated by passage through a solution of an optically active substance is proportional to the concentration and the path length. The *specific rotation*, $[\alpha]$, is defined by

$$[\alpha] = \phi/cd \tag{9.2}$$

where c is the concentration in g cm^{-3} and d is the path length in dm. The *molecular rotation*, $[\phi]$, is defined by

$$[\phi] = [\alpha]M/100 \tag{9.3}$$

where M is the molecular weight. The measurement of $[\alpha]$ is termed *polarimetry*, and instruments for this purpose are known as *polarimeters*.

The phenomenon of optical activity was observed and correctly interpreted before the development of X-ray crystallographic analysis. It was thus possible for early workers in the field to draw structures such as those in Fig. 9.6 or 9.7 but not know which corresponded to the dextrorotatory and which to the laevorotatory isomer. However, as has been stated, many optically active organic substances can be shown to have the same absolute configuration as others by linking them through series of chemical reactions that do not involve a change in the absolute configuration. Thus, if one assignment is made in a purely arbitrary manner, a large number of other assignments follow. The first convention of this type stated that dextrorotatory and laevorotatory glyceraldehyde have the absolute configurations shown in Fig. 9.8. Any molecule derived from dextrorotatory glyceraldehyde was assigned to be of type D, and any molecule derived from the laevorotatory isomer was assigned to be of type L, whatever their actual rotation might be. The actual rotation is indicated by (+) for dextro- and (−) for laevo-rotatory. For example, oxidation of D(+)-glyceraldehyde gives laevorotatory lactic acid without breaking any of the bonds to the asymmetric carbon atom. This isomer is thus termed D(−)-lactic acid. This convention has proved very useful in the elucidation of the structures of the sugars, all of which can be derived from glyceraldehyde. Subsequent X-ray crystallographic analysis has shown that D(+)-glyceraldehyde does indeed have the configuration assigned to it by this convention. It was, of course, an even chance that this would be so.

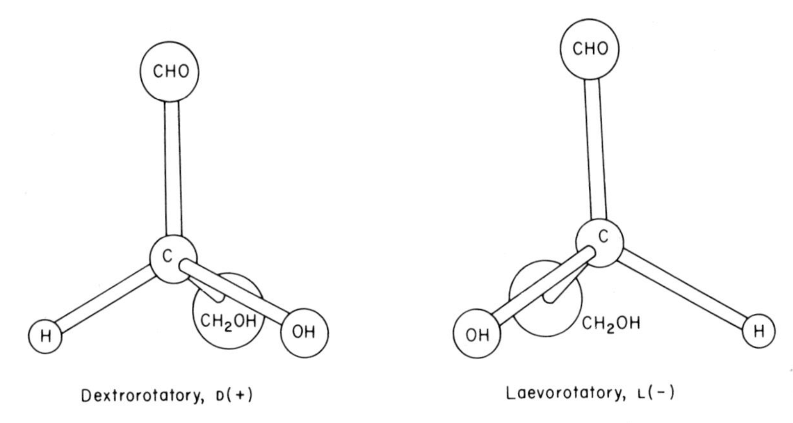

FIG. 9.8 The optical isomers of glyceraldehyde.

The D,L convention is ambiguous in some applications. A more recent convention employs the following set of rules to assign an absolute configuration. If a molecule contains an asymmetric carbon atom, the four atoms attached to it are assigned priority in order of decreasing atomic number. If two or more atoms have the same atomic number, one adds on the atomic numbers of the atoms attached to each. Multiple bonds count as separate single bonds, so a doubly bonded substituent atom is counted twice. A three-dimensional model of the molecule is viewed along the axis of the substituent of lowest priority, looking out from the central carbon atom. The sequence in decreasing priority of the other three substituents is then either clockwise, giving a configuration denoted by R (from the Latin for right, rectus), or anticlockwise, giving a configuration denoted by S (from the Latin for left, sinister). For example, glyceraldehyde has groups in descending priority order: $OH > CHO > CH_2OH > H$. Thus, as can be seen from Fig. 9.8, D(+)-glyceraldehyde is R and L(−)-glyceraldehyde is S.

9.A.2 Anomalous dispersion

The argument in Section 9.A.1 considers a set of atoms to be polarised such that the direction or magnitude of polarisation is time-dependent, fluctuating at a frequency equal to the frequency of the incident radiation. If this frequency is close to some natural frequency of the set of atoms, further effects may be seen. This natural frequency may be that of a real vibration of the atoms, i.e. a normal vibrational mode, or it may be that of a hypothetical oscillation related to some permitted transition between two electronic energy levels by the equation

$$\Delta E = h\nu_0 \tag{9.4}$$

where ΔE is the gap between the energy levels, and ν_0 is the frequency of the hypothetical oscillation.

It was stated in Section 1.A that the refractive index of a transparent medium is the ratio of the velocity of light in a vacuum to its velocity in that medium. This definition is incomplete, since light has more than one type of velocity. The most familiar velocity of light is the *signal velocity*, that is the velocity at which energy or information is transmitted by the light. It is this velocity that forms the foundation of Einstein's Theory of Relativity. However, the velocity that is relevant to considerations of refractive index is the *phase velocity*, which is defined as follows. Consider two points along the path of propagation of a ray of monochromatic light, distance d apart. At some arbitrary time the light has a certain phase at the first point, that is to say that the electric and magnetic vectors have a certain magnitude and direction. At a time t after this, the light has the same phase at the second point. The ratio d/t is the phase velocity. In a vacuum, the signal and phase velocity are equal, but in any medium they are not. The signal velocity has its maximum value in a vacuum, but the phase velocity can be greater than this under some conditions, as will be seen.

Consider radiation of frequency ν passing through a set of atoms with but one natural oscillation frequency, ν_0. If ν is very much less than ν_0, the atoms oscillate in phase with the incident radiation. Each oscillating atom acts as a source of secondary radiation. Because the amplitude of the electric vector of the secondary radiation is proportional to the rate of change of polarisation of the atom, the secondary radiation has a phase 90° behind that of the atomic oscillation, and hence 90° behind that of the incident radiation. In this limiting case, no interference occurs between the incident and secondary radiation. However, if ν is increased to a value approaching ν_0, the atomic oscillation can no longer keep up with the fluctuation of the applied electric field, and lags behind by a phase angle θ°. The secondary radiation then has a phase lag of $(90+\theta)$° relative to the incident radiation, and interference occurs.

This interference causes the phase velocity of the transmitted radiation, which is a composite of the primary and secondary radiation, to be reduced. The refractive index of the medium is thus greater than unity. When ν is nearly equal to ν_0, and θ approaches 90°, another effect occurs. Energy is transferred from the radiation to the oscillating atoms by the phenomenon of resonance, as discussed in Section 8.A.1. Were it not for the damping effect of this energy transfer, which causes loss of energy from the incident radiation, the amplitude of the atomic oscillation would increase to infinity when ν equals ν_0. When ν equals ν_0, θ has the value 90°, and the secondary radiation is exactly out of phase with the incident radiation. If ν is increased to a value slightly greater than ν_0, the secondary radiation lags behind the incident by a phase angle of more than 180°. The symmetry of a sinusoidal wave form is such that a phase lag of $(180+\phi)°$ is equivalent to a phase lead of $\phi°$. Thus, if ν is slightly greater than ν_0, the secondary radiation leads the incident radiation by a phase angle of nearly 180°, which falls to 90° at extremely high values of ν. The phase velocity of the transmitted radiation is thus extremely high for ν slightly greater than ν_0, falling to the value for a vacuum at high values of ν. The refractive index is thus less than unity for values of ν greater than ν_0. A detailed mathematical investigation, assuming no interatomic interactions, leads to equations of the type

$$n_r - 1 = \tfrac{1}{2}\nu_p{}^2(\nu_0{}^2 - \nu^2)/[(\nu_0{}^2 - \nu^2)^2 + \nu^2\tau^{-2}] \tag{9.5}$$

$$n_i = \tfrac{1}{2}\nu\,\nu_p{}^2\tau^{-1}/[(\nu_0{}^2 - \nu^2)^2 + \nu^2\tau^{-2}] \tag{9.6}$$

The effect on radiation of passage through a medium is summarised in general by the complex refractive index n, where

$$n = n_r + in_i \tag{9.7}$$

The real component n_r is the observed refractive index, and the imaginary component n_i is a measure of the extent of absorption of the radiation by the medium. The absorption index κ, defined by

$$I = I_0 \exp(-4\pi\kappa/\lambda) \tag{9.8}$$

is equal to n_i. The quantity ν_p is of dimensions s^{-1}, being related to the polarisability, and τ^{-1} is the width at half-height of the absorption peak. Plots of Eqns. 9.5 and 9.6 are shown in Fig. 9.9. For the purpose of evaluation, ν_0 was chosen to be 10^{17} Hz, corresponding to maximum absorption in the near ultraviolet, τ^{-1} was chosen to be 10^{16} Hz, and $\tfrac{1}{2}\nu_p{}^2$ was chosen to be $10^{33}\ s^{-2}$, so as to give a maximum value of unity for n_i. It can be seen from Fig. 9.9 that n_r has the form of the differential of n_i. This is an example of the *Kronig–Kramer transformation*, which relates the frequency-variation of refraction with the frequency-variation of absorption for any medium. The model for which Eqns. 9.5 and 9.6 are derived is a dilute gas. A good experimental example is provided by sodium vapour (Bevan, 1910) for which the refractive index is less than unity for a short wavelength range below 589 nm, the wavelength for absorption and emission. The simple model does not apply exactly to condensed systems, in which the induced dipole in one atom affects the dipoles in its neighbours, but the salient features of the curves in Fig. 9.9 are still seen.

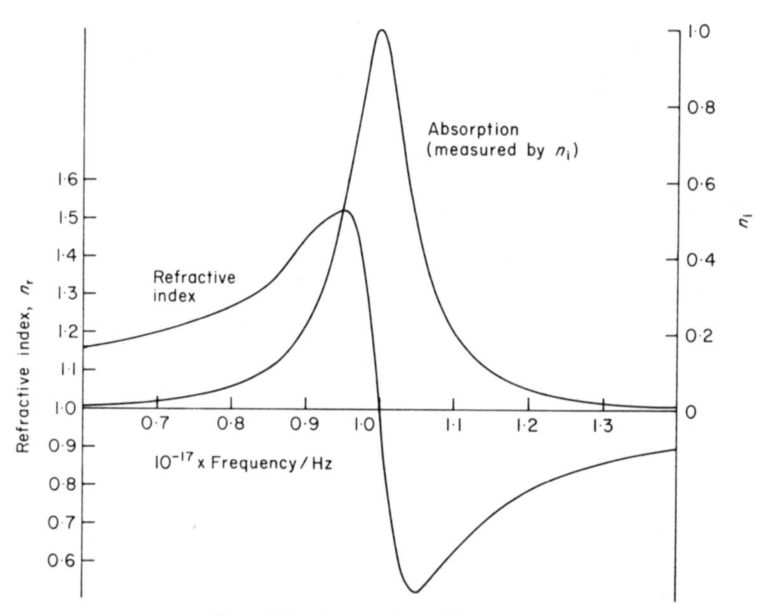

FIG. 9.9 Anomalous dispersion.

One important difference between the equations for gases at low pressures and condensed phases is that the absorption peaks in the latter are described by equations of the form

$$\kappa = \kappa_{\max} \exp[-(\nu_0 - \nu)^2/\Delta^2] \tag{9.9}$$

where 2Δ is the width of the peak at 1/e of the maximum height. Equation 9.6 is said to describe a *Lorentzian peak*, and Eqn. 9.9 is said to describe a *Gaussian peak*. In general, a system has many natural frequencies, and although ν may be greater than ν_0 for one of those, it is less than ν_0 for another. The observed refractive index is the total for all the different interactions, and, for all systems except gases at low pressures, this total is always greater than unity. However, the characteristic decrease in refractive index with increasing frequency may be observed on the high-frequency side of an absorption region. Early measurements of refractive index were made on substances transparent in the visible region of the spectrum, with absorption bands in the ultraviolet. It was found that the refractive index increased with decreasing wavelength of the light, as mentioned in Section 4.B.3, this variation being termed dispersion. It was then found that coloured substances, e.g. $KMnO_4$ (Taylor and Glover, 1933), showed a sharp decrease of refractive index in the absorption region, and this phenomenon was termed *anomalous dispersion.* It can be seen that "anomalous" is a misnomer; the phenomenon is quite general if looked for in the correct frequency region.

9.A.3 Circular dichroism and optical rotatory dispersion

Just as a substance may have a different refractive index for right and left circularly polarised radiation, so also it may have a different extinction coefficient. Figure 9.1(c) shows the effect if the right circularly polarised radiation is absorbed to a greater extent as well as being retarded. The vector sum of the two components traces out an ellipse, and the emergent radiation is said to be *elliptically polarised.* A substance that behaves in this way is said to exhibit circular dichroism, often referred to by the initials, CD. Circular dichroism can be measured either by the difference in extinction coefficients between the two hands of circularly polarised radiation

$$\Delta\varepsilon = \varepsilon_L - \varepsilon_R \tag{9.10}$$

or by the ellipticity, which is an angle whose tangent is the ratio of the minor to the major axis of the ellipse traced out by the vector sum. In practice the ellipse is very narrow and elongated, and $\tan\theta$ may be approximated to θ in radians. The *molecular ellipticity*, $[\theta]$ is defined in an analogous way to the molecular rotation, $[\phi]$. It can be shown (Eyring *et al.*, 1968) that

$$\begin{aligned}[\theta] &= 180 \times 100 \times \log_e 10 \times \Delta\varepsilon/4\pi \\ &= 3300\Delta\varepsilon\end{aligned} \tag{9.11}$$

if ε is in the conventional units of $\mathrm{l\ mol^{-1}\ cm^{-1}}$.

The rotation of the plane of polarised radiation by an optically active substance is in general found to vary with wavelength; this variation is termed optical rotatory dispersion, often referred to by the initials, ORD. Optical rotatory dispersion is related to circular dichroism in the same way that dispersion is related to absorption. Figure 9.10 shows the anomalous dispersion curves for right and left circularly polarised radiation passing through an optically active substance with an absorption maximum at 300 nm. Suppose, firstly, that the molecules interact more strongly with left

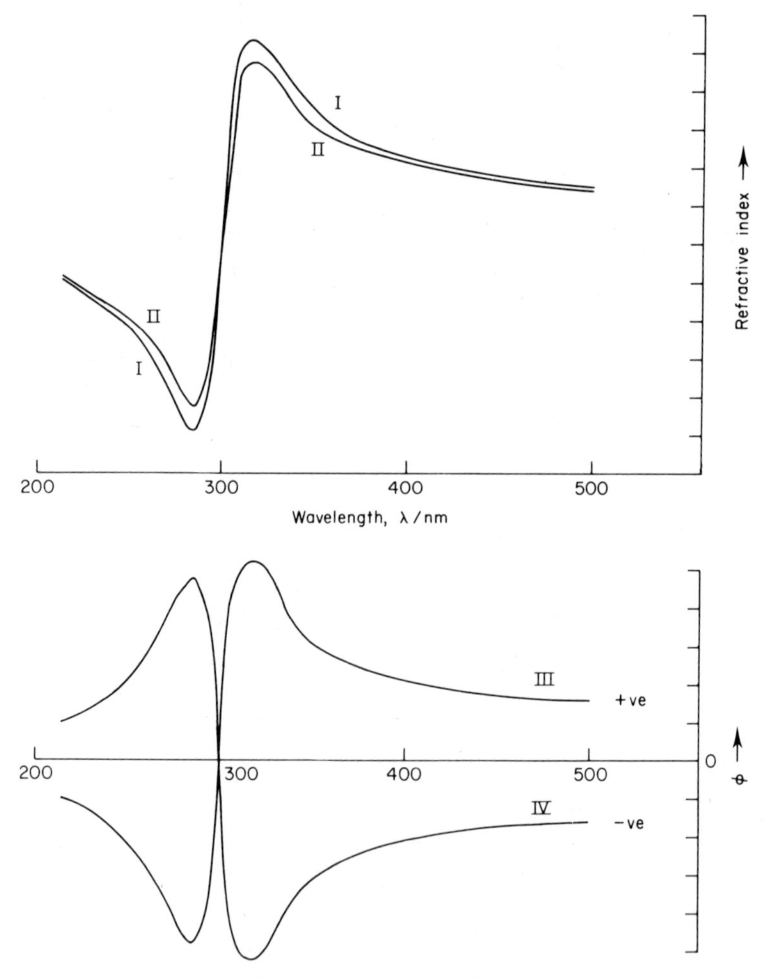

FIG. 9.10 Optical rotatory dispersion.

circularly polarised radiation than with right. Curve I then refers to left and curve II to right circularly polarised radiation. The observed rotation is, as shown by Eqn. 9.1, proportional to the difference in refractive index, and has the form shown by curve III. It can be seen that the substance is dextrorotatory for radiation of wavelength greater than 300 nm, but laevorotatory for radiation of wavelength less than 300 nm. The cross-over at 300 nm is termed the *Cotton effect*, and is said to be positive for curve III. If, on the other hand, curve I refers to right and curve II to left circularly polarised radiation, then the rotation is described by curve IV, which shows a negative Cotton effect.

If left circularly polarised radiation interacts more strongly with the medium, it will be absorbed more strongly as well as being retarded to a greater extent, so that

$$(n_L - n_R)_r > 0; \qquad (n_L - n_R)_i > 0$$

and hence

$$\varepsilon_L - \varepsilon_R > 0 \tag{9.12}$$

Thus a positive Cotton effect implies a positive $\Delta\varepsilon$, and vice versa.

The dependence of the sign of optical rotation on whether the wavelength used is greater or less than that for absorption introduces an ambiguity into the terminology "dextrorotatory" and "laevorotatory". It is the convention to use the (+) and (−) signs to refer to rotation at 589 nm. This is the wavelength of light emitted by the sodium vapour lamp. The sodium vapour lamp, in the form of a salt-impregnated asbestos pad in the flame of a bunsen burner, was the first and cheapest source of monochromatic light, and in its more modern form of an electric discharge lamp it is still widely used for polarimetry. The absorption/emission line of sodium vapour was originally labelled by Fraunhofer, in the early days of spectroscopy, as the D line (it is actually a doublet). The signs signifying sense of rotation may be clarified by the use of a subscript D to denote rotation at 589 nm, as $(+)_D$, $(-)_D$. Most electronic absorption bands of organic molecules occur at wavelengths shorter than 589 nm, so the sign of the rotation at 589 nm is often also the sign of the Cotton effect. It is this fortunate occurrence that enables so much stereochemical information to be obtained from values of $[\alpha]_D$, the specific rotation at 589 nm.

An interesting set of examples is provided by the helicenes, which are a series of polynuclear aromatic hydrocarbons in which each benzene ring is fused to the next at an angle, so that the molecule has an overall helical structure (Brown *et al.*, 1971; Brickell *et al.*, 1971). Pentahelicene, or 5-helicene, for example has the structure

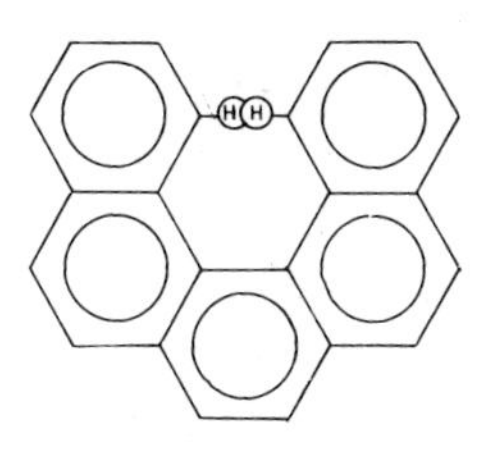

Because of steric hindrance between the two hydrogen atoms shown, the molecule is not planar but has a helical form. If the hydrogen atom on the right is uppermost, as in the figure, the molecule has the shape of a right-handed helix. 5-Helicene is thus found as a pair of optical isomers. The

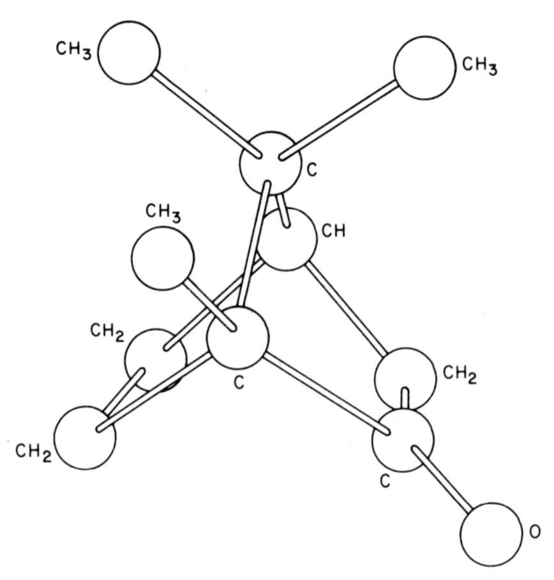

FIG. 9.11 The structure of D(+)$_D$-camphor; the hydrogen atoms are omitted for clarity.

isomers can be *resolved,* i.e. each prepared in a state free from the other isomer, but interconversion, or *racemisation,* is comparatively easy, and occurs rapidly in solution at room temperature because the energy barrier preventing the hydrogen atoms from pushing past each other is low. $(-)_D$-Pentahelicene is a left-handed helix, and the $(+)_D$-isomer of the 9,10-dicarboxylic acid derivative is a right-handed helix. The π electron system extends over the whole molecule, and is thus a helical chromophore. The first absorption maximum at 310 nm for the $(+)_D$-9,10-dicarboxylic acid derivative has an extinction coefficient of 5×10^4 l mol^{-1} cm^{-1} and exhibits a positive circular dichroism peak at this wavelength of +28 l mol^{-1} cm^{-1}, showing that the Cotton effect is positive for the π–π^* transition of lowest energy. At lower wavelengths absorption occurs due to other π–π^* transitions for which the Cotton effect is negative. The $(+)_D$-isomers of 6- and 7-helicene are similarly of right-handed helical form.

It is not necessary for the chromophore to be intrinsically helical or asymmetrical for circular dichroism to be observed. The carbonyl group, a much-studied chromophore, absorbs by virtue of a forbidden n–π^* transition. Absorption is made possible by molecular orbitals in neighbouring parts of the molecule. Permitted σ–σ^* transitions occur between these neighbouring molecular orbitals. Although the transitions themselves are of such high energy that they cause absorption only below 200 nm, they couple sufficiently

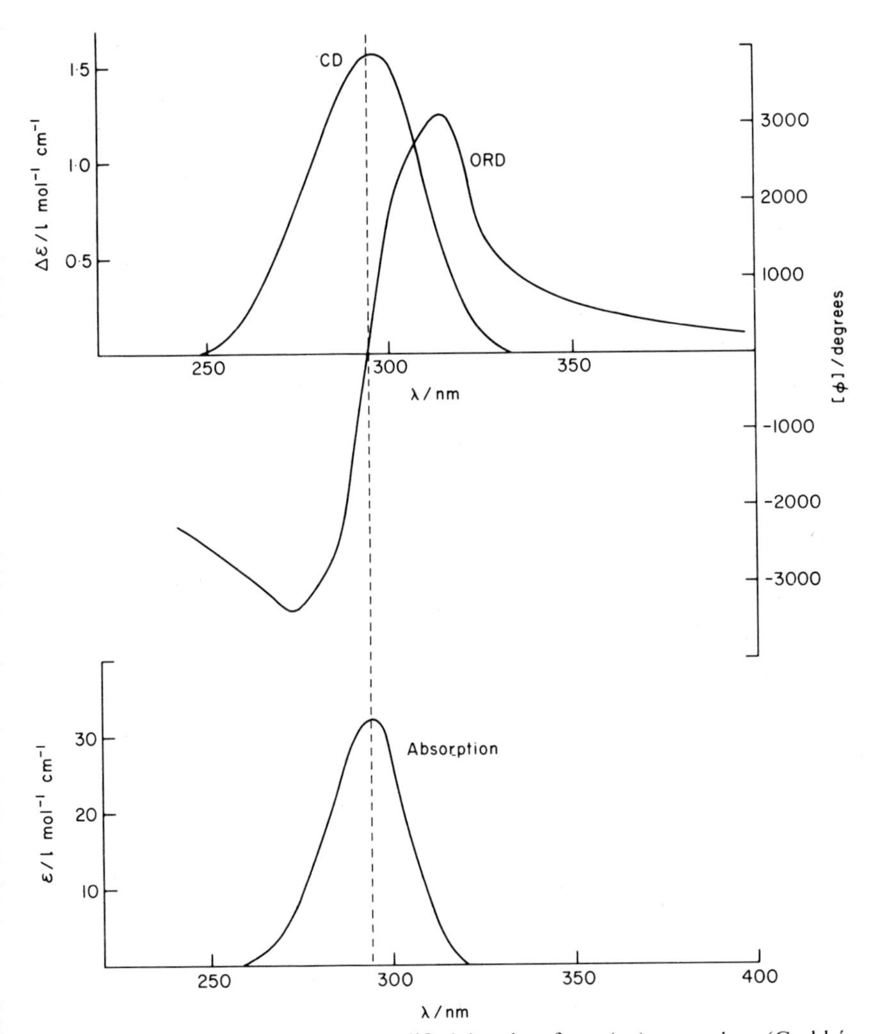

FIG. 9.12 The Cotton effect, exemplified by that for D(+)$_D$-camphor (Crabbé, 1972).

to the n–π^* transition to relax the selection rule and enable the carbonyl group to absorb weakly at around 270 nm. If the neighbouring parts of the molecule are asymmetric, the absorption of the carbonyl chromophore exhibits circular dichroism. A simple example is provided by camphor. The configuration of the D(+)$_D$-isomer is given in Fig. 9.11. The experimentally observed absorption spectrum, optical rotatory dispersion, and circular dichroism are shown in Fig. 9.12, and the similarities between these curves and those in Figs. 9.9 and 9.10 can be seen.

9.B Apparatus

9.B.1 Polarising devices

The measurement of optical activity requires a source of plane polarised radiation. Lasers provide such a source, but all other sources of radiation in the visible and ultraviolet, such as hot-filament and arc lamps, emit unpolarised radiation. The production of polarised radiation thus requires the use of a polarising device. Three types of such devices are in use, and are discussed below.

Crystals of the highest order of symmetry, the cubic class, e.g. NaCl, are *isotropic*, i.e. their physical properties are identical whether measured along the x, y, or z directions of the crystal axes. Other crystals of lower symmetry are *anisotropic*, i.e. the magnitude of some physical property depends along which axis it is being measured. Uniaxial crystals have the same magnitude of some physical property as measured along the x and y axes, but a different magnitude as measured along the z axis. This unique axis is termed the *optic axis*. One of the physical properties that has a unique value along this axis is the polarisability, which, as discussed in Section 9.A.1, determines the velocity of propagation of radiation through the crystal, and hence its refractive index. Uniaxial crystals thus have two values of the refractive index. A ray of unpolarised radiation directed along the optic axis is transmitted unchanged, since the electric vector, which is perpendicular to the direction of propagation, oscillates in a medium whose polarisability is the same for all orientations of the vector. However, for rays passing through the crystal in any other direction, the polarisability is a function of the orientation of the electric vector. The effect is to split the ray into two rays, plane polarised at right angles to each other, which travel through the crystal at different velocities and, in general, in different directions. The ray dependent on the polarisability along the x and y axes is termed the *ordinary ray*, and the beam dependent on the polarisability along the z axis is termed the *extraordinary ray*. The extraordinary ray is plane polarised in the plane containing the direction of propagation and the optic axis, and its electric vector is parallel to the optic axis. Crystals whose polarisability varies with orientation of the electric vector are said to exhibit *linear birefringence*, whereas crystals whose polarisability differs for the two senses of rotation of the electric vector in circularly polarised radiation are said to exhibit *circular birefringence*. Crystals in the latter class are much rarer than those in the former class, and the difference between the two values of the refractive index is usually much greater for linear birefringence than for circular birefringence. A commonly used linearly birefringent crystal is calcite, a transparent form of $CaCO_3$, for which the refractive index at 589 nm for the ordinary ray is 1.65836, and for the extraordinary ray is 1.48641. Thus, in calcite, the extraordinary ray is transmitted more rapidly than the ordinary ray and calcite is said to exhibit *negative birefringence*.

The magnitude of the change in direction of a ray of radiation passing from air into a refracting medium depends on the index of refraction, according to Snell's Law:

$$\sin i/\sin r = \mu \tag{9.13}$$

where i is the angle between the incident ray and the normal, i.e. a line perpendicular to the refracting medium, r is the corresponding angle for the refracted ray, and μ is the refractive index. Snell's Law does not apply to the extraordinary ray, which is why it is termed extraordinary. This is because, whereas the normal to the wavefront of an electromagnetic wave in an isotropic medium is along the direction of propagation, the corresponding normal for the extraordinary ray is always perpendicular to the plane of incidence. The ordinary ray behaves as if it were in an isotropic medium, and Snell's Law applies. The refractive index for the extraordinary ray is only given by Eqn. 9.13 if r is taken to be the angle between the normal to the wavefront and the normal to the plane of incidence. The consequence of the different values of μ for the ordinary and extraordinary rays is that, in general, they follow different paths through the crystal. In calcite, the ordinary ray is deviated more than the extraordinary ray. This difference is used to separate the two rays and so obtain plane polarised radiation. A crystal of calcite is cut into such a shape that the ordinary ray within the crystal meets a face at such an angle that it undergoes total reflection, and is lost, whereas the extraordinary ray passes through. There are many ways in which the crystal may be cut and two pieces cemented together to give a net zero deviation for the extraordinary ray and a total loss of the ordinary ray. The most commonly used design is the Nicol prism, in which the two pieces are cemented together by Canada balsam, which has a refractive index of 1.55, intermediate between the values for the ordinary and extraordinary rays. Birefringent polarising devices are often referred to as *nicols*, even though they may be constructed according to other designs. For use in the ultraviolet the Glan–Thompson–Taylor design, in which the two pieces are separated by an air gap, is more suitable. Calcite polarisers cannot be used below about 240 nm, since calcite in the required thickness becomes opaque to radiation below this wavelength. Ammonium dihydrogen phosphate may be used to construct polarisers functioning down to 185 nm.

Some substances are known that absorb radiation to a degree dependent on the orientation of the electric vector. This phenomenon is termed *linear dichroism*, and bears the same relation to circular dichroism as linear birefringence does to circular birefringence. The passage of unpolarised radiation through such a substance causes the absorption of all radiation except that polarised in one plane, so the emergent radiation is plane polarised. Although a few such substances occur naturally, e.g. tourmaline, the most familiar example is the synthetic material Polaroid, which contains oriented microcrystals of dyestuffs embedded in a transparent plastic film. Polarising elements based on linear dichroism are much cheaper than those based on linear birefringence, but the plane of polarisation is defined much less closely, and they are usually unsuitable for work in the ultraviolet.

Radiation reflected from isotropic surfaces is partially polarised, the plane of polarisation being normal to the plane of incidence. If the angle of incidence is Brewster's angle, $\tan^{-1} \mu$, the reflected radiation is 100% polarised. Reflection-type polarisers may be used in regions of the spectrum, such as the infrared and far ultraviolet, in which birefringent polarisers are opaque. Their principal disadvantage is the low intensity of the polarised radiation, and it is for this reason that birefringent polarisers are used when possible.

The electric vector of plane polarised radiation may be resolved into two components, perpendicular to each other, each at 45° to the original vector. If one of these components is caused to lag in phase by 90° behind the other, the resultant electric vector is that of circularly polarised radiation. Such a lag can be produced by passing the plane polarised light through an anisotropic crystal of the correct thickness. This device is called a *quarter-wave plate*. It suffers from the drawback that it functions only for monochromatic light, since the magnitude of the retardation depends on the wavelength. The *Fresnel rhomb* is another device for producing circularly polarised light. A Fresnel rhomb is a block of crown glass or silica, cut so that opposite sides are parallel, and adjacent sides are not perpendicular but at such an angle that incident light is subject to two internal reflections before emerging. The phase retardation is 45° at each reflection.

9.B.2 Electro-optic and magneto-optic devices

A substance that is intrinsically isotropic may become anisotropic in the presence of an electric field. This phenomenon is known as the *Kerr effect*. Liquids are intrinsically isotropic owing to their random molecular motion. However, if the molecules are polar they tend to line up along the lines of force of an applied electric field, and the liquid then acts as a uniaxial crystal and exhibits linear birefringence. The effect is most marked for liquids of high dielectric constant and low conductivity; nitrobenzene provides the best example.

The use of liquids is experimentally inconvenient, and a more popular device is the *Pockel cell*, which consists of a thin slice of a crystal of ammonium hydrogen phosphate across which an electric field can be applied. Ammonium hydrogen phosphate crystals are intrinsically uniaxial, so the crystal is cut to the optic axis to eliminate the intrinsic anisotropy. Light travelling along the optic axis then sees the crystal as intrinsically isotropic, of refractive index n_0. If an electric field is applied across the crystal, along the optic or z axis, the electron clouds in the ions in the crystal are polarised in such a way that the crystal becomes anisotropic in the xy plane. Two new axes are set up in the xy plane, the x′ and y′ axes, at 45° to the original x and y crystallographic axes. The difference in refractive indices for two rays of plane polarised light whose electric vector is along each of these new axes is given by

$$n_{x'} - n_{y'} = n_0^3 r_{eo} V_z / d \tag{9.14}$$

where r_{eo} is the electro-optic constant, V_z is the applied electric field along the z axis, and d is the thickness of the slice. The difference in velocity of propagation of the two rays means that they get out of phase during their passage through the crystal. The phase difference δ is given by

$$\delta = (2\pi/\lambda) n_0^3 r_{eo} V_z \tag{9.15}$$

and so is independent of the thickness of the crystal slice but varies inversely with λ, the wavelength of the light. There is thus a value of V_z for each value of λ for which the Pockel cell acts as a quarter-wave plate, having a value of 90° for δ. The Pockel cell can be used to produce circularly polarised light over a wide range of wavelengths by varying the applied voltage as the wavelength varies. The magnitude of r_{eo} for ammonium dihydrogen phosphate is such that the value of V_z required ranges from 1350 V at 220 nm to 5390 V at 600 nm.

The sense of the circularly polarised light is reversed by reversing the polarity of the applied voltage. The main difficulty with the construction of the Pockel cell is the design of the electrodes. These must be transparent, since they are in the light path. A thin layer of glycerol, in which ammonium hydrogen phosphate has been dissolved to give adequate electrical conductivity, has been successfully used.

Any substance may be made to exhibit circular birefringence, i.e. optical activity, by the application of a magnetic field. This phenomenon is known as the *Faraday effect*. A free electron moves in a circular path in a magnetic field, since it is constantly subjected to a force perpendicular to its direction of motion at any instant. The electrons of the molecules of any substance in a magnetic field thus tend to move in circles. The plane of motion is perpendicular to the lines of force of the field, and the sense of the rotation depends on the direction of the field. The circular motion of the electrons causes their polarisability to be different for right and left circularly polarised radiation, and hence the substance is rendered optically active. If the magnetic field is set up by a solenoid, the sense of the optical rotation is the same as the sense of the magnetising current for most substances. If the positions of light source and observer are interchanged, the sense of the observed optical rotation is reversed, since the sense of the magnetising current has been reversed. Thus optical rotation due to the Faraday effect differs from optical rotation due to intrinsic optical activity, since the sense of the latter is independent of the relative positions of light source and observer. It follows that, if a mirror is put at the opposite side of the sample to the light source, and if the observation is made on the same side of the sample as the light source, the rotation due to the Faraday effect is doubled whereas that due to intrinsic optical activity is cancelled out. This affords a useful technique for the separation of Faraday rotation from that due to intrinsic optical activity. The magnitude of Faraday rotation is small compared with that due to intrinsic optical activity; it is rare to obtain a rotation as great as 0.5° for a field of 1 T.

9.B.3 The measurement of optical rotation

Every chemistry student is familiar with the simplest device for the measurement of optical rotation, the polarimeter, in which the light source is a sodium lamp and the detector is the eye of the observer. Light from the lamp is made plane polarised by a nicol before passing through the sample cell, which is a cylindrical tube, 10 cm long, with windows at each end. After leaving the sample cell, the light passes through a second nicol, the analyser, before being observed in a telescope eyepiece. If the sample cell is empty, a minimum intensity of light is observed if the analyser nicol has an orientation 90° to the polariser nicol. If the sample cell is now filled with an optically active liquid giving a dextrorotation of $\alpha°$, the analyser must be rotated through an angle of $\alpha°$ in a clockwise direction to restore the condition of minimum intensity of observed light. The magnitude of the rotation is measured by a circular scale fitted to the analyser. Because it is much easier for the eye to detect whether two adjacent luminous regions are of the same brightness than it is to detect a minimum brightness, modern polarimeters have a more elaborate optical arrangement giving a half-shade effect. The field of view is split into two and the correct orientation of the analyser is shown by the equality of brightness of the two parts of the field.

The design of a recording *spectropolarimeter*, a device that displays optical rotation as a function of wavelength, could in principle be based on that of a simple polarimeter. The sodium lamp would be replaced by a source of white light, e.g. a xenon arc, and a monochromator whose wavelength drive would be linked to the wavelength drive for the recorder chart. A photomultiplier would replace the eye, controlling a servomotor which would rotate the analyser until the electric signal from the photomultiplier was at a minimum. However, such a design would suffer from the fundamental disadvantage that the optical signal disappears at the point of measurement, just when it should be as large as possible. Shot noise would be significant, and the displayed trace on the chart would be very noisy. It is far preferable to have some analogue of the half-shade technique of visual polarimetry. This is provided by modulating the orientation of the plane polarised light incident on the sample. The modulation can be provided by rapidly oscillating the polariser through a small angle by mechanical means, or by applying an alternating current to a Faraday cell, consisting of a cylinder of quartz within a powerful solenoid, in the optical path. If the analyser is in the correct orientation, the signal from the photomultiplier is a simple sine wave at twice the modulation frequency. If the analyser is not in the correct orientation, the sinusoidal signal is distorted in that the amplitudes of successive peaks are not equal. This is equivalent to the superposition of a second sine wave on the original sine wave found for the correct analyser orientation. This second sine wave is at half the frequency of the original, i.e. its frequency is equal to the modulation frequency. The signal from the photomultiplier is electronically filtered so that only the signal at the modulation frequency is observed, and a servomotor rotates the analyser until this signal is zero.

In general, the samples studied are solutions of fairly low concentration, since the absorbance at the wavelength of maximum absorption must not be so great that insufficient light reaches the photomultiplier. In practice a value of 2.0 is taken as the maximum usable absorbance. The rotations observed for these dilute solutions are low, typically one or two degrees at most, so it is necessary to measure rotations to an accuracy of a few millidegrees. The change in light intensity corresponding to a change in rotation of the analyser of a few millidegrees is extremely small, so an intense light source and a sensitive low-noise detector are required. As a rough rule-of-thumb guide, it may be taken that the required photoelectric signal-to-noise ratio for spectropolarimetry is a hundred times greater than for ordinary spectrophotometry.

9.B.4 The measurement of circular dichroism

The earlier commercial instruments for the measurement of circular dichroism were adaptations of spectropolarimeters. The passage of plane polarised radiation through a circularly dichroic sample causes the radiation to become elliptically polarised as shown in Fig. 9.1(c). A measurement of the ellipticity

θ gives a value for the circular dichroism $\Delta\varepsilon$ by use of Eqn. 9.10. θ is measured by introducing an achromatic quarter-wave retarder, such as a Fresnel rhomb, into the optical path of a spectropolarimeter. The rotation displayed by the spectropolarimeter is then the ellipticity θ, not the optical rotation α. The main drawback to this technique is that elliptical polarisation need not be solely due to circular dichroism. Reflections of stray light will also be elliptically polarised and give rise to error.

A more satisfactory technique for the measurement of circular dichroism is the direct measurement of the difference in absorbance by the sample of right and left circularly polarised light. The light from a monochromator is first plane polarised then sent through a Pockel cell to which an alternating voltage in a sinusoidal waveform is applied, so that pulses of right and left circularly polarised light are produced alternately. The magnitude of the applied voltage is controlled by the monochromator setting so that the phase retardation, as given by Eqn. 9.15, is exactly the required value of 90°. The light then passes through the sample and thence to a photomultiplier. Because of the difference in absorbance for the right and left circularly polarised light, the signal from the photomultiplier is modulated at the Pockel cell modulation frequency, varying about a mean that corresponds to the mean absorption of the sample. The ratio of the magnitudes of the alternating to the steady signal that is obtained electronically gives a measure of $\Delta\varepsilon$. The minimum difference in absorbance between right and left circularly polarised light that can be observed is of the order of 0.0002.

9.C Information Available from Optical Rotatory Dispersion and Circular Dichroism

9.C.1 Organic molecules

The discussion in Section 9.A.3 shows that in principle the sign of the Cotton effect can be used to determine whether an asymmetrical molecule is right- or left-handed. The Cotton effect is a much more reliable guide to configuration than the sign of the optical rotation at 589 nm, since the latter is the algebraic sum of the rotation due to all the chromophores in the molecule at a wavelength far removed from their maximum effect. An example is provided by the steroid 16β-hydroxymethyl-17α-progesterone. The observed rotation at 589 nm is +12°, but the Cotton effect at 340 nm, associated with the Δ^4-3-keto chromophore is negative, whereas that at 300 nm associated with the saturated keto chromophore is positive as shown in Fig. 9.13. The Cotton effects of the steroids have been much studied, because the steroid nucleus is

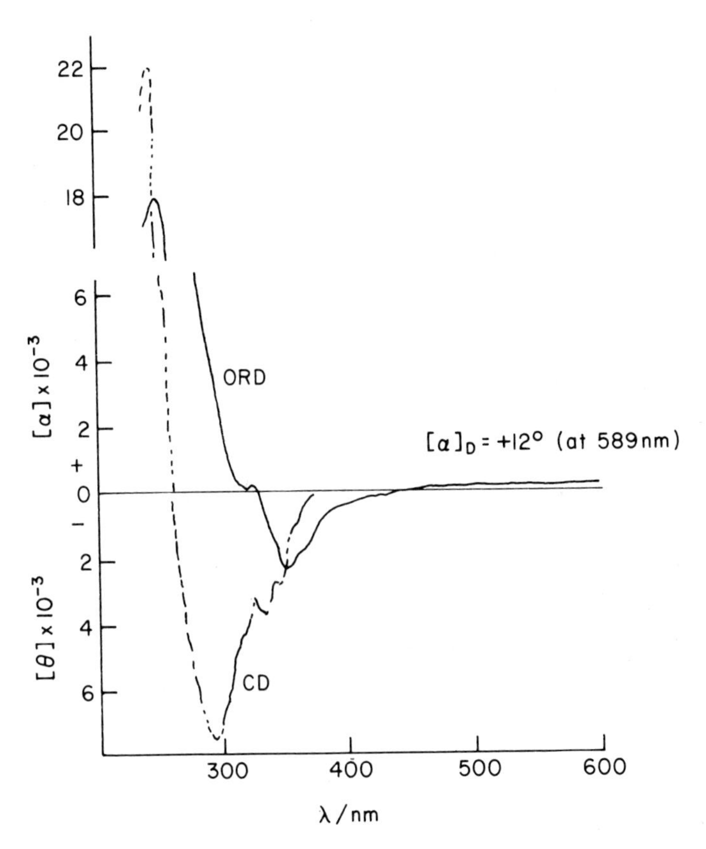

FIG. 9.13 The Cotton effect for 16β-hydroxymethyl-17α-progesterone (Crabbé 1971).

of well-characterised stereochemistry. Many empirical rules relating the sign of the Cotton effect to absolute configuration have been established using steroids whose absolute configurations have been established by other methods. The first rule of this type to be formulated was the *octant rule* for the carbonyl chromophore. The molecule is visualised as being oriented such that an observer is looking along the axis of the carbonyl chromophore, with the oxygen atom nearest to the observer, as shown in Fig. 9.14. The molecule is divided into eight sections by three orthogonal planes which intersect at the centre of the oxygen atom. The atoms in each section make a contribution to the sign of the Cotton effect as shown in Fig. 9.15. For the molecule chosen as example in Fig. 9.14, D(+)$_D$-camphor, all the other atoms lie behind the oxygen atom, so only the back quadrant need be considered. Carbon atoms 1, 2, 3, 4, 5, and 10 lie on the dividing planes, and so make no

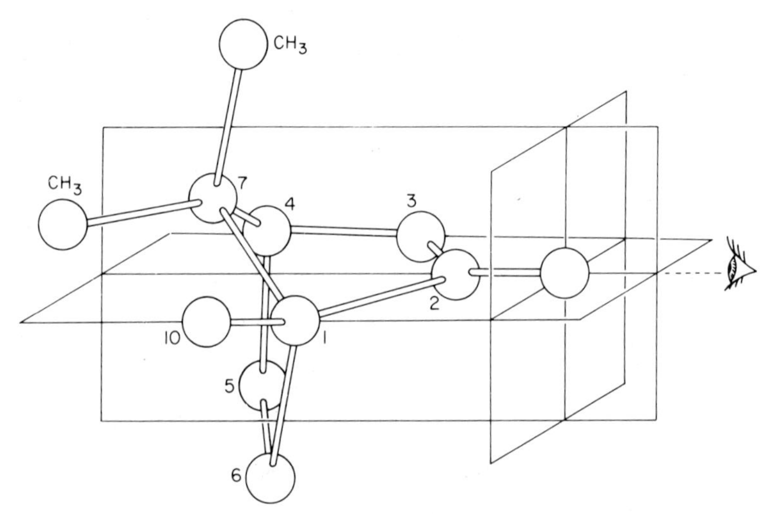

FIG. 9.14 The application of the octant rule for ketones to D(+)$_D$-camphor.

contribution, and the effect of atom 6 in the lower left-hand quadrant is exactly counterbalanced by the effect of atom 7 in the upper left-hand quadrant. The chirality of the molecule is determined by the two methyl groups in the upper left-hand quadrant, and the octant rule thus predicts that, as observed, D(+)$_D$-camphor has a positive Cotton effect, as shown in Fig. 9.12. The L-isomer has the two methyl groups in the upper right-hand quadrant and so has a negative Cotton effect. Unfortunately the octant rule is not invariably true, and sometimes "antioctant" behaviour is observed. Octant and quadrant rules have been suggested for a large number of chromophores.

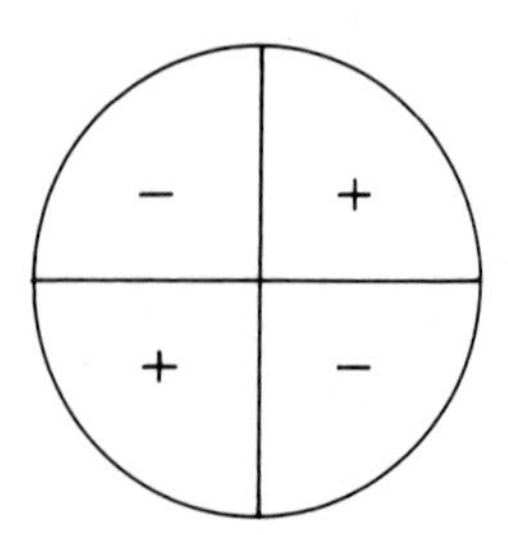

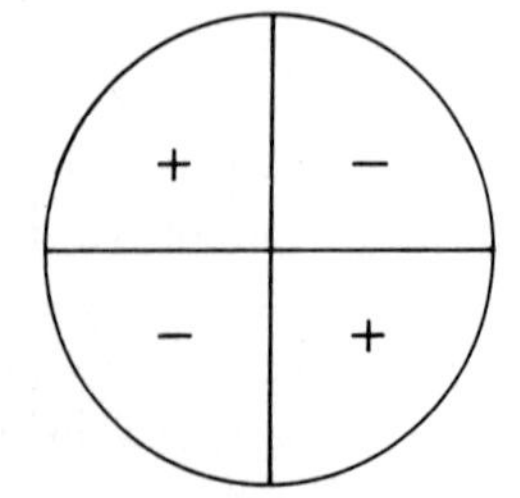

FIG. 9.15 The octant rule for carbonyl compounds. Contributions to the sign of the Cotton effect made by atoms in the eight octants.

Cyclohexane rings are flexible, and may undergo internal rotation so that a substituent may be either axial or equatorial to the plane of the ring. The different shapes that a flexible molecule can assume are known as *conformations*. For example, *trans*-2-chloro-5-methylcyclohexanone might have the chloro and methyl substituents both axial or both equatorial as shown in Fig. 9.16. The octant rule predicts that the axially substituted conformation, with substituents in the upper right quadrant will have a negative Cotton effect. Since a negative Cotton effect is observed at room temperature, the molecule does have the axially substituted conformation. However at -192 °C the Cotton effect is positive (Djerassi, 1964), showing that the equatorially substituted conformation is more stable at -192 °C. The preferred conformation of a flexible molecule is also often dependent on the solvent as well as on the temperature, so the sign of the Cotton effect may be dependent on both solvent and temperature. Much information about conformational equilibria may thereby be obtained. Even if no conformation change can occur, as in steroids which have rigid molecules, the sign of the Cotton effect may be solvent-dependent since different chromophores interact to a different extent with each solvent. Optically active solvents may induce Cotton effects in symmetrical solute molecules.

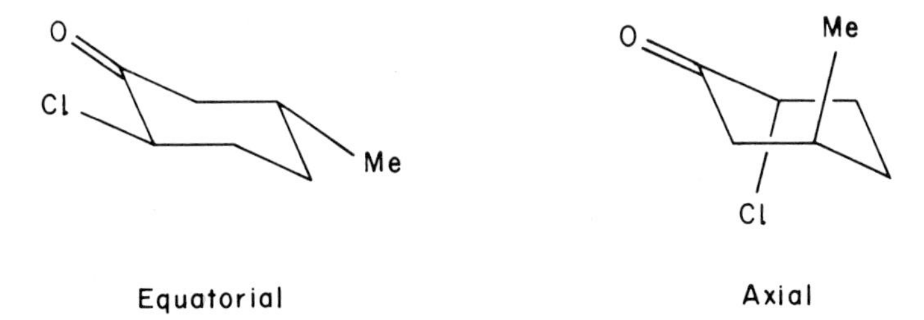

FIG. 9.16 The conformations of *trans*-2-chloro-5-methylcyclohexanone.

Since the observation of circular dichroism, often referred to as CD, and of optical rotatory dispersion, often referred to as ORD, lead to the measurement of the same quantity, the sign of the Cotton effect, the question arises as to whether CD or ORD is the more useful technique. The answer depends on the application. Since CD effects are localised at the wavelength region of absorption by the chromophore in question, it is easier to distinguish the effect of one chromophore from that of another that absorbs in a neighbouring region. ORD spectra, on the other hand, show background effects, that is the rotational contribution of absorption bands at some wavelengths considerably different from that under observation. Background effects are a nuisance if one is interested in the use of the Cotton effect to determine molecular stereochemistry, or in the theoretical interpretation of the Cotton effect for a given chromophore. However, background effects are helpful for fingerprint purposes, since they give each ORD spectrum an individual character. The ORD spectra of closely similar steroids may differ to such an extent that the spectra are of use in qualitative analysis. It is also sometimes necessary to obtain the sign of the Cotton effect for ORD if the chromophore responsible absorbs below the lower wavelength limit (typically 200 nm) at which the apparatus can be operated.

9.C.2 Transition-metal complexes

Transition-metal complexes may exhibit optical activity. This may be due either to inherently asymmetrical ligands, such as amino-acids, or to a helical configuration of chelating ligands around the central metal ion. The complexes absorb radiation in the visible and ultraviolet regions by excitation of electrons from filled to empty d orbitals in the metal ion, and these transitions are perturbed by the asymmetrical environment set up by the ligands, so as to give rise to Cotton effects. It is common for several transitions to be important in the spectra of transition-metal complexes. The sign of the Cotton effect for each transition depends on the symmetry types of the orbitals between which the electronic transition is occurring. Some generalisations, of the type of the octant rule for the carbonyl chromophore, have been made, but a knowledge of group theory is required for their comprehension. This is an active field of research, and many of the conclusions reached are the subject of disputation.

The complex ion $[Co(H_2NCH_2CH_2NH_2)_3]^{3+}$ is optically active because the three chelating ethylenediamine groups are arranged in helical fashion around the central cobalt ion, in a manner reminiscent of the blades of a marine propeller. The structure of the dextrorotatory isomer, as determined by X-ray diffraction, is shown in Fig. 9.17. The CD spectrum (McCaffery *et al.*, 1965) shows a pair of Cotton effects at 493 and 428 nm of $+1.89$ and -0.17 l mol^{-1} cm^{-1} respectively, corresponding to the absorption around 469 nm, and a single Cotton effect of $+0.25$ l mol^{-1} cm^{-1} at 351 nm, corresponding to the absorption at 340 nm. The dextrorotatory complex ion has a left-handed helical configuration if viewed along the three-fold symmetry axis, since each N–C bond is twisted counterclockwise relative to the adjacent N–C bond nearer to the observer. It is noteworthy that the ion has a right-handed helical configuration if viewed along the two-fold symmetry axis.

The Faraday effect, as discussed in Section 9.B.2, is also used in the study of transition-metal complexes. The magnetically induced optical activity is accompanied by circular dichroism at the absorption maxima. The study of

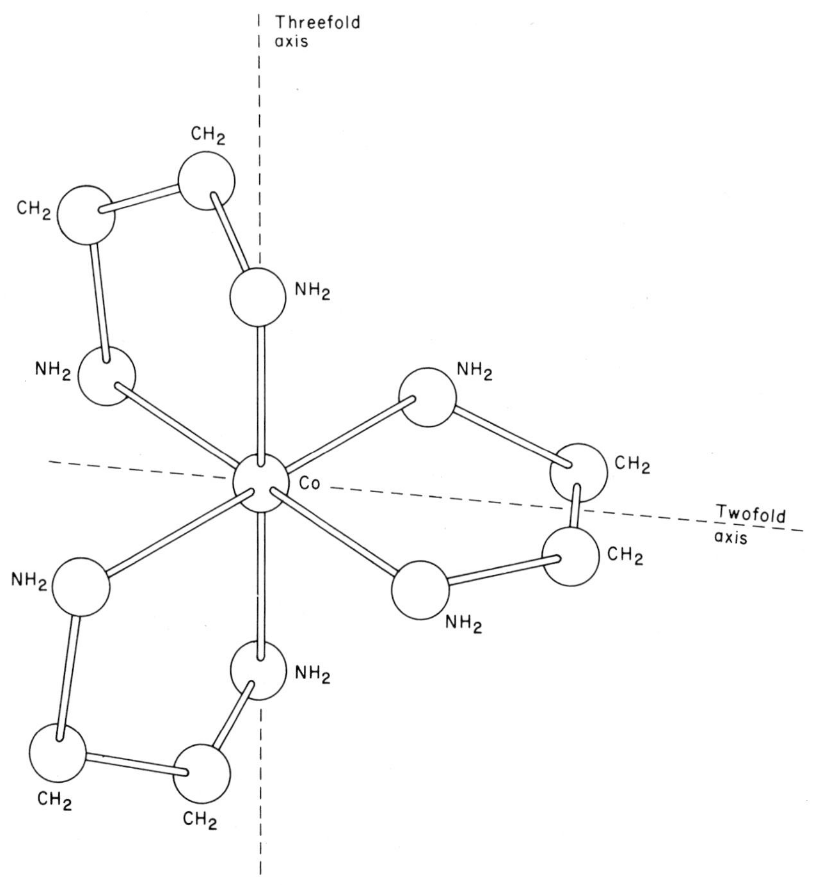

FIG. 9.17 The configuration of $(+)$-$[Co(H_2NCH_2CH_2NH_2)_3]^{3+}$.

MORD (magnetically induced optical rotatory dispersion) and MCD (magnetically induced circular dichroism) is experimentally difficult, since the effect is small. It is rare to observe a rotation greater than 0.5° for a 1 T magnetic field, and the CD effects are correspondingly small. Typical instrumentation for the observation of these effects employs magnetic fields of up to 10 T and is sensitive to 1 millidegree of rotation and 10^{-6} of an absorbance unit.

REFERENCES

Bevan, P. V. (1910) *Proc. Roy. Soc.* **84**, 209.

Brickell, W. S., Brown, A., Kemp, C. M. and Mason, S. F. (1971) *J. Chem. Soc. A*, 756.

Brown, A., Kemp, C. M. and Mason, S. F. (1971) *J. Chem. Soc. A*, 751.
Djerassi, C. (1964) *Proc. Chem. Soc.*, 314.
Eyring, H., Han-Chung Liu and Caldwell, D. (1968) *Chem. Rev.*, **68**, 525.
Jones, W. M. and Walbrick, J. M. (1968) *Tetrahedron Letters*, 5229.
McCaffery, A. J., Mason, S. F. and Ballard, R. E. (1965) *J. Chem. Soc.*, 2883.
Taylor, A. M. and Glover, A. M. (1933), *J. Opt. Soc. Amer.*, **23**, 206.
Waters, W. L. and Caserio, M. C. (1968) *Tetrahedron Letters*, 5233.

BIBLIOGRAPHY

Caldwell, D. J. and Eyring, H. (1971) "The Theory of Optical Activity", Wiley-Interscience, New York and London.

Caldwell, D. J. and Eyring, H. (1972) in "Techniques of Chemistry, Vol. I. Physical Methods of Chemistry Part IIIC. Optical, Spectroscopic and Radioactivity Methods" (A. Weissberger and B. W. Rossiter, eds.), pp. 1–51, Wiley-Interscience, New York and London.

Ciardelli, F. and Salvadori, P. (eds.) (1973) "Fundamental Aspects and Recent Developments in Optical Rotatory Dispersion and Circular Dichroism" Heyden, London and New York.

Crabbé, P. (1971) in "Determination of Organic Structures by Physical Methods", Vol. III (F. C. Nachod and J. J. Zuckerman, eds.), pp. 134–206, Academic Press, New York and London.

Crabbé, P. (1972) "ORD and CD in Chemistry and Biochemistry", Academic Press, New York and London.

Crabbé, P. and Parker, A. C. (1972) in "Techniques of Chemistry, Vol. I, Physical Methods of Chemistry, Part IIIC. Optical, Spectroscopic and Radioactivity Methods" (A. Weissberger and B. W. Rossiter, eds.), pp. 183–271, Wiley-Interscience, New York and London.

Djerassi, C. (1960) "Optical Rotatory Dispersion", McGraw-Hill, New York.

Hawkins, C. J. (1971) "Absolute Configuration of Metal Complexes", Wiley-Interscience, New York and London.

Heller, W. and Curmé, H. G. (1972) in "Techniques of Chemistry, Vol. I, Physical Methods of Chemistry, Part IIIC, Optical, Spectroscopic and Radioactivity Methods" (A. Weissberger and B. W. Rossiter, eds.), pp. 51–182, Wiley-Interscience, New York and London.

Neville-Jones, D. (1973) in "MTP International Review of Science, Organic Chemistry", Series One, Vol. I (W. D. Ollis, ed.), pp. 85–122, Butterworths, London.

Velluz, L., Legrand, M. and Grosjean, M. (1965) "Optical Circular Dichroism", Academic Press, New York and London.

10 Electron Spectroscopy

10.A Photoelectron Spectroscopy

10.A.1 Basic principles

The energy, I, required to expel an electron from a molecule is usually in the range 5–20 eV. Thus, if a molecule is irradiated by electromagnetic radiation in the vacuum ultraviolet, an electron will be expelled. Absorption of the photon may cause an increase in the vibrational and rotational energy of the molecule as well as ionisation, so the kinetic energy E of the expelled electron is given by

$$E = h\nu - I - \Delta E_{\text{vib}} - \Delta E_{\text{rot}} \tag{10.1}$$

Typically, ΔE_{vib} is of magnitude a few hundred meV and ΔE_{rot} is of magnitude a few meV. The energy of the expelled electron is also reduced by the kinetic energy imparted to the ion, but since the ion produced is so much more massive than the electron, this effect is negligible. In photoelectron spectroscopy, or PES as it is sometimes termed, the sample is irradiated with monochromatic radiation of fixed frequency ν. The variation in the number of expelled electrons of kinetic energy in some narrow range of fixed width as a function of the mean kinetic energy gives the photoelectron spectrum. The spectrum shows a peak for the first ionisation potential, I_1, where E is close to $(h\nu - I_1)$, with fine structure due to variations in ΔE_{vib}. The hyperfine structure due to variations in ΔE_{rot} is usually beyond the resolving powers of the apparatus, and is not seen. A molecule has a series of ionisation potentials, corresponding to electron loss from various molecular orbitals, so the photoelectron spectrum shows a series of bands, i.e. clusters of peaks, each corresponding to a particular ionisation potential.

10.A.2 Apparatus

The successful development of photoelectron spectroscopy in the vacuum ultraviolet rests, to a considerable extent, on the use of the low-pressure helium discharge lamp (Al-Joboury and Turner, 1963). This provides an intense monochromatic source of radiation at 58.4 nm (21.22 eV), of sufficient energy to ionise from at least one energy level of all molecules except helium and neon, and from several energy levels in most polyatomic molecules. The line-width is a few meV, offering the potential of considerable resolving power. There is great difficulty in finding a window material transparent to radiation of this wavelength, so the apparatus is designed to expose the sample to the radiation without any intervening solid material. Figure 10.1 is a schematic diagram of the apparatus. The helium discharge occurs in a quartz capillary (0.5 mm bore) open at both ends. The radiation from the open end of the capillary passes through another capillary, which acts as collimator and prevents mixing of the helium in the lamp with sample vapour, and irradiates the sample chamber. The sample chamber contains the sample as a gas at low pressure, in the range 2–20 N m^{-2} (0.015–0.15 mm Hg). The sample can

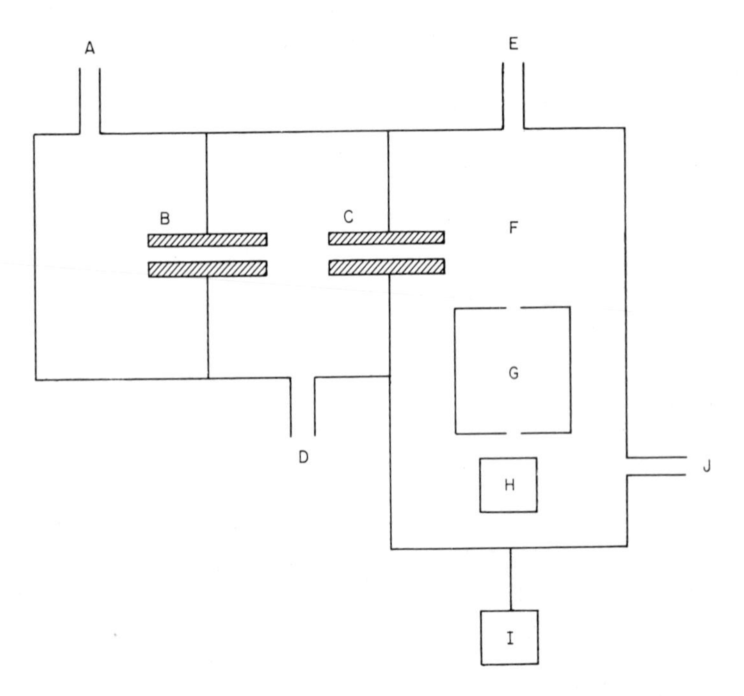

FIG. 10.1 Schematic diagram of a photoelectron spectrometer. A, Inlet for helium; B, lamp capillary tube; C, collimating capillary tube; D, exit for helium and sample vapour to pump; E, inlet for sample vapour; F, collision chamber; G, electron analyser; H, electron multiplier; I, counting equipment; J, exit for sample vapour to pump.

be liquid or solid at room temperature and pressure, provided that it has a sufficiently high vapour pressure. As the photons hit the molecules, electrons are expelled in all directions. The yield of electrons depends on the effective area presented by a molecule to the incident photons, termed the *photoionisation cross-section.* It is necessary for the production of a spectrum that the electrons of a given range of energies be collected at any one time. This function is performed by the electron analyser, which is for electrons what a monochromator is for photons. In the early studies, a grid analyser was used. A cylindrical grid, to which an adjustable retarding voltage could be applied, was placed round the ionisation region. For a given value of the grid voltage, only electrons of energy above a certain value could pass through to the collector, a cylinder coaxial to, and outside, the grid. The spectrum was scanned by varying the grid voltage, and was obtained as a plot of collector current against grid voltage. This spectrum was an integrated spectrum, since the current was proportional to all the electrons above a certain energy, not to the electrons of energy in a certain narrow range. Thus steps appeared in the spectrum where peaks would appear in the more familiar conventional spectrum. Electronic differentiation could be used to give a conventional display. Modern photoelectron spectrometers use focusing deflection analysers (Wannberg *et al.*, 1974). In these the electrons are forced to travel in curved paths under the influence of electrostatic or magnetic fields, just as in the classic experiments on the determination of the charge/mass ratio of the electron. Only electrons travelling with velocities in a certain narrow range pass through the exit slit at the end of the analyser. The radius of curvature of the path of an electron of given velocity (a parabola for electrostatic deflection, a circle for magnetic deflection) is set by the field intensity, so the spectrum is scanned by varying the field intensity. Deflection analysers are capable of greater resolving power than grid analysers, and can separate peaks 15 meV apart. Electrostatic analysers are more convenient than magnetic analysers in that the energy scale is linear, whereas magnetic analysers discriminate by momentum, which is proportional to the square-root of the energy. Both types of analyser give a conventional spectrum. The main disadvantage of focusing deflection analysers is that only a small proportion of all the electrons emitted pass through the entry slit into the analyser. An electron multiplier must be used to amplify the current. The electron multiplier is very similar to a photomultiplier, without the photosensitive cathode, and produces almost noiseless amplification by a factor of up to 10^9. Each electron emitted by photoionisation thus produces a measurable pulse of current. The pulse rate is measured by standard counting techniques, as used for radioactivity measurements, so the vertical axis of the photoelectron spectrum is in counts per second.

10.A.3 Interpretation of spectra

Figure 10.2 shows a schematic diagram for the energy levels of a hypothetical diatomic molecule, X, and two ions, $\tilde{X}$ and $\tilde{A}$, derived from it. Ion $\tilde{X}$ is similar to X in mean interatomic distance, binding energy, and force constant; that is to say, $\tilde{X}$ is produced from X by the loss of an electron from a non-bonding orbital in X. The energy, E, of the electron emitted is equal to the energy of the incident photon $h\nu$ less the energy required for ionisation, i.e.

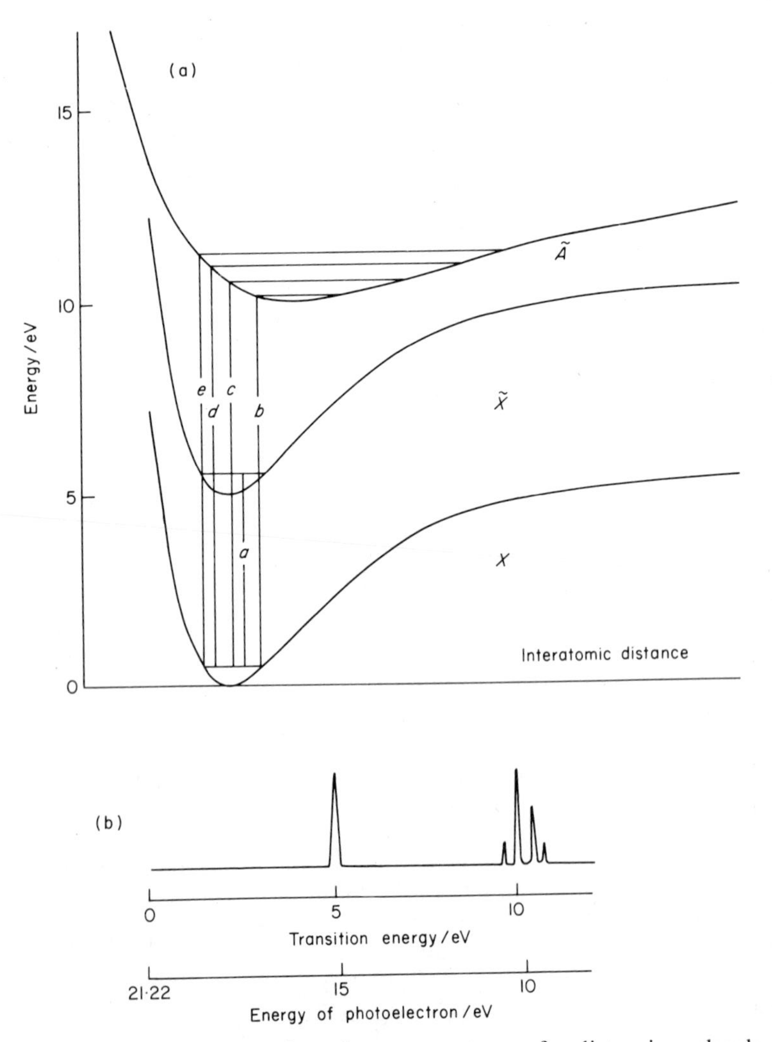

FIG. 10.2 The origin of the photoelectron spectrum of a diatomic molecule. (a) Energy levels. (b) Observed photoelectron spectrum.

the first ionisation potential, I_1, so $(h\nu - E)$ is the energy of transition *a*. The interatomic distance is unchanged during this transition, by virtue of the Franck–Condon principle discussed in Section 6.A, so the line on the diagram is vertical. The photoelectron spectrum is plotted as counts per second as a function of $(h\nu - E)$, so transition *a* gives a peak where $(h\nu - E) = I_1$. Transition *a* is from the zeroth vibrational level of X to the zeroth vibrational level of $\tilde{X}$. Such a transition is termed *adiabatic*. Because of the similarity between $\tilde{X}$ and X, and because only the lowest vibrational level of X is populated to any appreciable extent at room temperature, the Franck–Condon principle requires that only the adiabatic transition from X to $\tilde{X}$ appears in the spectrum, so a single sharp peak is observed. By contrast, $\tilde{A}$ differs from X, in that the interatomic distance is greater and the force constant and binding energy less; that is to say, $\tilde{A}$ is produced from X by the loss of an electron from a bonding molecular orbital. The Franck–Condon principle permits transitions *b*, *c*, *d*, and *e* to occur; in transitions *c*, *d*, and *e* the photon has not only produced the ion but raised it to an upper vibrational level, so the emitted electron is of reduced energy. Transitions *b*, *c*, *d*, and *e* thus give rise to four peaks in the observed spectrum, the separation between them giving the spacing of the vibrational levels in $\tilde{A}$. The adiabatic transition *b*, of energy equal to the second ionisation potential I_2, gives rise to a comparatively low peak, since the probability of molecule X having the right interatomic distance for this transition to occur is low. It may happen that the adiabatic transition is not observed, being completely forbidden by the Franck–Condon principle. The true, or adiabatic, ionisation potential then cannot be measured directly, but only by extrapolation. The ionisation potential taken from the largest peak, i.e. that corresponding to the transition of greatest probability, is termed the *vertical ionisation potential.*

Photoelectron spectroscopy thus gives values for one or more ionisation potentials and also values for vibration frequencies for ions in ground or upper excited states. Although Fig. 10.2 resembles that for the interpretation of an ultraviolet absorption spectrum, e.g. Fig. 6.2, the information obtained is somewhat different. Ultraviolet spectroscopic data are used to calculate energy differences between the ground state and electronically excited states of molecules, and evaluate vibration frequencies of molecules in ground and excited states, as discussed in Chapter 6. One type of transition is sometimes observed in the vacuum ultraviolet region in which a valency electron is excited into a very high-energy molecular orbital. These transitions give a series termed a *Rydberg series* of lines, tending to a limit at complete ionisation, and this affords another way of measuring ionisation potentials which predates photoelectron spectroscopy. This ultraviolet absorption method shares with the other earlier methods, electron impact and photoionisation, the limitation of being a threshold technique; i.e. the observed ionisation potential is the minimum required to ionise the molecule, so only the first ionisation potential can be measured. Photoelectron spectroscopy is also not usually complicated by autoionisation processes, i.e. ionisation from an electronically excited state of the molecule, which cause threshold methods to give spurious results.

A considerable proportion of photoelectron spectroscopic studies is concerned with the correlation between observed ionisation potentials and those calculated by molecular orbital theory (Worley, 1971). The link between theory and experiment is *Koopman's theorem*, which states that each ionisation potential is equal to the negative calculated value of each molecular orbital energy. This is an approximation, founded on the assumptions that there is no change in molecular geometry on ionisation and that the wave-functions describing the other electrons are unaltered. This approximation means that orbital energies should be more similar to vertical than to adiabatic ionisation potentials. However, the differences between adiabatic and vertical ionisation potentials are usually in the range 0–0.3 eV, which is within the error limits of current theoretical calculations.

A simple photoelectron spectrum, that of nitrogen, is shown in Fig. 10.3. This spectrum is plotted so that the ionization energy, $(h\nu - E)$, increases from right to left in accordance with the convention for ultraviolet and infrared spectra. Three adiabatic ionisation potentials, of values 15.58, 16.69, and 18.75 eV for the production of ions $\tilde{X}$, $\tilde{A}$, and $\tilde{B}$ respectively, can be seen.

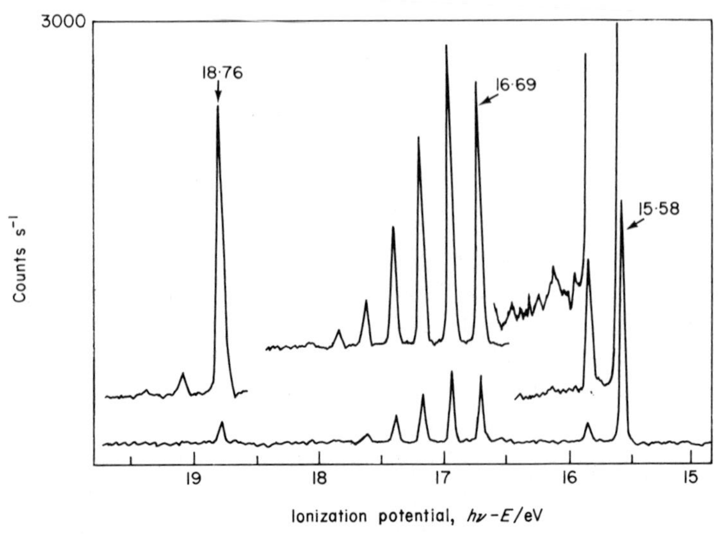

FIG. 10.3 The photoelectron spectrum of nitrogen, N_2 (Turner *et al.*, 1970).

These correspond to ionisation from the $\sigma_u 2p$, $\pi_u 2p$, and $\sigma_u{}^*2s$ molecular orbitals respectively. The molecular orbital energy diagram for N_2 is given in Fig. 10.4, together with the energy levels for the orbitals in atomic nitrogen from which the molecular orbitals are derived. Energy zero for both atomic and molecular nitrogen in Fig. 10.4 is taken for the state in which an electron is completely removed from the atom or molecule. The spacing of the peaks in each band in the spectrum gives the vibration frequencies for each $N_2{}^+$ ion, namely 2100, 1810, and 2340 cm^{-1} for $\tilde{X}$, $\tilde{A}$, and $\tilde{B}$ respectively. (Note that 1 eV $\equiv$ 8065.7 cm^{-1}.) The fundamental vibration frequency for N_2 in the ground state is 2345 cm^{-1}. As discussed in Chapter 4, the lower the vibrational frequency, the weaker the bond. It thus seems that the orbitals $\sigma_u 2p$, $\pi_u 2p$, and $\sigma_u{}^*2s$ are respectively weakly bonding, more strongly bonding, and very weakly antibonding. The sparse amount of vibrational fine structure in the peaks due to ionisation from $\sigma_u 2p$ and $\sigma_u{}^*2s$ is a further indication of the non-bonding character of these orbitals. However, this does not mean that σ bonding is ineffective in holding together the nitrogen molecule, since the bonding strengths deduced in this way apply only to the ground state molecule with equilibrium bond length. Consider a hypothetical attempt to pull apart the nitrogen molecule. At first the molecule resists stretching by virtue of its π bonds, but at greater internuclear separation σ bonding becomes more significant, so the total amount of energy required to pull the molecule apart is very large.

The photoelectron spectrum of methane is of interest from the light shed on the concept of sp^3 hybridisation (Turner *et al.*, 1970). Evidence from organic stereochemistry, as well as from Raman and ultraviolet spectroscopy and diffraction studies, shows that the CH_4 molecule is symmetrical, the four hydrogen atoms being at the corners of a tetrahedron with the carbon atom at the centre. All four C–H bonds are the same length, and the angle between any pair is the same, 109°. However, the carbon atom has one full 2s orbital and two half-full and one empty 2p orbitals, which might be expected to give two σ bonds at an angle of 90°. The tetrahedral tetravalency of carbon is often explained by postulating that one electron has been excited to the empty p orbital, and the four orbitals rearranged to give four equivalent orbitals at 109° to each other, the sp^3 hybrid. The evidence of the photoelectron spectrum of methane shows that this concept does not present the full picture. The 58.4 nm helium-excited photoelectron spectrum only has one band which is very broad, and contains a double maximum, extending from 12.7 to 16.0 eV. The shape of this band is caused by the lack of symmetry in the $CH_4{}^+$ ion, which is not exactly tetrahedral. Excitation by 30.4 nm radiation gives another peak near 23 eV (Hamrin *et al.*, 1968). The band around 13 eV is attributed to ionisation from the molecular orbital formed from the 2p

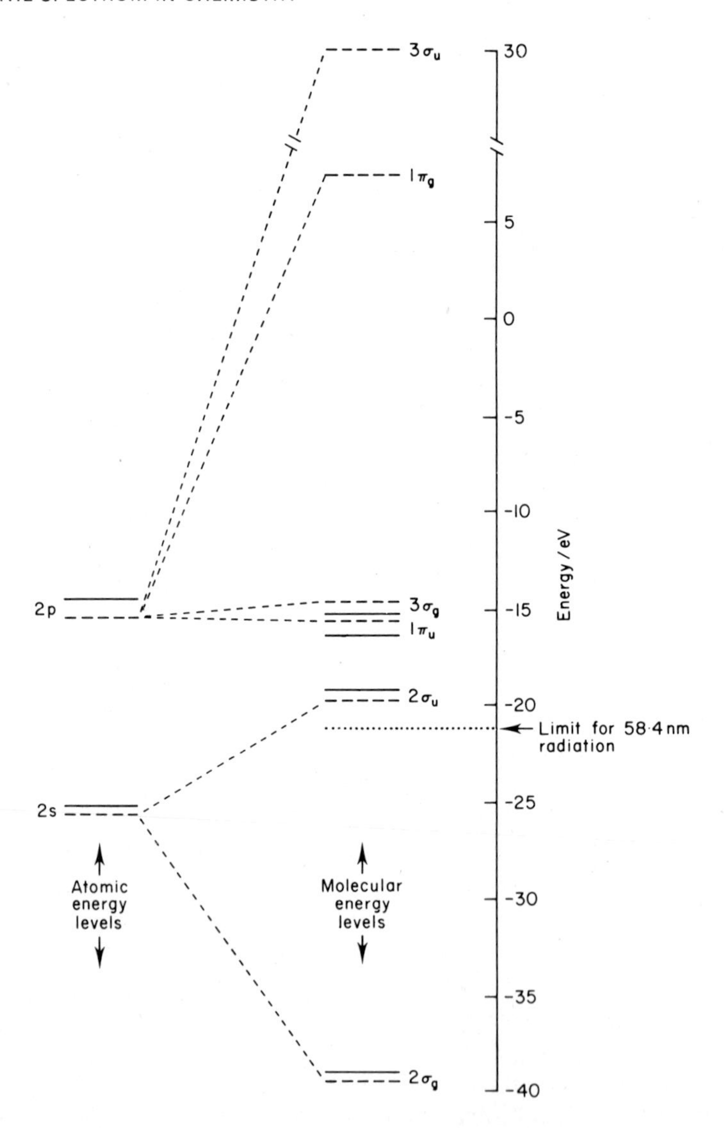

FIG. 10.4 Energy levels in atomic and molecular nitrogen. The molecular orbital energy diagram for N_2 is rather more complicated than that usually given in the elementary text-books, in which the 2s atomic orbitals give the $\sigma 2s$ molecular orbital, and so forth. Sophisticated calculations give energy levels as in the diagram, in which orbitals $2\sigma_g$, $2\sigma_u$, $1\pi_u$, $3\sigma_g$, $1\pi_g$, and $3\sigma_u$ are shown. These orbitals are combinations of the simple molecular orbitals, and only roughly correspond to $\sigma_g 2s$, $\sigma_u{}^*2s$, $\pi_u 2p$, $\sigma_g 2p$, $\pi^*{}_g 2p$, and σ^*2p respectively. The molecular orbital energy levels denoted by broken lines are those calculated by Scherr (1955). The unbroken lines represent experimentally determined energy levels. That at −39 eV was found

orbital on carbon and the 1s orbital on hydrogen, whereas the band at 23 eV is attributed to ionisation from the molecular orbital formed from the 2s orbital on carbon and the 1s orbital on hydrogen. There is no band attributable to ionisation from an sp^3 orbital. It must be emphasised, however, that the hybrid orbital treatment is not wrong, merely incomplete. It provides a satisfactory description of molecular shapes, but not molecular energies. More complex spectra are shown in Fig. 10.5, in which the effect of halogen substitution in ethylene can be seen. The lowest band, at 10.51 eV for ethylene, is due to ionisation from the C═C π orbital. Interaction between the π orbital and one p orbital (whose axis is perpendicular to the plane of the molecule) on the substituent halogen lowers the ionisation potential of the C═C π orbital and raises the ionisation potential of the interacting p orbital. This type of interaction has long been recognised in organic chemistry, and is termed *mesomerism*. It is often discussed in terms of structures such as $^{-}CH_2—CH{=}Cl^{+}$. Thus, electron loss from the C═C bond is made easier because of partial compensation by transfer from one lone pair on Cl. The other lone pair of electrons on the halogen, in the p orbital whose axis is in the plane of the molecule, does not interact with the C═C bond, and is truly non-bonding. Ionisation from this orbital is responsible for the sharp peaks at 11.8 eV for Cl and 10.95 eV for Br. The corresponding peak for F, at around 17.4 eV, is hidden by other peaks.

10.A.4 Analytical applications

At present, photoelectron spectroscopy is not used extensively for either qualitative or quantitative analysis, although it has potential for both applications. The photoelectron spectrum of an unknown may be used as a "fingerprint", in the same way that infrared spectra are used. However, the infrared spectrum is more informative in that many characteristic group frequencies are observed, whereas group features are not so clearly defined in photoelectron spectra. A few such features are observed; e.g. fine structure in the 8–11 eV region suggests π bonds, whereas a peak around 13 eV suggests an alkyl group. Sharp peaks indicate the presence of lone pairs. The use of

---

by excitation by high-energy radiation, as described in Section 10.B. The atomic energy levels denoted by broken lines are those calculated by Clementi *et al.* (1962). The experimental value for the energy of 2p is taken as the first ionisation potential for atomic nitrogen, and the experimental value for the gap between 2s and 2p is taken as the energy for the spectroscopically observed transition from the $2s^22p^3$ state to the $2s^12p^4$ state.

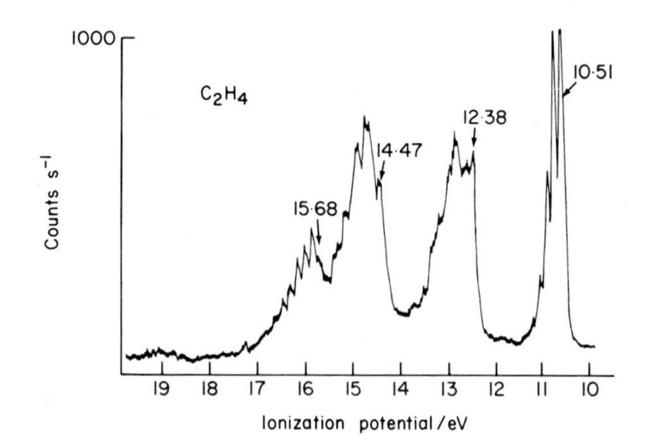

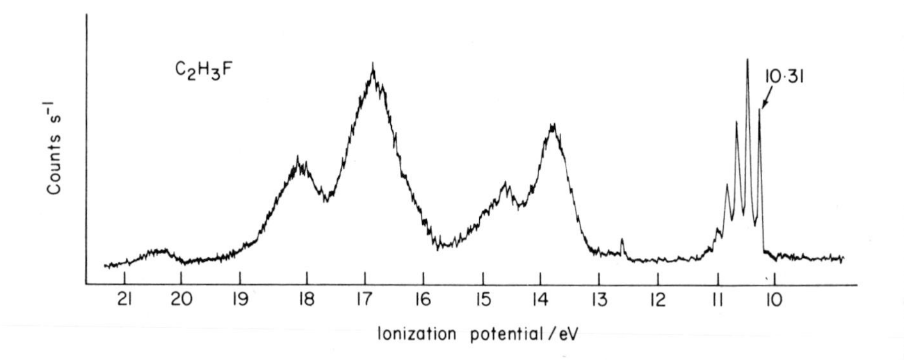

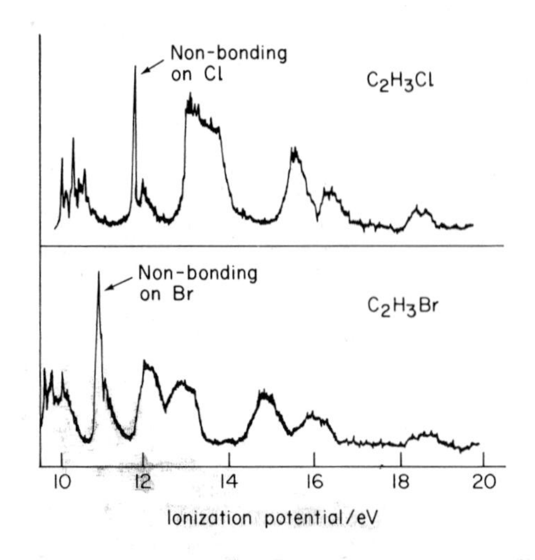

FIG. 10.5 Photoelectron spectra of alkenes (Turner *et al.*, 1970).

photoelectron spectroscopy for quantitative analysis suffers from the problem of uncertainty in sample size. Adsorption of sample on the walls of the ionisation chamber may occur to an unknown extent.

10.B X-Ray Photoelectron Spectroscopy

10.B.1 Basic principles

If X-rays rather than vacuum ultraviolet radiation are used to irradiate molecules, electrons are emitted from inner, or core, orbitals rather than from valency or molecular orbitals. This technique is frequently called *ESCA* (*Electron Spectroscopy for Chemical Analysis*). The generation and properties of X-rays will be discussed in detail in Chapter 11. The wavelengths and photon energies for X-rays from some commonly used sources are listed in Table 10.1. The apparatus used for ESCA resembles that for photoelectron spectroscopy illustrated in Fig. 10.1, with the difference that an X-ray source replaces the helium lamp. A beryllium window, beryllium being transparent to X-rays, separates the source from the sample chamber.

TABLE 10.1

Monoenergetic X-ray sources (Baker and Betteridge, 1972)

Source	Energy/eV	Wavelength/pm
Mg (K_α)	1253.6	989
Al (K_α)	1486.6	834
Cu (K_α)	8048	154.1
Ag (K_α)	22 163	55.9

The binding energies of electrons in the 1s orbitals of some light atoms are listed in Table 10.2, and comparison with the values in Table 10.1 shows that 1s electrons may be expelled from molecules exposed to X-rays. For heavier atoms, e.g. Cl or S, other, less tightly bound inner orbitals such as 2p may be used instead of 1s. Except in the case of hydrogen the 1s or other core electrons do not take part in chemical binding, so to a first approximation the energy of the emitted electrons does not depend on the nature of the bonding between atoms. For example, a peak around 400 eV may be confidently attributed to emission from nitrogen atoms, whereas a peak around 530 eV may be attributed to emission from oxygen atoms. However, the inner electrons are affected by the bonding to a slight extent, so small differences exist between

TABLE 10.2

Binding energies of 1s electrons (Baker *et al.*, 1972)

Atom	Energy/eV	Atom	Energy/eV
Li	50	Na	1070
Be	110	Mg	1305
B	190	Al	1560
C	280	Si	1840
N	400	P	2150
O	530	S	2470
F	690	Cl	2823
Ne	867	Ar	3203

the energies of electrons emitted from atoms of the same element in dissimilar molecules, or in different parts of the same molecule. These small changes are termed *chemical shifts*. Other types of chemical shift will be discussed in later chapters, in connection with Mössbauer and n.m.r. spectroscopy. A simple example of chemical shift can be seen in the ESCA spectrum of sodium

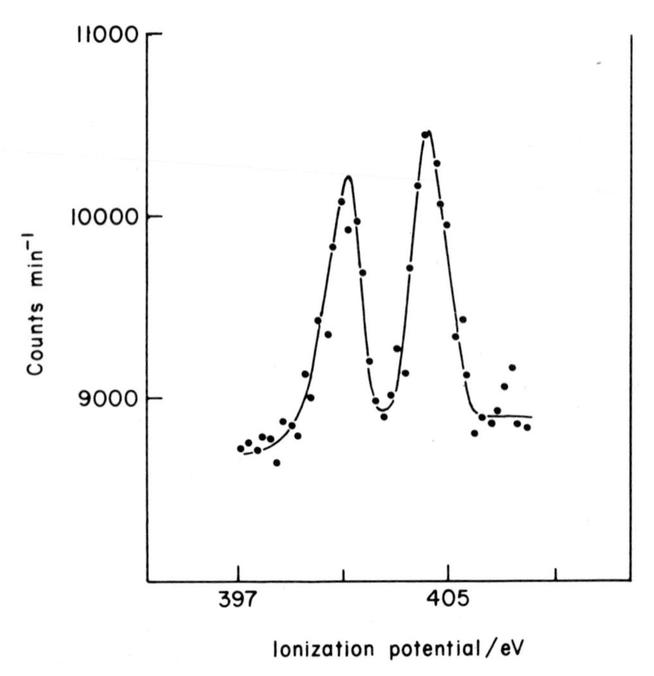

FIG. 10.6 The ESCA spectrum of $Na_2N_2O_3$; N 1s region (Hollander and Jolly, 1970).

oxyhyponitrite, shown in Fig. 10.6. There are three possible structures for the $N_2O_3^{2-}$ ion:

$$O{=}N{-}O{-}\bar{N}{-}O \qquad O{=}N{-}N\begin{matrix} \diagup O^- \\ \diagdown O^- \end{matrix} \qquad {}^-O{-}N{=}N{-}O{-}O^-$$

(1) (2) (3)

The ESCA spectrum shows that there are nitrogen atoms in two different locations in the ion, so the symmetrical structure (1) cannot be correct. Detailed molecular orbital calculations suggest that (3) is the correct structure. The ESCA spectrum of ethyl trifluoroacetate (Fig. 10.7) shows four peaks of equal intensity, one for each of the four different carbon atoms in the molecule.

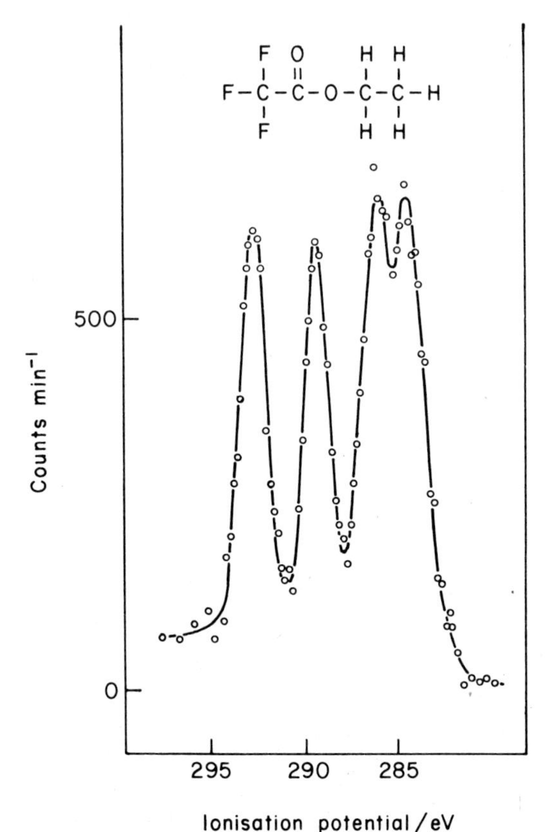

FIG. 10.7 The ESCA spectrum of $CF_3CO_2C_2H_5$; C 1s region (Hollander and Jolly, 1970).

As can be seen from Figs. 10.6 and 10.7, the resolving power of ESCA at present is much less than that of photoelectron spectroscopy in the vacuum ultraviolet. This is due more to the frequency spread of the X-ray emission from the source, which gives a multiplet of lines, than to deficiencies in the electron analyser. At present it is not possible to separate ESCA peaks less than about 1 eV apart, although attempts have been made to resolve more closely spaced peaks by mathematical curve-fitting techniques. Further advances in ESCA will probably be in the development of more truly monochromatic X-ray sources.

10.B.2 Interpretation of chemical shifts

The magnitude of the chemical shift is a measure of the extent to which the adjacent atoms affect the core electrons of the emitting atom. The more electronegative the adjacent atoms are, the lower is the energy of the emitted electron, and the higher the value of $(h\nu - E)$, as shown in Table 10.3. This

TABLE 10.3

Chemical shifts for carbon compounds (Shirley, 1973)

Compound	Observed 1s binding energy/eV	Calculated energy/eV
CH_4	290.8	292.9
CH_3F	293.6	297.6
CH_2F_2	296.4	300.6
CHF_3	299.1	304
CF_4	301.8	—

effect can also be seen in Fig. 10.7. The emitted electron has been held in the molecule not only by the field of its own nucleus but also, to a small extent, by the fields of the adjacent nuclei. This interaction can be studied theoretically by the methods of quantum mechanics, and chemical shifts can be thus calculated from first principles (Shirley, 1973). Some calculated values are listed in Table 10.3. The agreement with observed values is not too bad considering that the chemical shift is only a small proportion of the total energy of the atom. Theoretical calculations of ESCA chemical shifts offer a better chance of success than those for Mössbauer or n.m.r. spectra which will be discussed in Chapters 12 and 13 respectively. ESCA chemical shifts are found to be additive to quite a high degree of accuracy; in other words, the total shift can be calculated as the sum of the individual shifts for each adjacent atom.

10.B.3 The Auger effect

ESCA spectra may be complicated by peaks due to emission of electrons from higher orbitals in the core. The initial expulsion of an electron from the 1s orbital caused by impact of an X-ray photon leaves an empty hole in the 1s orbital, which may be filled by an electron falling from the 2s orbital. This electron must lose energy in so doing, and may do so in one of two ways. The atom may emit radiation, X-ray fluorescence, as will be discussed in Chapter 11, or the energy may be lost in the form of an electron expelled from a higher core orbital, say 2p. Thus one incident photon may cause the emission of two electrons of different energies. This is termed the Auger effect. It is possible to decide whether a given peak in an ESCA spectrum is due to Auger emission by changing the X-ray source. This causes the energy of the electrons emitted by the primary process to change but not the energy of Auger electrons, since the latter is determined solely by the energy levels within the atom.

10.B.4 Analytical applications

As the name implies, ESCA has great potential usefulness as an analytical technique. ESCA can, in principle, detect and identify all the elements present in a sample, give a measure of the relative numbers of the various atoms, show whether the atoms of each element are all in the same environment or not, and indicate the oxidation state of each element. An important figure-of-merit for any chemical-shift technique is the ratio of the total range of shifts observed to the individual peak width, and unfortunately ESCA does not score very highly on this basis. The total range of chemical shifts for C 1s, for example, is only 11 eV, whereas the peak widths commonly obtained at present are about 1 eV. Thus not all the carbon atoms in an organic molecule can be distinguished.

ESCA can be used for quantitative analysis, since the peak area is proportional to the relative number of atoms, for emission from a specified orbital in a specified atom. The apparatus must be calibrated using samples of known composition. The accuracy is not very great; typical discrepancies between ESCA and other analyses are of the order of 5%, but a comprehensive analysis can be carried out for all the elements in a very small sample. A notable success of ESCA was the demonstration that "polywater", a viscous liquid formed by the condensation of water vapour in fine glass capillary tubes, was not a highly polymeric allotrope of water, as previously believed, but a concentrated solution, mostly sodium silicate, containing less than 5% water (Davis, 1971).

ESCA differs from vacuum ultraviolet photoelectron spectroscopy in that it is quite common for the sample to be in the solid rather than in the vapour phase. The solid is spread on a plate at an angle of 45° to the incident X-rays, and the emitted electrons are analysed at 45° to the plate. The binding energy for electrons in atoms in the solid phase includes a term for the field of the solid as a whole, so observed ionisation potentials for atoms in the solid phase are a few eV greater than for atoms in the vapour phase. Build-up of electrostatic charge on the sample during the experiment also affects the observed ionisation potential. An important feature of the ESCA analysis of solids is that the observed electrons are only emitted from a layer 10 nm or so deep on the surface of the solid. Electrons emitted from atoms further within the solid are lost by collisions with other atoms on their way out. From one point of view this is a disadvantage, since care must be taken to prepare the sample in such a way that the surface composition is representative of the bulk. From another point of view this sensitivity to surface effects is of great value in the study of heterogeneous catalysis and surface absorption. For example, the differences between ESCA spectra of fresh and used catalyst may be used to gain understanding of the mode of action of the catalyst. Auger spectroscopy is of similar value in the study of surfaces.

REFERENCES

Al-Joboury, M. I. and Turner, D. W. (1963) *J. Chem. Soc.*, 5141.
Baker, A. D., Brundle, C. R. and Thompson, M. (1972) *Chem. Soc. Rev.* **1**, 355.
Clementi, E., Roothaan, C. C. J. and Yoshimine, M. (1962) *Phys. Rev.*, **127**, 1618.
Davis, R. E. (1971) Electron Spectroscopy Proc. Int. Conf., 909.
Hamrin, K., Johansonn, G., Gelius, U., Fahlman, A., Nordling, C. and Siegbahn, K. (1968) *Chem. Phys. Letters*, **1**, 613.
Hollander, J. M. and Jolly, W. L. (1970) *Accounts Chem. Res.* **3**, 193.
Scherr, C. W. (1955) *J. Chem. Phys.* **23**, 569.
Shirley, D. A. (1973) *Adv. Chem. Phys.* **23**, 85.
Wannberg, B., Gelius, U. and Siegbahn, K. (1974) *J. Phys., E.* **7**, 149.
Worley, S. D. (1971) *Chem. Rev.* **71**, 295.

BIBLIOGRAPHY

Albridge, R. G. (1972) in "Techniques of Chemistry, Vol. I, Physical Methods of Chemistry, Part IIID, Optical Spectroscopic and Radioactivity Methods" (A. Weissberger and B. W. Rossiter, eds.), pp. 307–364, Wiley-Interscience, New York and London.
Allan, C. J. and Siegbahn, K. (1973) in "International Review of Science: Physical Chemistry", Series One, Vol. 12. Analytical Chemistry—Part I (T. S. West, ed.), pp. 1–44, Butterworths, London.

Baker, A. D. and Betteridge, D. (1972) "Photoelectron Spectroscopy", Pergamon, Oxford.
Eland, J. H. D. (1974) "Photoelectron Spectroscopy", Butterworths, London.
Turner, D. W., Baker, A. D., Baker, C. and Brundle, C. R. (1970) "Molecular Photoelectron Spectroscopy", Wiley, London and New York.

11 X-ray Emission and Absorption Spectroscopy

11.A The Origin and Properties of X-Rays

11.A.1 X-ray emission spectra

When an atom is bombarded with particles of the energies of tens of kilo-electron volts, an electron may be ejected from a core orbital, i.e. an orbital near the nucleus, as discussed in Section 10.B. As well as electrons or protons, the bombarding particles may be photons, that is to say, quanta of electromagnetic radiation of such short wavelength that the radiation has some of the properties of a stream of particles. The link between the wave and particle properties is the de Broglie relationship,

$$\lambda = h/mv \tag{11.1}$$

where λ is the wavelength of the radiation and mv is the momentum of a quantum of radiation considered as a particle. If λ is 10 pm, mv is about 6.6×10^{-23} kg m s^{-1} which is similar to the momentum of an electron accelerated by a field of 12 kV. The ionised atom is in a high energy state, and is then stabilised by the falling of an electron from an outer orbital into the vacancy in the core orbital. The excess energy is emitted as a quantum of radiation. The magnitudes of the transitions involved are such that the wavelength of the emitted radiation, X-rays, is in the region of 10 pm to 1 nm.

A typical set of core electronic energy levels, those for the silver atom, are shown in Fig. 11.1. The energy of each state shown is that of an atom lacking an electron from the level corresponding to that state. The energy of an atom lacking an electron from the innermost state is arbitrarily taken as zero. Since all the other states are more stable than this, their energies are all negative on this scale. The energy levels are labelled by their spectroscopic terms, as discussed in Section 6.A.2, with the modification that the prefix giving the value of $(2S+1)$ has been omitted, and the term is prefaced by the principal quantum number of the orbital. The terms are worked out on the electron-hole principle, since all the orbitals are full except for the lack of a single electron. The configuration of the valency electrons is ignored in establishing the spectroscopic term for this purpose, since the valency electrons are well separated from the core electrons. The S levels are singlets, since L has the value zero, but the P and D levels are doublets, since the electron

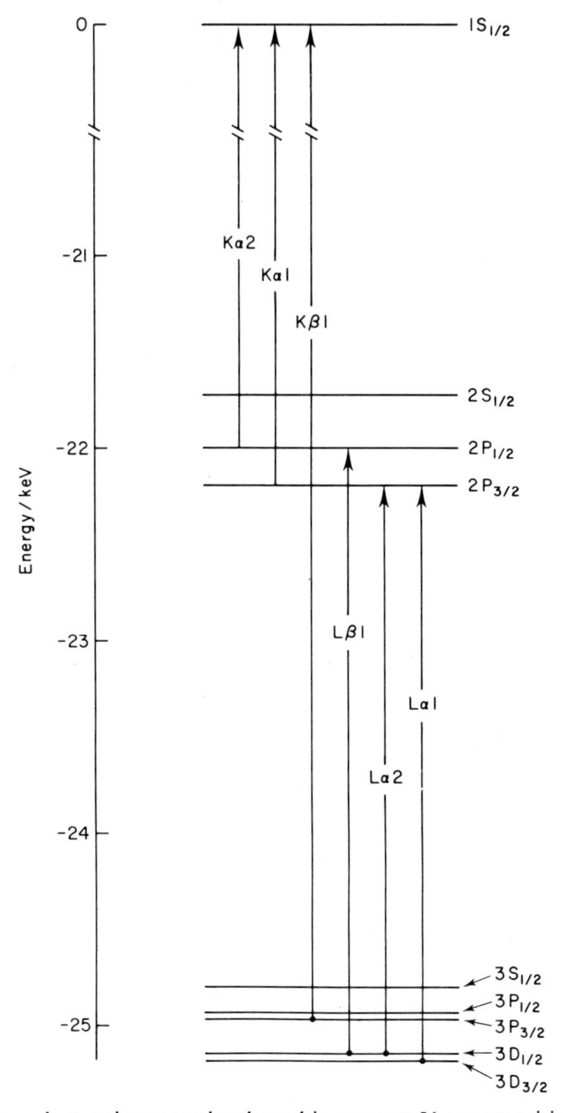

FIG. 11.1 Core electronic energy levels and important X-ray transitions for silver.

spin can couple with or against the orbital spin. The selection rule for a permitted transition is that $\Delta L = \pm 1$. Emission lines due to electrons falling into a hole in the $1S_{\frac{1}{2}}$ state are known as K lines, and the innermost orbital is thus sometimes known as the K shell. If the electron has come from a level of principal quantum number 2, the line is a K_α line; if from a level of principal quantum number 3, the line is a K_β line. Lines due to transitions to a hole in a level of principal quantum number 2 are known as L lines, and so on. The more important transitions in the silver atom are shown in Fig. 11.1. The student will note that the arrows in Fig. 11.1 representing emission point upwards, rather than downwards as in, say, Fig. 7.1. It may be helpful to think of X-ray spectra as being caused by the movement of an electron hole leaving behind it a full orbital rather than the movement of an electron leaving behind it an empty orbital as is customary in the consideration of other emission spectra. The energy levels in Fig. 11.1 are electron-hole energy levels, whereas the energy levels in Fig. 7.1 are electron energy levels.

The energy of the transition causing X-ray emission depends on the nuclear charge, which is given by the atomic number. The greater the nuclear charge, the greater the energy required to eject an electron from a core orbital, and the greater the energy released when an electron falls from an outer to an inner orbital. Moseley's pioneering work on X-rays showed that the wavelength of the emitted radiation is inversely proportional to the square of the atomic

TABLE 11.1

Characteristic X-ray emission lines

		$K_{\alpha 1}$		$L_{\alpha 1}$	
Element	Atomic number	λ /pm	Energy /keV	λ /pm	Energy /keV
Na	11	1190.9	1.041	—	
Si	14	712.5	1.739	—	
S	16	537.2	2.307	—	
Cr	24	229.0	5.412	2171.4	0.571
Fe	26	193.6	6.401	1760.2	0.704
Cu	29	154.1	8.042	1335.7	0.928
As	33	117.6	10.541	967.1	1.281
Mo	42	70.9	17.480	540.6	2.292
Sn	50	49.1	25.24	360.0	3.442
W	74	20.9	59.3	147.6	8.396
Pb	82	16.5	75.1	117.5	10.55
U	92	12.6	98.4	91.1	13.60

number, after applying a small correction for the effect of the other electrons screening the nuclear charge. This square-law dependence is that predicted by elementary electrostatics theory. The energy levels of the core orbitals are little affected by the outer orbitals or environment of the atom, so the wavelength of the emitted radiation is practically independent of the valency state or chemical environment of the atom. Small effects, the chemical shifts discussed in Section 10.B, are observed but are only of the order of a few hundred parts per million. The effects are much smaller than for ESCA, which measures the core orbital energy relative to complete ionisation, because there is a cancellation effect; both of the orbitals involved in the transition are subject to a chemical shift in the same direction. Although some work has been done in the past on the study of chemical bonding by X-ray chemical shifts, ESCA is now the preferred technique.

The significance of X-ray emission spectroscopy for the chemist is that each element has only a small number of characteristic emission lines, the wavelengths of which are in practice independent of the state of chemical combination of the element, and which differ markedly from the wavelengths of lines of other elements by virtue of the square-law dependence on atomic number. The rare-earth metals, or lanthanides, for example, give a range of well-separated emission lines, since they differ widely in their atomic numbers although their chemical properties are so similar. It is thus comparatively easy to determine what elements are present in a sample by examination of its X-ray emission spectrum. Furthermore the intensity of emission by an element depends on the number of atoms of that element present in the sample, so the concentration of each element can also be determined. X-ray spectra are much simpler than visible/ultraviolet spectra in that there are no complicating effects due to vibrational or rotational energy levels.

It is common practice to use X-rays to supply the energy necessary to ionise the sample in X-ray emission spectroscopy. There is thus a resemblance between this technique and that of fluorescence, as discussed in Chapter 7, and the technique is therefore often termed *X-ray fluorescence analysis*. This term is somewhat of a misnomer. The essential feature of fluorescence is that the excited state lasts for a finite period of time, during which it loses energy by non-radiative processes before emitting a photon. There is no such delay in X-ray emission.

11.A.2 The absorption of X-rays

Atoms absorb X-rays by the ejection of electrons from core orbitals, as discussed in Section 11.A.1. Because the ejected electron has a non-quantised kinetic energy, the absorption of X-rays differs from that of radiation of longer wavelengths, which only occurs if there is an exact correspondence between the energy of the quantum of radiation, $h\nu$, and the gap between two quantised energy levels in the atom or molecule. There does not have to be an exact coincidence between the energy of the X-ray photon and the energy gaps between the core orbitals. However, the efficiency of the absorption process is a maximum if there is such a coincidence, and decreases with increasing proportion of the energy of the incident photon which must be carried away by the ejected electron. The absorption spectrum thus shows a maximum at a wavelength corresponding to some emission line, an L line for example, followed by a smooth decrease with decreasing wavelength, which follows a cube law, the absorption being proportional to the cube of the wavelength. At a wavelength corresponding to the next emission line, a K line for example, there is a discontinuity leading to a sharp maximum in the absorption. These discontinuities are known as *absorption edges.* A typical absorption spectrum, that of lead, is shown in Fig. 11.2. The absorption edges K, L_I, L_{II}, and L_{III} correspond to transitions to the $1S_{\frac{1}{2}}$, $2S_{\frac{1}{2}}$, $2P_{\frac{1}{2}}$, and $2P_{\frac{3}{2}}$ levels respectively. It can be seen that the absorption edges are superimposed on a background which continuously increases at increasing wavelength. This absorption is due to ejection of electrons from orbitals further out from the core, which, since they are more numerous, gives rise to greater absorption than the ejection of electrons from core orbitals. Light elements are comparatively transparent to X-rays because even the ejection of an electron from the innermost orbital requires so little energy that an excessive amount of the energy of the absorbed X-ray photon must be transferred to kinetic energy of the ejected electron. Organic material, which is principally composed of carbon, hydrogen, and oxygen, is thus quite transparent to X-rays. The high nuclear charge in heavy atoms such as lead means that the core ionisation energies are similar to the energies of X-ray photons, and absorption is comparatively efficient. The large number of electrons available for ionisation in each atom also favours absorption of X-rays by heavy atoms.

Because core energy levels are little affected by the valency electrons, the proportion of X-rays absorbed by an element depends only on the mass of the element present in the path of the radiation, and not on its physical state or state of chemical combination. The form of Beer's Law which is appropriate to X-ray absorption is thus, for monochromatic radiation,

$$\ln(I/I_0) = \mu m \tag{11.2}$$

where I_0 is the intensity of the incident radiation, and I is the intensity after passing through a mass m. μ is the mass absorption coefficient, a quantity analogous to the extinction coefficient defined by Eqn. 4.27. X-ray absorption

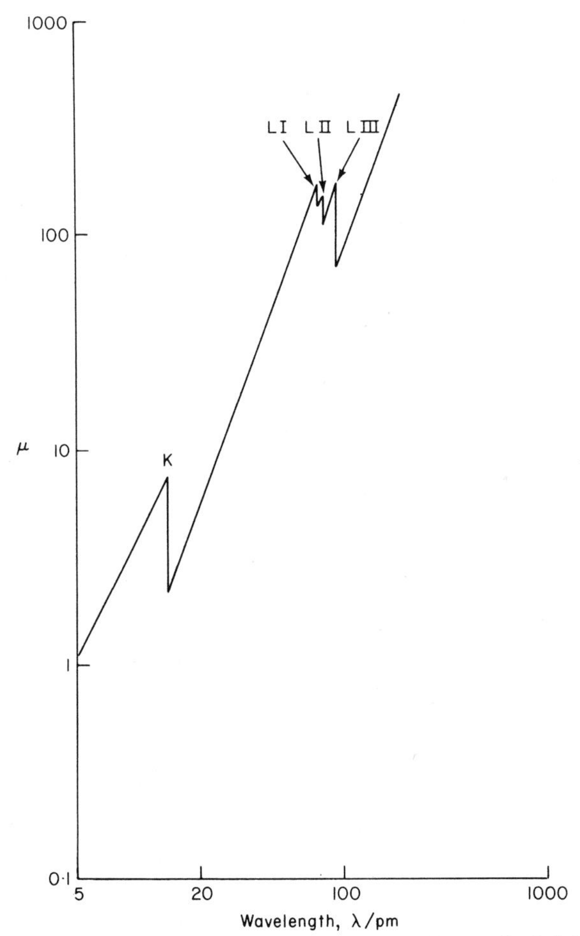

FIG. 11.2 X-ray absorption spectrum of lead; log–log scale (Liebhafsky, 1951).

may thus be used as a technique for quantitative elemental analysis. Although this technique has been very popular in the past, it has now been largely superseded by X-ray emission spectrometry. Emission occurs over a much narrower range of wavelengths than absorption, so interference from other elements is less in X-ray emission spectrometry. Furthermore, X-ray emission spectrometry is more suitable for the analysis of elements present in low concentration in the sample. These produce only a small attenuation of the transmitted beam, which is much more difficult to measure accurately than the intensity of emitted radiation, low though this be. A similar comparison between absorption and fluorescence in the visible/ultraviolet region has been made in Section 7.C.2.

11.B Apparatus

11.B.1 Excitation sources

The excitation source used in most commercial X-ray emission spectrometers is a Coolidge X-ray tube. X-rays are produced by the bombardment of a target of some heavy inert metal, such as molybdenum or tungsten, by electrons accelerated through a high vacuum by a potential of several tens of kilovolts. As the electrons hit the target they slow down, and lose energy by emitting radiation over a wide range of wavelengths. This radiation is sometimes referred to as *Bremsstrahlung*, a German word literally meaning "braking radiation". There is an abrupt lower limit to the wavelength of the radiation emitted, corresponding to complete conversion of all the energy of the electrons into radiation. If the accelerating potential is V kV, this minimum wavelength is λ_0 pm, where

$$\lambda_0 = hc/eV = 1239.3/V \tag{11.3}$$

where c is the velocity of light and e is the charge on the electron. The conversion of the kinetic energy of the electrons into radiation is inefficient, and more than 99% is lost as thermal energy. Some electrons lose more energy in this way than others, so the emission spectrum shows a broad distribution of energy at wavelengths greater than λ_0. Typically the maximum intensity is found at about $3\lambda_0/2$. Some of the radiation produced by this primary process causes ionisation of atoms within the target, so the X-rays produced contain the characteristic emission lines of the target element superimposed on the polychromatic or "white" background. The intensity of X-ray emission increases with the atomic number of the target element, which is why targets are made of heavy metals. The X-rays leave the tube through a window of beryllium (atomic number 4) which is very transparent to X-rays for reasons discussed in Section 11.A.2.

It would seem most efficient to generate X-rays from a sample by direct bombardment with electrons rather than via the prior generation of excitation X-rays. The energy conversion for the direct process is much more efficient, and this technique is favoured by some workers. One disadvantage is that the emitted radiation contains the white background of Bremsstrahlung which is missing from the emission spectrum produced by X-ray excitation. There are also practical disadvantages. The sample must be within an evacuated enclosure, and an electrostatic charge builds up on the sample if it is non-conductive. A special application for which electron activation has a unique advantage is the study of regions of small area on a sample. The electron beam can be focused down so as to concentrate all its energy on an area of 1 μm^2 which could not be done for a beam of X-rays.

A source of X-rays, which, although less intense than the Coolidge tube, has the great practical advantage of not requiring a high-voltage supply, is provided by various artificial isotopes. ^{55}Fe, for example, is a radioactive isotope of iron produced in a nuclear reactor. The ^{55}Fe nucleus is unstable, but may achieve stability by extracting an electron from the innermost, i.e. K, orbital to form the stable ^{55}Mn nucleus. This process is known as *K-capture*. An electron then falls from the L shell to fill the vacancy thus created in the K shell, emitting a photon of K_α or K_β radiation characteristic of manganese at a wavelength of 210.2 or 210.6 pm respectively. This source will ionise elements up to and including vanadium. Other radioactive sources useful for the analysis of elements of higher atomic number are ^{238}Pa and ^{147}Pm.

There has been much recent interest in the use of a beam of protons for ionisation. The attractive feature of proton excitation is the high efficiency of conversion of the energy of the protons into X-ray emission. Proton beams of energies of the order of 10 MeV, as produced by linear accelerators of the type used in nuclear physics laboratories, can produce detectable X-ray emission from as little as 10^{-12} g of an element present in a sample. The background contribution from Bremsstrahlung is much lower than that observed for excitation by an electron beam. This technique is thus of value for the estimation of trace quantities as required for pollution research (Campbell *et al.*, 1974). Unfortunately linear accelerators for the production of 10 MeV proton beams are very expensive, so the technique is not likely to find widespread acceptance. Proton beams of lower energies, of the order of 100 keV, can be produced by the less expensive Cockroft–Walton type of accelerator. These low-energy beams only penetrate the sample to a depth of a few tens of nanometres, and are thus of interest in the study of surface layers.

Radioactive sources which emit α-particles (an α-particle consists of two protons and two neutrons) offer a cheap and convenient alternative to linear accelerators for the production of a beam of heavy charged particles. The efficiency of X-ray production is proportional to the square of the charge of the bombarding particle; so, other things being equal, α-particles are four times as efficient as protons. ^{244}Cm has been suggested as the radiation source for an X-ray emission spectrometer for *in situ* analysis of rocks in the surface of other planets (Franzgrote, 1971).

11.B.2 Dispersion

The X-rays emitted from a sample containing several elements cover a wide range of wavelengths, and some dispersive device is needed to separate the emission for each element. Monochromators for use in the infrared and visible/ultraviolet regions of the spectrum commonly employ mechanically ruled diffraction gratings as the dispersive element, as discussed in Section 4.B.3. However, the line-spacing must be of the same order of magnitude as the wavelength of the radiation to be dispersed, and it is not possible to rule lines 100 pm apart. Fortunately Nature provides ready-made diffraction gratings with a suitable line-spacing, in the form of ionic crystals. The LiF crystal, for example, has a cubic lattice with a unit cell dimension of 201.34 pm. The crystal is cleaved so that one face of the unit cell lies on the cleavage plane. Diffraction then occurs as if the cleavage plane were acting as a mirror, the phenomenon of *Bragg reflection*. X-rays of wavelength λ are diffracted as if reflected at angle θ to the cleavage plane, where θ is given by *Bragg's Law*:

$$n\lambda = 2d \sin \theta \qquad (11.4)$$

where d is the unit cell dimension perpendicular to the cleavage plane. The resolution of a dispersion device based on diffraction by a crystal is of the order of 10 eV. As with diffraction gratings for use in the infrared and visible/ultraviolet, higher-order diffracted beams may cause interference. These can be eliminated by the use of filters, made of thin sheets of the appropriate metal, which exploit the phenomenon of the absorption edge. The spectrum is scanned by rotating the crystal, the detector being simultaneously moved in a circle, centred on the crystal, at an angular velocity twice that of the crystal. An alternative arrangement is to have a separate detector fixed at each point around the circle to which is diffracted the most intense characteristic line of the element to be analysed for. Commercial apparatus is available in which twelve detectors are in simultaneous use, enabling the simultaneous determination of twelve elements in a sample. The work can be further speeded up by the attachment of a small on-line computer to the apparatus, so that a full analysis can be printed out without any computation on the part of the operator.

The instrument which measures the intensity of the X-rays diffracted at any angle by a crystal may be used to investigate the structure of the crystal. If it is used for this purpose, it is known as an X-ray diffractometer. If a monochromatic X-ray source is used, the angles at which a diffracted beam is observed give a measure of the dimensions of the unit cell, and the intensities of the diffracted beams provide information about the distribution of electron density within the unit cell, from which the molecular structure can be deduced.

The student may notice a circular argument, in which X-ray wavelengths are calculated from crystal lattice dimensions, and crystal lattice dimensions are calculated from X-ray wavelengths. The wavelengths of a few well-characterised, intense, X-rays have been determined absolutely by the use of mechanically ruled gratings at grazing angles of incidence. Although the actual distance between the lines on the most finely ruled gratings is about 560 nm, the effective spacing as seen by the X-rays is reduced to less than 1 nm by the small angle of incidence. The relationship between the angle of diffraction and the wavelength is given by

$$n\lambda = 2d \sin 2\psi + \omega/2 \sin \omega/2 \qquad (11.5)$$

where ψ is the angle between the incident radiation and the plane of the grating, and $(\psi + \omega)$ is the angle between the diffracted radiation and the plane of the grating. ψ is typically less than 1°.

11.B.3 Detection

There are three main types of detector used for the quantitative detection of X-radiation, namely the gas-ionisation cell, the scintillation counter, and the semiconductor detector. The relative merits of these have been discussed by Israel *et al.* (1971).

The gas-ionisation detector consists of a cylindrical cell, about 1 cm radius, containing an inert gas, such as xenon, at a pressure of a few kN m^{-2}. A fine wire runs down the axis of the cylinder. The wire is charged to a positive potential of a kilovolt or so, the walls of the cylinder being held at earth potential. X-ray photons enter the cell at one end, through a thin beryllium window, and collide with gas molecules until all the energy of the X-ray photons has been used to ionise the gas. The ionisation energy of Xe, 11 eV, is only a small fraction of the energy of an X-ray photon. The process is about 50% efficient, so an X-ray photon of energy 1.2 keV (corresponding to radiation of wavelength 100 pm) produces about 600 electrons. These are accelerated by the electrostatic potential and gain energy. They collide with more gas molecules, causing further ionisation. There is thus a multiplier effect, each X-ray photon causing the arrival of a large number of electrons at the axial wire anode. The amplification factor depends on the applied voltage; factors of the order of 10^6 are typical. The current pulse so produced is large enough to be counted by standard electronic instrumentation. If the electrostatic field is constant, the magnitude, or height, of the pulse depends on the energy of the X-ray photon, and hence on the wavelength of the X-radiation. The relationship between the pulse height and the photon energy is linear, so the detector is thus often called a *proportional counter*. An error in this linear relationship is caused by the phenomenon of *escape peaks*. The number of electrons produced in the primary ionisation process has been calculated as if the X-ray photons caused ionisation from the outermost orbital of the Xe atom. Of course, as discussed in Section 11.A.1, X-ray photons may cause emission of electrons from the core orbitals. To a large extent, this does not affect the argument. The hole in the core orbital is filled by an electron falling from an outer orbital, leaving a hole which in turn is filled by an electron falling from another orbital still further out, until eventually the hole is in the outermost orbital. This rearrangement is accompanied by the emission of photons of various energies, lower than that of the original photon, which then ionise other Xe atoms. As long as none of the secondary photons escapes from the cell without losing all its energy in ionisation processes, the net effect is the same as if the original photon had directly ionised all the atoms involved. However, such escape does occur to a limited extent, being most serious for X-ray photons of energy equal to or just greater than that required for ionisation from a core orbital. The consequence is that some of the pulses are of lower height than they would be if pulse height were always proportional to photon energy. Another possible cause of error is *Compton scattering*, the collision between a photon and an electron in which the photon loses energy to the electron. This is only significant for photons of energies about 100 keV. There is a limit to the counting rate of a gas-ionisation detector, set by the need for the ionised atoms to re-acquire electrons before the arrival of the next X-ray photon. The time required for this process, the *dead-time*, is of the order of 1 microsecond, so the maximum count-rate is of the order of $10^6\ s^{-1}$.

If the applied voltage in a gas-ionisation detector is sufficiently high, say 1.5–2 kV, the amplification is so great that practically all the gas molecules in the region around the axial wire anode become ionised on the arrival of a single X-ray photon. The pulse height reaches a maximum value, which is independent of the energy of the original X-ray photon. The gas-ionisation detector operated in this mode is called a *Geiger–Müller* counter. It is useful for the detection of low-energy X-ray photons, but suffers from the drawback of a long dead-time, of the order of 1 millisecond.

The photomultiplier, as described in Section 6.B, is sensitive to X-radiation. However, it is experimentally more convenient for the photomultiplier to monitor light in the visible region of the spectrum emitted by luminescent materials irradiated by the X-rays. Single crystals of sodium iodide, activated by trace quantities of thallium halides, are often used. An X-ray photon creates an ionised track through the crystal. The ionisation energy for a site in a sodium iodide crystal is about 500 eV, so the number of ionisation processes is an order of magnitude smaller than the number of primary ionisations in a gas-ionisation detector. The electron-deficient sites are filled by electron transfer from electron-rich sites by a process that involves the loss of excess energy by the emission of photons of light in the visible region of the spectrum. About 20 photons are emitted as the result of the impact of a single X-ray photon of energy 1 keV. The number of photons depends on the energy of the X-ray photon, so the device has a proportional response. This minute pulse of light is detected by the photomultiplier, which generates a pulse of current of measurable size. The pulses are counted by the usual electronic devices. A stream of X-ray photons causes the crystal to scintillate, the scintillation being just perceptible to the human eye if properly accommodated to darkness. The combination of scintillator crystal, photomultiplier, and pulse-counting apparatus is known as a *scintillation counter*. The dead-time of the scintillation counter depends on the rate of electron transfer within the crystal, which is usually slower than the photomultiplier response. Sodium iodide crystals have a dead-time of 250 ns; anthracene, dissolved in styrene which is then polymerised, has a dead-time of about 5 ns. The geometrical arrangement of a scintillation counter is simple, since the scintillators are transparent to light. A slab of scintillator crystal is placed directly in front of the photomultiplier window so as to cover it completely. This arrangement optimises the collection of the emitted light by the photomultiplier.

Solid-state detectors are also used for the detection of X-rays. Silicon or germanium, doped with lithium, are the favoured semiconductor materials. The principle of operation is similar to that described in Section 6.B, with the important difference that each X-ray photon causes a large number of electrons to jump the gap between full and empty bands. Solid-state detectors also have a proportional response, since the higher the X-ray photon energy the greater the number of excited electrons. The main drawback of the semiconductor detector is the comparatively long dead-time. The limiting count-rate is of the order of $4 \times 10^4\ s^{-1}$ at present, although there seems no fundamental reason why this could not be improved.

11.B.4 Energy-dispersion spectrometers

The detectors described in Section 11.B.3 have the common feature that the height of the pulses produced depends on the energy of the X-ray photon. Thus, if the pulses are stored and sorted, such that the number of pulses in a given narrow range of heights is displayed as a function of pulse height, the display is an energy spectrum. This sorting and display can be performed by a multichannel analyser, which has been described in Section 7.B.4. It is thus possible to do away with a dispersion device, at the expense of introducing more complex electronics. An X-ray spectrometer of this type is termed a *non-dispersive* or *energy-dispersion spectrometer*. The resolution of an energy-dispersion spectrometer depends on the energy resolution of the detector.

This depends on many factors, one of which is the magnitude of the energy quantum needed to generate an electron. The greater this is, the less fine the discrimination of the detector for variations in X-ray photon energy. The scintillation counter thus offers particularly poor resolution: about 15 keV for 100 keV photons, and as great as 5 keV for 10 keV electrons. The gas-ionisation detector is better, as expected from the smaller size of the energy quantum for electron generation; although the resolution at 100 keV is not much higher than that of the sodium iodide scintillator counter, the resolution at 10 keV is 2 keV. The energy required for electron generation in silicon is low, being only 2.8 eV compared with the 22 eV required for xenon and 500 eV required for sodium iodide. The increase in resolution is not as great as one would hope; the resolution is about 2 keV at 100 keV but only 1 keV at 10 keV. Fluctuations due to thermal noise, i.e. the excitation of electrons by thermal energy, make a major contribution to the loss of resolution by the silicon semiconductor detector. Cooling in liquid nitrogen helps reduce this, and a practical limiting value of about 100 eV has been reached for the resolution. This resolution is much less than that obtained by crystal dispersion, for which values of a few eV are typical, but is often good enough for analytical X-ray spectrometry.

An energy-dispersion X-ray spectrometer with a radioactive source and

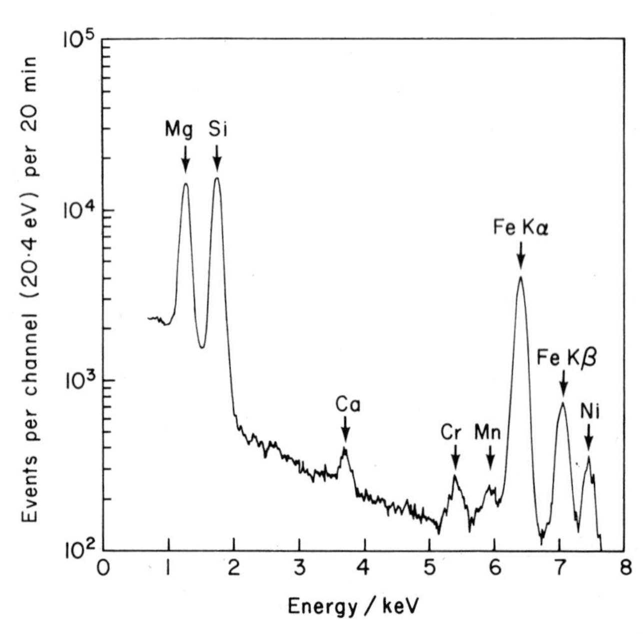

FIG. 11.3 Energy dispersion spectrum of a rock sample (Franzgrote, 1971).

semiconductor detector is a compact and rugged device. Since it does not require a high-voltage supply, it can be made portable, and used for *in situ* investigations of minerals in a geological survey. It can be lowered down a drill-hole for the analysis of minerals deep underground, the data processing equipment being at the surface. It can even be transported to the surface of another planet, the pulse counts being transmitted back to Earth for data processing. A spectrum obtained from an energy-dispersion X-ray spectrometer designed for *in situ* analysis of lunar rocks is shown in Fig. 11.3.

11.C Quantitative Analysis by X-ray Emission Spectrometry

X-ray emission spectrometry is a versatile and powerful technique for quantitative elemental analysis. It has been used for elements of as low atomic number as fluorine (atomic number 9), although the X-ray quanta emitted from low atomic-number elements are of such low energy that their detection is difficult and the count data are more liable to error. Concentrations from up to 100% to down to a few parts per million can be determined to a precision of 0.1 per cent of the value measured by a single determination. Most quantitative analytical methods based on spectroscopic techniques require the sample to be in solution or in the gaseous phase, and X-ray emission spectroscopy offers the advantage that the sample may be a solid mass or powder. The technique is thus especially indicated for the analysis of insoluble refractory materials, such as silicates, and is much used in the analysis of cements, ceramics, and geological samples.

The intensity of the emission of the characteristic radiation from a solid sample of a pure element depends on the thickness of the sample. If the sample is sufficiently thin, the intensity increases with increasing mass of material exposed to the radiation, i.e. with increasing thickness. Some losses by absorption occur for thicker samples. Samples of thickness greater than a certain limiting value, typically of the order of 100 μm may be considered as infinitely thick; only radiation emitted within the surface layer of limiting thickness is able to escape from the sample, so the intensity of emission is independent of sample thickness. Unless the sample is a thin film, e.g. an electrolytically deposited coating of one metal on another, the sample is usually infinitely thick.

If other elements are present in the sample, it is natural to assume that, to a first approximation, the intensity I_E of the characteristic radiation from element E is proportional to W_E, the weight fraction for that element in the sample, so that

$$I_E = W_E I_E^0 \tag{11.6}$$

where I_E^0 is the intensity for pure E. The proportionality only holds if other elements present in the sample, known collectively as the matrix, do not interfere, which they may do in various ways. If one of the other elements absorbs the characteristic radiation of E more intensely than does E itself, the I_E will be less than predicted by Eqn. 11.6; if less intensely, then I_E will be greater than predicted by Eqn. 11.6. These two possibilities are termed *negative* and *positive absorption* respectively. Excitation of the sample may cause one of the other elements present to emit X-radiation of shorter wavelength than that characteristic of E. This X-radiation then acts as a secondary source, exciting further characteristic radiation from E, so I_E is greater than $W_E I_E^0$. This is known as *enhancement*. Examples of these effects may be found in the quantitative determination of iron. Aluminium in the sample matrix causes positive absorption, lead negative absorption, as can be deduced from the relative positions of these three metals in the list of atomic numbers. Nickel present in the sample causes enhancement of the iron K_α radiation, which is excited by the nickel K_α radiation. These various effects must be allowed for in quantitative analysis, and a careful standardisation of emission intensity is necessary. This is best done using an internal standard; a small amount of the element to be analysed for is added so as to produce a known increase of concentration, of magnitude low enough not to perturb the matrix significantly.

REFERENCES

Campbell, J. L., Herman, A. W., McNelles, L. A., Orr, B. H. and Willoughby, R. L. (1974) *Adv. X-ray Analysis* **17**, 457.
Franzgrote, E. J. (1971) *Adv. X-ray Analysis* **15**, 388.
Israel, H. I., Lier, D. W. and Storm, E. (1971) *Nucl. Instrum. Methods* **91**, 141.
Liebhafsky, H. A. (1951) *Ann. New York Acad. Sci.* **53**, 997.

BIBLIOGRAPHY

Bertin, E. P. (1970) "Principles and Practice of X-ray Spectrometric Analysis", Plenum Press, New York and London.
Jenkins, R. (1973) in "MTP International Review of Science, Physical Chemistry Series One, Vol. 13, Analytical Chemistry, Part 2" (T. S. West, ed.), pp. 95–126, Butterworths, London.
Jenkins, R. (1974) "An Introduction to X-ray Spectrometry", Heyden, London and New York.
Liebhafsky, H. A., Pfeiffer, H. G., Winslow, E. H. and Zemany, P. D. (1972) "X-rays Electrons and Analytical Chemistry: Spectrochemical Analysis with X-rays", Wiley-Interscience, New York and London.

12 Mössbauer Spectroscopy

12.A The Mössbauer Effect

We have seen in Chapter 6 that an atom has a set of energy levels corresponding to the distribution of electrons in the orbitals around the nucleus, and transitions between these levels give rise to absorption and emission of electromagnetic radiation. The atomic nucleus also has a set of energy levels, transitions between which also give rise to absorption and emission of radiation. The lowest excited state of the ^{57}Fe nucleus, for example, has an energy of 14.4 keV (i.e. 2.3×10^{-15} J) above that of the ground state. A quantum of radiation of this energy, corresponding to a frequency of 3.48×10^{18} Hz and a wavelength of 86 pm, is emitted when the nucleus undergoes a transition from this excited state to the ground state. Electromagnetic radiation due to nuclear transitions is termed *γ-radiation*. Although γ-radiation is in general of shorter wavelength than X-radiation there is a considerable region of overlap.

If the photon emitted by an excited ^{57}Fe nucleus encounters a ^{57}Fe nucleus in the ground state, it might be expected that this second nucleus would be raised up to the excited level by the phenomenon of resonance absorption, as discussed in Section 6.D.2 for electronic transitions. This is not necessarily so. As for electronic transitions within atoms in the gas phase, the linewidth of the emitted radiation is very narrow. It follows from the Heisenberg Uncertainty Principle, as discussed in Section 6.D.2, that the linewidth, Γ, is inversely proportional to τ, the lifetime of the excited state:

$$\Gamma = (2\pi\tau)^{-1} \tag{12.1}$$

The lifetime of the excited state of the ^{57}Fe nucleus under consideration is 97.7 ns, so the linewidth is 1.63×10^{6} Hz. The linewidth is thus an extremely small proportion of the frequency, so only a small proportional difference in frequency between the absorbing and emitting nucleus will cause the photon to fail to be absorbed. One possible cause of such a difference is the motion of the emitting nucleus. If this is in the vapour phase or in solution, so that it is free to move, the momentum of the emitted photon is sufficient to cause the nucleus to recoil, just as a rifle recoils when a bullet is fired from it. There is a corresponding recoil in the absorbing nucleus, as it takes up the momentum of the fast-moving photon. Thus not all of the energy of the nuclear transition appears as energy of the photon; a quantity E_0^2/mc^2 is lost as recoil energy, where E_0 is the energy of the nuclear transition, m is the mass of the nucleus, and c is the velocity of light. For ^{57}Fe the recoil energy is 3.9×10^{-3} eV, sufficient to cause a decrease of 9.45×10^{11} Hz in the frequency of the γ-radiation. This shift is much greater than the linewidth, so the condition for absorption is not fulfilled. Since the relative magnitude of the recoil effect is proportional to the energy of the transition, the recoil effect is much smaller for emission in the visible/ultraviolet region. For example, the recoil energy of the photon emitted at 589 nm by the excited Na atom is 2×10^{-10} eV, causing a decrease in frequency of 5×10^4 Hz, which is negligible compared with the linewidth of 1×10^7 Hz. Recoil effects thus do not prevent resonant absorption in the visible/ultraviolet region. For a resonant absorption of γ-radiation to be observed, the emitting and absorbing nuclei must both be held firmly in a crystal lattice, so that the effective mass m is that of the crystal as a whole, or, at least, a substantial part of it. The importance of this was first realised by Mössbauer, so the phenomenon of recoilless emission and absorption of γ-radiation is termed the *Mössbauer effect*, and resonance γ-radiation spectroscopy is termed Mössbauer spectroscopy.

Even if recoil effects are eliminated by embedding the nuclei in a crystal lattice, the photon emitted by one ^{57}Fe nucleus cannot be absorbed by another if the energy gap between ground and excited states is different for the two nuclei. Such a difference may be produced by differences in the electronic states of the two ^{57}Fe atoms. s Orbitals have the distinctive characteristic that they penetrate through the other orbitals right to the nucleus, whatever their principal quantum number. The 4s valency orbital of ^{57}Fe thus affects the nucleus, and the extent to which this orbital is filled modifies the energy levels within the nucleus. The modification is proportionately very small; the difference in the energies of the nuclear transition for $^{57}Fe^{2+}$ and $^{57}Fe^{3+}$ is only 4×10^{-8} eV, corresponding to a difference in frequency of absorption of around 10^7 Hz. This difference is called the *isomer shift*. The magnitude of the isomer shift thus gives a measure of the relative degree of occupation of the 4s level. Although the isomer shift is so small, the linewidth is even less, so the isomer shift is measurable. It is necessary to scan through a frequency range of the order of 10^7 Hz, centred on a frequency of 3.48×10^{18} Hz. At first sight this presents a formidable problem, but an elegant solution is provided by the Doppler effect, discussed in Section 6.D.2. If the source of radiation, of frequency ν_0, is caused to move towards the absorbing sample at a velocity v, the apparent frequency as seen by the sample is ν, where

$$(\nu - \nu_0)/\nu_0 = v/c \tag{12.2}$$

The magnitudes of ν and ν_0 are such that velocities of the order of 1 mm s^{-1} are required to compensate for the isomer shifts observed for ^{57}Fe. These velocities are readily produced by mechanical means. A device that measures the absorption of γ-radiation as the frequency is scanned by the Doppler shift is known as a Mössbauer spectrometer. Some of the nuclei on which Mössbauer spectroscopy has yielded information of chemical interest are listed in Table 12.1.

TABLE 12.1

Some nuclei used in Mössbauer spectroscopy (Greenwood and Gibb, 1971)

Emitter nucleus	γ-Ray energy/keV	Half-life of emitter nucleus/ns	$10^{13} \times$ ratio of linewidth to total energy	Radioactive source	Half-life of source nucleus
^{57}Fe	14.4	98	4.7	^{57}Co	270 d
^{83}Kr	9.3	147	4.8	^{83}Rb	83 d
^{119}Sn	23.8	18.5	14.9	^{119m}Sn	245 d
^{129}I	26.8	16.3	15.1	^{129}Te	33 d
^{129}Xe	40	0.96	171	^{129}I	1.6×10^7 y
^{133}Cs	81	6.23	13.0	^{133}Ba	7.2 y
^{151}Eu	21.6	8.8	34.6	^{151}Gd	120 d
^{197}Au	77.3	1.8	47.3	^{197}Hg	65 h

12.B Apparatus

12.B.1 Sources

A source of the appropriate nuclei in an excited state is required. Most commonly this is provided by a comparatively long-lived radioactive isotope, which undergoes a nuclear reaction to give the excited nucleus required. For example, the ^{57}Co isotope of cobalt is unstable, having a half-life of 207 days. The ^{57}Co nucleus undergoes transformation to ^{57}Fe by electron capture, as described in Section 11.B.1. The ^{57}Fe nucleus is formed in an excited state, 137 keV above the ground state, which emits a γ-ray photon to form the excited state 14.4 keV above the ground state which is used for Mössbauer spectroscopy of Fe compounds. The source is made by the electrochemical deposition of a thin film of ^{57}Co on a diamagnetic support, e.g. platinum metal. The ^{57}Co is diffused into the bulk of the support by heating for some time. The minimum linewidth for emission is only obtained from nuclei in a diamagnetic matrix, so solid cobalt metal, which is ferromagnetic, would not provide a suitable source. Some examples of radioactive sources are listed in Table 12.1. Although electron capture is the most common route, it is not the only process by which the emitter nucleus may be formed. For example, ^{119m}Sn is a comparatively long-lived isotope which decays to the ^{119}Sn Mössbauer emitter by γ-emission. The radioactive source must have a half-life sufficiently long, so that replacement is not needed at inconveniently short intervals.

A limitation on the nuclei used as emitter is that the γ-ray photon energy be not so high that the condition for recoilless emission is not fulfilled. The practical limit for Mössbauer spectroscopy is about 150 keV. The emission of photons of greater energy than this is accompanied by significant loss of energy to the crystal lattice. Thermal vibration of the crystal lattice causes line-broadening, and the Mössbauer effect can only be observed at room temperature for low-energy emitters, such as ^{57}Fe. Most sources must be cooled in liquid nitrogen, or even, in extreme cases, liquid helium. The sample is often similarly cooled.

12.B.2 Detectors

Since γ-ray photons are similar in energy to X-ray photons, the detectors described in Section 11.B.3 are also used for Mössbauer spectrocopy.

12.B.3 Scanning the spectrum

In the simplest type of Mössbauer spectrometer, the source is moved towards or away from the sample for a fixed time at a constant velocity. The total distance moved must be not more than a small proportion of the total distance between source and sample, to keep constant the fraction of the total radiation from the source intercepted by the sample. The count rate from the detector, which is mounted on the other side of the sample, is measured. This is repeated for a succession of different velocities, and the spectrum is plotted, point by point, as a function of count rate against velocity. The calculation of frequency shift can be readily performed by application of Eqn. 12.2, but it is customary to leave the horizontal axis graduated in mm s^{-1}. Resonance corresponds to absorption, and is shown by a decrease in the count rate. If only a small proportion of the photons are absorbed, a higher signal-to-noise ratio can be obtained if the photons re-emitted from the sample are observed in a direction perpendicular to that of the incident photons. The reason for this improvement is as discussed in Section 6.D.3. For this geometry, resonance is indicated by a maximum in the count rate.

In more elaborate Mössbauer spectrometers, the spectrum is scanned by varying the velocity during a single sweep. The source is repetitively moved towards and away from the sample, and the count data are accumulated for a large number of cycles. The form of velocity variation that is the easiest to produce experimentally is sinusoidal; the source is coupled by a crank to a shaft rotating at constant angular velocity. However, it is now more common for the velocity variation to be linear, to give a linear scale to the displayed spectrum. This is achieved by coupling the source, mounted on a spring-loaded suspension, to a solenoid. The solenoid current is varied so as to give a constant acceleration in each cycle, and hence a linear variation of velocity. The repetition frequency is typically of the order of a few kHz, and the maximum displacement is a fraction of a millimetre. The count data from the detector are processed by a multichannel analyser, as described in Section 7.B.4. Each channel of the analyser displays the number of counts obtained for a certain narrow range of velocities. The change from one channel to the next is timed by a clock pulse, the clock frequency being an integral multiple of the source vibrator frequency. For example, suppose the spectrum is to be scanned over the range -2 mm s^{-1} to $+2$ mm s^{-1} at a repetition rate of 20 Hz, using a 400-channel analyser. At the beginning of each cycle, pulses from the detector, corresponding to the arrival of each γ-ray photon, are counted and displayed in the first channel, in which are accumulated the count data for the velocity range -2.00 to -1.99 mm s^{-1}. At the end of 2.5 ms, the pulses from the detector are diverted into the second channel in which are stored the count data for the velocity range -1.99 to -1.98 mm s^{-1}, and so on. The apparatus is allowed to cycle repetitively until sufficient pulses have accumulated in each channel to make the spectrum clear. Typically, 10^5 to 10^6 counts per channel are accumulated. A simple Mössbauer spectrum is shown in Fig. 12.1.

12.C Information Available from Mössbauer Spectroscopy

12.C.1 Isomer shift

The energy of the nucleus is reduced to a small extent by interaction with the s electrons that penetrate to it. The larger the nucleus, the greater is this stabilisation effect. In general a nucleus in an excited state is of different radius from that of its ground state, so the effect of the s electrons will be different. For example, the ^{119}Sn nucleus is larger in its first excited state than

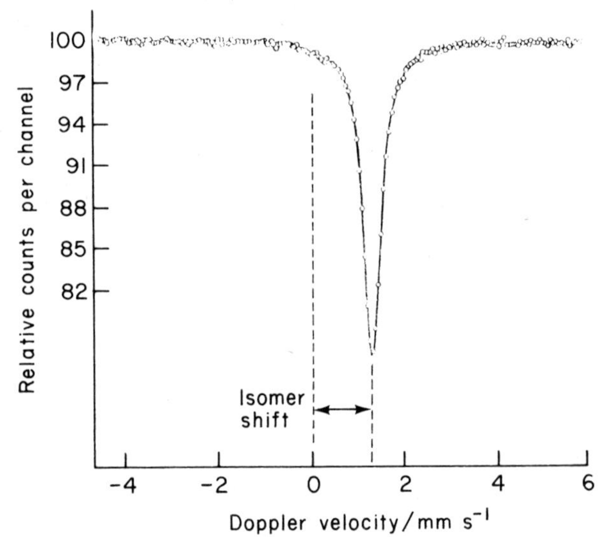

FIG. 12.1 Mössbauer spectrum of ^{57}Fe showing isomer shift; sample $RbFeF_3$ at 127 K (paramagnetic) (Wertheim *et al.*, 1967).

in its ground state. The effect of s electron interaction is thus to reduce the energy gap between these two states, and so reduce the Mössbauer resonance frequency. The difference between the resonance frequency in a sample and in a standard source, the isomer shift, conventionally measured in mm s^{-1}, thus gives a measure of the relative s electron density in the sample compared with the source. A positive isomer shift (i.e. the resonance frequency of the sample is greater than the source, so resonance occurs when the source is moving towards the sample) for a ^{119}Sn sample indicates a greater occupancy of the s orbitals in the sample than in the source. This can be seen for the Sn^{IV} halides. As shown in Table 12.2, the more covalent the Sn–Ha bond, the

TABLE 12.2

Isomer shifts in Sn^{IV} halides (Goldanskii *et al.*, 1964)

Compound	Isomer shift relative to β-Sn/ mm s^{-1}
SnF_4	−3.1 (mean value for doublet)
$SnCl_4$	−2.0
$SnBr_4$	−1.7
SnI_4	−1.0

greater is the occupancy of the 5s orbitals in Sn, and the greater the isomer shift. All the isomer shifts for $SnHa_4$ are negative with respect to the source, elemental grey Sn, in which the 5s orbitals are fully occupied. It can be seen that isomer shifts provide a useful tool for the study of chemical bonding.

The most widely studied nucleus is ^{57}Fe, since ^{57}Fe Mössbauer spectra are comparatively easy to obtain. The excited ^{57}Fe nucleus is smaller than the ground-state nucleus, so a positive isomer shift indicates a higher occupancy of s orbitals in the sample than in the source. As discussed in Section 6.C.4, Fe complexes may be classified as high-spin or low-spin. For high-spin complexes the isomer shift indicates increasing occupancy of the s orbitals with decreasing ligand electronegativity, as found for the Sn^{IV} halides. This is readily interpreted in terms of the increasing covalency of the metal–ligand bonds, since the ionic character of a bond is correlated with the different in electronegativity of the linked groups. However, for low-spin complexes the isomer shift increases with increasing ligand electronegativity, e.g. along the series $CN^- < CO < NO^+$. The isomer shift for $Na_2[Fe(CN)_5NO]2H_2O$ corresponds to the highest s orbital occupancy for any Fe compound in a common oxidation state. In these complexes π bonding occurs between the metal ion and the ligands, with donation of electrons form the 3d orbitals of the metal ion to the empty antibonding π orbitals on the ligand. The more electronegative the ligand, the more strongly this *back-donation* occurs. The 3d electrons shield the 4s electrons from the Fe nucleus, so the greater the back-donation the greater the s electron density at the nucleus, and the greater the isomer shift.

An early success of ^{57}Fe Mössbauer spectroscopy was the elucidation of the structures of the hitherto intractable insoluble cyanide complexes Prussian Blue and Turnbull's Blue. Insoluble Prussian Blue is formed by mixing aqueous solutions containing $Fe(CN)_6{}^{4-}$ and Fe^{3+} respectively, whereas Turnbull's Blue is formed by mixing aqueous solutions of $Fe(CN)_6{}^{3-}$ and Fe^{2+}. Both of these substances are of the type $Fe_4[Fe(CN)_6]_3$, but the oxidation states of the Fe ions were unknown; structures $Fe^{III}{}_4[Fe^{II}(CN)_6]_3$ and $Fe^{III}Fe^{II}{}_3[Fe^{II}(CN)_6]_3$ were postulated, and it was also considered possible that electron exchange within the complex was so rapid that the oxidation numbers were indeterminate. The Mössbauer spectra showed that Prussian Blue and Turnbull's Blue are identical. The spectra consisted of two peaks, one attributable to high-spin Fe^{III} and one to low-spin Fe^{II} (Duncan and Wigley, 1963). Both substances are thus $Fe^{III}{}_4[Fe^{II}(CN)_6]_3$.

12.C.2 Nuclear quadrupole coupling

As discussed in Section 2.E.2, nuclei with spin quantum number greater than $\frac{1}{2}$ have an electric quadrupole moment that may be positive or negative corresponding to oblate or prolate distortion of the nucleus from spherical symmetry. If the electric field set up by the environment of the nucleus lacks spherical symmetry, the quadrupole is aligned in one of a set of possible orientations to the field, so that a set of energy levels is established. For a nucleus of spin quantum number 3/2, such as the first excited state of ^{57}Fe in an electric field of axial symmetry, the nuclear quadrupole moment can couple with or against the field, giving a pair of energy levels. The ground-state ^{57}Fe nucleus has spin quantum number $\frac{1}{2}$, and so does not have a quadrupole moment. Thus the transition from the excited to the ground state of ^{57}Fe gives a pair of lines, i.e. a doublet, in the Mössbauer spectrum if the ^{57}Fe nucleus is in a local electric field gradient with axial symmetry. Such a field is produced if one ligand differs from the others which are identical, e.g. in the $[Fe(CN)_5NO]^{2-}$ ion. The Mössbauer spectrum of $Na_2[Fe(CN)_5NO]2H_2O$ is thus a doublet, as shown in Fig. 12.2. The magnitude of the splitting is related to the nuclear quadrupole moment by Eqn. 3.1. If the Mössbauer transition is from a state of spin quantum number 3/2 to one of spin quantum 1/2, e.g. as for ^{57}Fe, the splitting, Δ, in units J is given by

$$\Delta = eQq \tag{12.3}$$

where eQ is the quadrupole moment in units C m^2, and q is the electric field gradient in V m^{-2}. Transitions occurring between nuclear states of higher spin quantum numbers give more complex multiplet spectra. If the electric field gradient lacks axial symmetry, a further factor describing the asymmetry

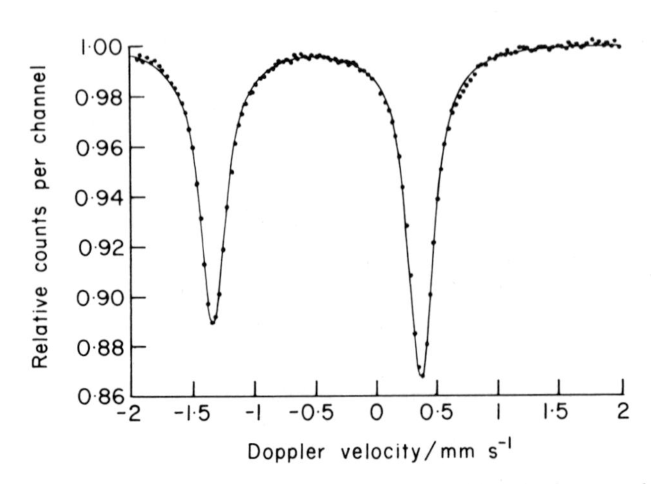

FIG. 12.2 Mössbauer spectrum of sodium nitroprusside (single crystal at 23°C) showing quadrupole splitting (Grant *et al.*, 1969).

is required. For the $3/2 \rightarrow 1/2$ type transition,

$$\Delta = eQq(1+\eta^2/3)^{\frac{1}{2}}/2 \qquad (12.4)$$

where η is a measure of the asymmetry. Since values of η must lie between zero and unity, the maximum effect asymmetry can have on the splitting is only 16%. For ^{57}Fe, a Doppler velocity of 1 mm s^{-1} corresponds to a value of Δ of 35 MHz, which, as discussed in Chapter 3, is an expected order of magnitude for a nuclear quadrupole transition. Values of Δ are thus of the same order of magnitude as isomer shifts, δ. Some illustrative values of Δ and δ are given in Table 12.3.

TABLE 12.3

Some isomer shifts (δ) and quadrupole splittings (Δ) (Greenwood and Gibb, 1971)

	Temp/K	δ/mm s^{-1}	Δ/mm s^{-1}	Zero δ for
$^{57}FeF_2$ (high-spin)	77	1.48	2.92	^{57}Fe metal
$^{57}FeI_2$ (high-spin)	77	0.99	0.82	^{57}Fe metal
$^{57}Fe(en)_3Cl_3$ (low-spin)	300	0.14	1.09	^{57}Fe metal
$^{57}Fe(phen)_3(ClO_4)_2$	77	0.34	0.23	^{57}Fe metal
$^{119}SnF_2$	80	1.60	1.80	Sn metal
$^{119}SnCl_2py$	80	1.19	0.98	Sn metal
$^{125}TeO_2$	77	0.72	6.54	^{125}I
^{129}ICl	77	1.73	62.7	^{129}Te
$^{151}EuSO_4$	0.057	−14.64	1.15	$^{151}Eu_2O_3$

en = ethylenediamine; phen = phenanthroline.

The field gradient causing quadrupole splitting may be due either to charges on adjacent ions, either on adjacent sites in the crystal lattice or as ligands, or to electrons of the atom itself if one of the orbitals is incompletely filled, as in transition-metal ions. The existence of quadrupole splitting thus affords useful evidence as to molecular structure. A simple example is provided by the Sn^{IV} halides. Whereas the ^{119}Sn Mössbauer spectra for $SnCl_4$, $SnBr_4$, and SnI_4 show only single peaks, as expected for a symmetrical tetrahedral disposition of the ligands around the central Sn atom, the spectrum of SnF_4 shows a doublet. There are thus two types of Sn–F bond in the molecule. Each Sn^{4+} ion is coordinated to six F^- ions, of which four are coordinated to other Sn^{4+} ions and two are coordinated solely to the central Sn^{4+} ion. (Goldanskii *et al.*, 1964).

12.C.3 Magnetic hyperfine interaction

Nuclei, being in effect tiny magnets, are oriented by a magnetic field, as will be discussed in more detail in Chapter 13. Only certain angles of orientation are permitted as a consequence of the quantisation of energy, so a nucleus of spin quantum number I has $(2I+1)$ magnetic energy levels. The energy gaps between these levels are such as to cause noticeable splitting in Mössbauer spectra, and this effect is known as magnetic hyperfine interaction. The magnetic field responsible for this splitting may be due to unpaired electrons around the nucleus, but splitting of the energy levels by this interaction is not observed for all paramagnetic species, since the electrons usually exchange spins so rapidly that the time-averaged effect is zero. Magnetic hyperfine interaction is observed only for ferromagnetic materials, or for samples in which paramagnetic ions, e.g. Fe^{3+}, are held well-separated from each other in a dilute solid solution in a diamagnetic matrix, e.g. of Al_2O_3. A spectrum showing magnetic hyperfine interaction is shown in Fig. 12.3. Below a certain temperature, the *Curie temperature*, thermal agitation is insufficient to prevent the alignment of paramagnetic ions to form a ferromagnetic crystal, so a crystal containing paramagnetic ions will show magnetic hyperfine splitting below the Curie temperature. Magnetic hyperfine splitting may also be induced by the application of an intense external magnetic field, of the order of several tesla.

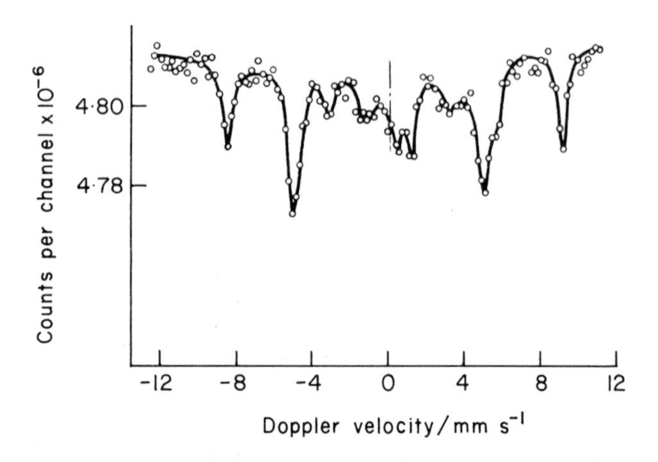

FIG. 12.3 Mössbauer spectrum of ^{57}Fe in an Al_2O_3 matrix (0.08 atom % ^{57}Fe in Al_2O_3 at 78 K) showing magnetic hyperfine interaction (Wertheim and Remeika, 1964).

REFERENCES

Duncan, J. F. and Wigley, P. W. R. (1963) *J. Chem. Soc.*, 1120.

Goldanskii, V. I., Makarou, E. F., Stukan, R. A., Sumarokova, T. N., Trukhtanov, V. A. and Khrapov, V. V. (1964) *Dokl. Akad. Nauk SSSR* **156**, 400.

Grant, R. W., Housley, R. M. and Goner, U. (1969) *Phys. Rev.* **178**, 523.

Wertheim, C. K., Guggenheim, H. J., Williams, H. J. and Buchanan, D. N. E. (1967) *Phys. Rev.* **158**, 446.

Wertheim, G. K. and Remeika, J. P. (1964) *Phys. Lett.* **10**, 14.

BIBLIOGRAPHY

Bancroft, G. M. (1974) "Mössbauer Spectroscopy: An Introduction for Inorganic Chemists and Geochemists", McGraw-Hill.

Greenwood, N. N. and Gibb, T. C. (1971) "Mössbauer Spectroscopy", Chapman and Hall, London.

Herber, R. H. and Hazony, Y. (1972) in "Techniques of Chemistry, Vol. I, Physical Methods of Chemistry, Part IIID, Optical Spectroscopic and Radioactivity Methods" (A. Weissberger and B. W. Rossiter, eds.), pp. 215–306, Wiley-Interscience, New York and London.

Wertheim, G. K. (1964) "Mössbauer Effect, Principles and Applications", Academic Press, New York.

13 Nuclear Magnetic Resonance Spectroscopy

13.A The Principles of Nuclear Magnetic Resonance

13.A.1 Nuclear magnetic energy levels

As discussed in Section 2.E.1, the atomic nucleus has a magnetic moment. In an applied magnetic field the nucleus orients itself so as to set up a set of $(2I+1)$ energy levels, where I is the spin quantum number. In the simplest instance, for nuclei of $I = \frac{1}{2}$, there are two energy levels, separated by a gap ΔE, given by

$$\Delta E = 2\mu B \tag{13.1}$$

where μ is the resolved part of the magnetic moment $\boldsymbol{\mu}$ in the direction of the applied magnetic field, whose flux density $\boldsymbol{B}$ has magnitude B. If μ has units $\mathrm{J\,T^{-1}}$, which are the same as units $\mathrm{A\,m^{-2}}$, and B has units T, then ΔE has units J. The student will note that the use of H rather than B, as is often found in the older literature, renders Eqn. 13.1 dimensionally inhomogeneous. The two energy levels do not correspond to alignment of the magnetic dipole exactly along the lines of magnetic force, as would be found for a macroscopic magnet, because of quantisation effects. The angular momentum of the nucleus is quantised so that the resolved part along the lines of force is an integral or half-integral multiple of $\hbar$. However, as shown by Eqn. 2.30, the total angular momentum $\boldsymbol{I}$ for a nucleus of spin quantum number $\frac{1}{2}$ is $\hbar(3/4)^{\frac{1}{2}}$, so the nucleus must be oriented at an angle of $\cos^{-1}(3^{-\frac{1}{2}})$ to the field. The two possible orientations for a nucleus of spin quantum number $\frac{1}{2}$, each labelled by a particular value of M_I, are illustrated in Fig. 13.1. The orientation in which the moment is oriented with the field is of lower energy. Absorption of electromagnetic radiation of frequency $\Delta E/h$ causes excitation to the other orientation. This phenomenon is termed nuclear magnetic resonance or n.m.r. for short.

It can be seen from Fig. 13.1 that the observable value of $\boldsymbol{\mu}$ is $g_N\mu_N/2$. In general, the maximum observable value of $\boldsymbol{\mu}$ is $g_N\mu_N I$, because no nucleus

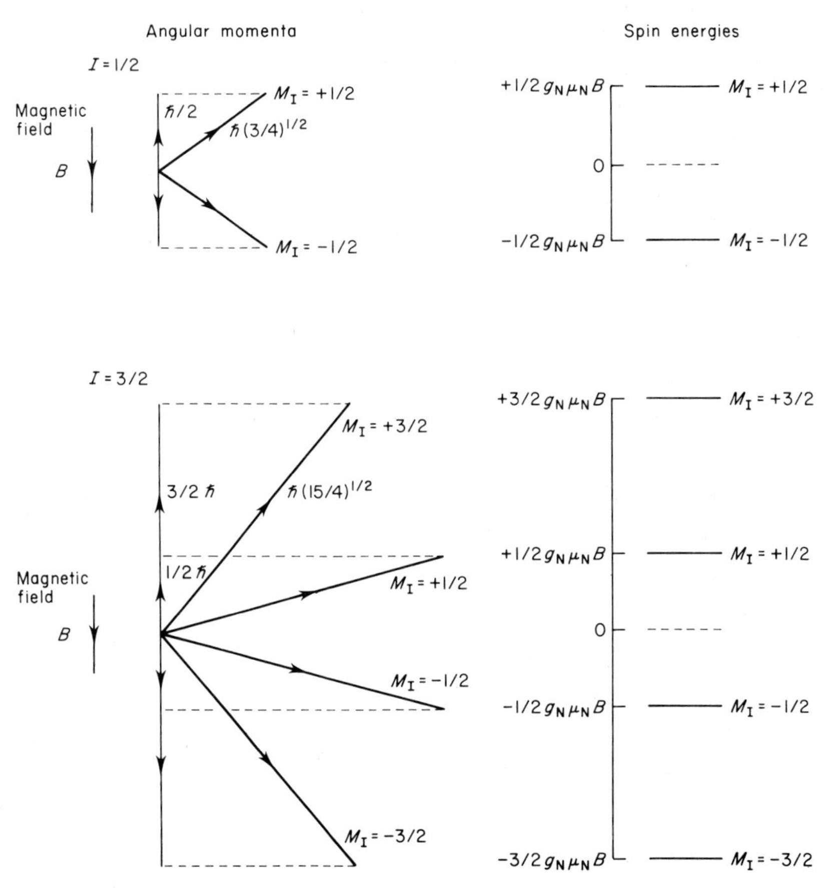

FIG. 13.1 Nuclear magnetic energy levels; magnetic moment vectors are obtained by replacing $\hbar$ by $g_N\mu_N$.

can be oriented exactly along the lines of force of the magnetic field. The four possible orientations for $I = 3/2$ are also shown in Fig. 13.1, corresponding to observable values of $\boldsymbol{\mu}$ of $3g_N\mu_N/2$ and $g_N\mu_N/2$ respectively.

Experiments with gyroscopes show that if a couple is applied to a spinning object, acting on its axis of rotation, the axis rotates, or precesses about an axis parallel to the direction of the applied couple. Since the spinning nucleus had its axis of rotation at an angle to the magnetic field, the field exerts a couple on the nucleus, and the nucleus may be thought of as precessing in this fashion. This phenomenon is known as *Larmor precession*. Calculation of the rate of the Larmor precession shows that the frequency is equal to $\Delta E/h$, where ΔE is given by Eqn. 13.1. The Larmor precession provides a mechanism by which electromagnetic radiation may interact with the spinning nucleus. If a secondary magnetic field is applied perpendicular to the principal field, with its vector rotating at the Larmor frequency, the nucleus will be induced to flip over into the other orientation, absorbing or emitting energy to the secondary field as it does so, according to the direction of the flip. Such a rotating magnetic field is provided by circularly polarised electromagnetic radiation. In practice, linearly polarised radiation is used, the nucleus selecting the circularly polarised component that it requires (cf. Section 9.A.1). Substitution of numerical values into Eqn. 13.1 shows that, for the ^1H nucleus, often referred to as the proton,* ΔE corresponds to a frequency of 60 MHz if B is of magnitude 1.41 T. Magnets giving this flux density are commonly employed in n.m.r. spectrometers, but a few instruments have been constructed in which the magnets give flux densities of up to 7 T. N.m.r. is thus located in the radio-frequency part of the electromagnetic spectrum.

The discussion so far has treated the nucleus as if it were isolated in space. Nuclei are of interest to chemists only if they form parts of atoms, ions, or molecules, and are thus surrounded by clouds of electrons. These electrons modify the local magnetic field around the nucleus, so nuclei in different chemical environments precess at slightly different Larmor frequencies. The differences are known as *chemical shifts*. The maximum observed variation in the magnetic field around ^1H nuclei is of the order of 20 p.p.m. If either the frequency of radiation incident on a sample or the magnetic field is scanned through a range of the order of 20 p.p.m., absorption will be observed in various regions of the scan due to ^1H nuclei in various environments. Since the chemical shift to be expected for a ^1H nucleus in a particular environment has been determined by measurements on standard samples, the n.m.r. spectrum can be interpreted to give the molecular structure of the sample. N.m.r. spectroscopy is thus an extremely powerful analytical technique, and is one of the two most widely used spectroscopic techniques in organic chemistry, the other being infrared spectroscopy. Chemical shifts are discussed in more detail in Section 13.C.1.

13.A.2 Relaxation and linewidths

The energy gap between the nuclear magnetic energy levels is extremely small, much less than kT. For example, for the ^1H resonance at 60 MHz, ΔE is 4×10^{-26} J, whereas kT at 300 K is 4×10^{-21} J. The ratio of the populations of upper and lower levels, n_u and n_l respectively, is given at

* The student may be accustomed to think of protons as ionisable hydrogen atoms in a molecule. In n.m.r. spectroscopy all hydrogen atoms are termed protons, whatever their binding, since it is their nuclei that are being studied.

equilibrium, by adaptation of Eqn. 2.12, by

$$\begin{aligned} n_u/n_l &= \exp(-\Delta E/kT) \\ &\approx 1-(\Delta E/kT) \text{ (for small } \Delta E/kT) \end{aligned} \tag{13.2}$$

Substitution of numerical values for ^1H at 60 MHz, at room temperature, shows that the population of the lower level is greater than that of the upper by a factor of only 1×10^{-5}. A nucleus flipping from the upper to the lower state emits as much energy as one absorbs flipping from the lower to the upper state, so the net energy absorbed by the sample is proportional to the difference in populations, and is thus small. If energy is supplied to the sample at such a rate that there is no time for it to be lost in other ways, i.e. converted into thermal motion, the population of the upper state rises to become equal to that of the lower. The system then emits as much energy as it absorbs, and the net absorption is zero. A system in such a state is said to exhibit *saturation*.

The value of n_u/n_l given by Eqn. 13.2 will not be that observed for a system exposed to a moderate amount of radio-frequency power at the Larmor frequency. Equation 13.2 may be used to define a *spin temperature*, T_s, which fits the equation for the actual value of n_u/n_l observed. T_s has the value infinity at saturation, when $n_u = n_l$. If $n_u > n_l$, as can be caused by, for example, a rapid reversal of the polarity of the applied field, T_s is negative. This provides an answer to a question sometimes asked by students as to the meaning of negative absolute temperature; such a temperature may be thought of as hotter than infinity!

The mechanism by which nuclei lose spin energy, so that the spin temperature falls to the thermodynamic temperature, is known as *spin–lattice relaxation*. The surroundings of the nucleus are known as the lattice because early experiments were carried out on crystalline samples. The most important mechanism for spin–lattice relaxation involves the interaction between the magnetic dipole of the nucleus in question and the magnetic dipoles of adjacent nuclei, and is thus known as *dipole–dipole relaxation*. Adjacent molecules or ions move in a random fashion owing to thermal agitation, setting up fluctuating magnetic fields at the nucleus. A component of the motion is at the Larmor frequency, so the nucleus flips from the upper to the lower state, transferring energy to the thermal motion of the surrounding molecules. The process is much more efficient in liquids than in solids owing to the greater thermal agitation. Each excited nucleus in the sample has the same chance of flipping down as any other, so the rate of energy transfer to the lattice at any instant is proportional to the number in the excited state. Energy transfer from the lattice to nuclei in the lower spin state, causing them to flip up, is similarly a first-order process. Spin–lattice relaxation is the net effect of two first-order processes, and thus is itself first-order, having a rate constant of T_1^{-1} where T_1 is the *spin–lattice relaxation time*. T_1 is of the order of a few seconds in liquids at room temperature, but is several orders of magnitude greater in solids. Since spin–lattice relaxation is dependent on thermal agitation, T_1 is temperature-dependent. T_1 is much reduced by the presence of paramagnetic species, such as O_2 or transition-metal ions. These produce fluctuating magnetic fields which couple very efficiently to the nuclear spin.

An individual nucleus may change its spin orientation by interaction with the magnetic field due to the magnetic dipole of an adjacent nucleus. One nucleus flips up as the other flips down, and there is no net loss of spin energy. This process, termed *spin–spin relaxation*, may be thought of as the process by which the spin temperature is maintained uniform throughout the sample. The relaxation process is first-order, and is characterised by the *spin–spin relaxation time*, T_2. The more efficient the magnetic coupling, the shorter is T_2. Values of T_2 are much less for solids than for liquids. Values of T_2 for liquids are similar to the values of T_1, whereas T_2 for solids is of the order of 10^{-6} s. Spin–spin relaxation is also facilitated by the presence of paramagnetic ions or molecules.

An n.m.r. absorption peak is Lorentzian in shape, i.e. is described by an equation of the type of Eqn. 9.6. The peak width at half-height, $2\delta\nu$, is related to the lifetime of the upper state by virtue of the Heisenberg Uncertainty Principle, as has been discussed for other types of spectroscopic transitions, e.g. in Section 12.A. The lifetime of a nucleus in an excited state is given by T_2, so that

$$T_2^{-1} = 2\pi\delta\nu \tag{13.3}$$

For a value of 1.6 s for T_2, $\delta\nu$ is 0.1 Hz, so linewidths of the order of 0.2 Hz are to be expected.

TABLE 13.1

Some nuclei commonly studied by n.m.r.

Nucleus	100 × Relative natural abundance	Spin quantum number	Frequency/MHz for $B = 1.000$ T	Chemical shift range/p.p.m.	Relative sensitivity*
^{1}H	99.98	1/2	42.58	~20	1.0000
^{11}B	81.17	3/2	13.66	~140	0.165
^{13}C	1.108	1/2	10.71	~350	0.0159
^{19}F	100	1/2	40.01	~300	0.834
^{31}P	100	1/2	17.24	~300	0.0664

* Referred to 100% abundance.

13.B Apparatus

13.B.1 Magnets

The quality of the magnet employed is of great importance. The magnetic flux density must be constant to less than one part in 10^8 both in time and over the volume of the sample. If the field is slightly inhomogeneous, the resonance lines are broadened. Both permanent magnets and electromagnets are in common use for n.m.r. spectroscopy. Permanent magnets have the advantage that no electrical power is required to maintain the field, but they suffer from various disadvantages. The field is temperature-dependent, so accurate thermostatting is required; an increase in temperature of only 5×10^{-5} K is sufficient to decrease the flux density by a factor of 10^{-8}. Permanent magnets of flux density greater than about 1.5 T are excessively expensive; magnets of 1.5 T represent a useful and economic compromise, which accounts for the adoption of 60 MHz as a standard operating frequency for many commercial proton magnetic resonance spectrometers. Electromagnets, i.e. solenoids with ferromagnetic cores, can be used to give fields of up to 2.2 T, the limiting value for saturation of ferromagnetic alloys. High currents, of the order of 50 A are required, and these must be accurately stabilised. These currents produce considerable Joule heating, so the coils must be cooled by a rapid flow of thermostatted water. Whether the magnet is a permanent magnet or an electromagnet, the pole pieces must be highly polished and accurately parallel to ensure field homogeneity. Coils of one or two turns of wire, called *shim coils*, are placed on the pole-pieces to increase field homogeneity. When properly adjusted, these shim coils improve the field homogeneity by a factor of 10^3. The sample cell, a precisely-made cylindrical glass tube, of the order of 3 mm internal diameter, is held midway between the pole-pieces, which are about 3 cm apart.

The development of superconducting electromagnets was a major breakthrough in n.m.r. instrumentation. The coils are made of a Nb–Ti alloy wire which is superconducting, i.e. has zero electrical resistance, at 4.2 K, the temperature of boiling helium. Once a current has been established in the coil, the magnetic field is maintained indefinitely without further input of energy. Since even the best ferromagnetic materials become magnetically saturated at a flux density of around 2.2 T, there is no point in having a ferromagnetic core for magnets of higher flux density, and superconducting magnets are thus air-cored. The sample is conveniently placed in the core, so larger sample volumes can be used than if a metal-cored magnet is used. Since the intensity of the signal is proportional to the number of resonating nuclei, the larger the volume of the sample, the greater the signal-to-noise ratio. Superconducting magnets of up to 7 T are now in use in n.m.r. spectrometers. The principal drawback is the need for a supply of liquid helium. The consumption of liquid helium is kept to a minimum by surrounding the liquid helium bath with another of liquid nitrogen, boiling at 77 K.

The magnitude of the flux density provides a figure-of-merit for n.m.r. spectrometers; the higher the flux density the better. The separation between the lines for nuclei of the same type in different environments is linearly proportional to the flux density, whereas the perturbation that one nucleus has on another adjacent to it is independent of the flux density. The higher the flux density, the less is the proportional effect of these perturbations which add to the complexity of the spectrum, so the spectrum becomes easier to interpret. This point is discussed further in Section 13.C.2. The other beneficial effect of increased flux density is an increase in the signal-to-noise ratio. The signal intensity is proportional to the difference in population between the lower and upper energy levels, which is linearly dependent on the energy gap, ΔE, for small values of $\Delta E/kT$, as shown by Eqn. 13.2. The value of ΔE is in turn linearly dependent on the flux density B, as shown by Eqn. 13.1. Thus, doubling the flux density doubles the signal-to-noise ratio.

13.B.2 Samples

The sample usually consists of 1 cm^3 or so of liquid in a precisely-made cylindrical glass tube, of the order of 3 mm internal diameter. Only a length of about 5 mm is in the detector coil, but a column of liquid about 2 cm long is required to obviate end-effects. To reduce flux inhomogeneity, the sample tube is rotated about its axis of symmetry at about 100 Hz. Since n.m.r. absorption is weak, samples consist of pure liquids or concentrated solutions, the lower concentration limit being typically of the order of 10^{-2} mol l^{-1}. The solvent must not have a spectrum overlapping that of the solute. C^2HCl_3 is often used for organic compounds whose ^1H spectrum is required. The sample must be in the liquid phase if chemical shifts are to be observed. T_2 is so short for solid samples that the peaks are far too broad for the structure due to the various chemical shifts to be seen.

13.B.3 Continuous-wave detection

An n.m.r. spectrometer differs in design from most of the spectrometers previously described in that there is no clearly defined path in space from source to detector through the sample. The wavelengths of the radiation are so long (e.g. 60 MHz radiation has a wavelength of 50 m) that the whole apparatus extends over less than a tenth of the wavelength. There are two modes in which nuclear magnetic resonance may be observed. In the single-coil mode the sample is surrounded by a coil, with its axis perpendicular to the applied magnetic field, which forms part of an alternating-current bridge. The bridge is balanced at a frequency at which absorption by the sample does not occur. As the frequency is scanned through the resonant frequency, power is absorbed by the sample, thereby throwing the bridge out of balance. The out-of-balance signal is monitored and displayed by a chart recorder as a function of frequency. In the double-coil mode, a second coil is placed around the sample, with its axis perpendicular to both the applied magnetic field and the axis of the first coil. Radiation is supplied by the first coil, and the signal from the second coil is monitored. At resonance the nuclei flipping down from the excited state emit radiation which is detected by the second coil. The signal from the second coil is displayed on a chart recorder after appropriate electronic processing.

The spectrum may be scanned by changing either the magnetic flux density or the radiation frequency, and both methods are in common use. The changes required are small; a change in flux density of 14 μT is sufficient to scan the whole range of proton resonances at 60 MHz, and a change of 600 Hz is sufficient to scan this range at a flux density of 1.41 T. The requisite change in flux density is conveniently produced by changing the current in a pair of small secondary coils attached to the pole pieces of the magnet. The requisite change in frequency is produced by modulating the fundamental frequency ν_0 at a frequency ν. The resulting radiation has a spectrum that consists of the original *centre-band*, and a pair of *side-bands* at $(\nu_0 + \nu)$ and $(\nu_0 - \nu)$.

Since linewidths may be as little as 1 Hz, frequency stabilisation to ± 0.2 Hz is necessary. This is achieved by locking the frequency to the natural vibration of a piezoelectric quartz crystal, a technique also used in microwave spectroscopy as discussed in Section 2.B. It is not necessary to stabilise flux density and frequency independently, as long as the ratio is kept constant, since they are linked by Eqn. 13.1. This can be achieved by locking the frequency to the resonance frequency of a standard substance. Tetramethylsilane, $Si(CH_3)_4$, usually referred to as TMS, is used for this purpose. The ^1H nuclei in TMS are equivalent, i.e. all in the same environment, and give a single sharp resonance peak. Since TMS is a volatile inert liquid, miscible with organic solvents, it can be added to the contents of the sample cell and distilled off afterwards if the sample is to be preserved.

13.B.4 Signal averaging and Fourier transform techniques

The signal-to-noise ratio of an n.m.r. spectrum is improved if the spectrum is scanned repetitively and the data are accumulated in a *computer of average transients*, or *CAT*. The frequency scan is divided into a large number of narrow ranges, and the signal for each one of the ranges is added to that for the same range observed during the preceding scan. The signal accumulates as the scans are repeated, but the noise, being random, tends to cancel itself out. The improvement in signal-to-noise ratio is proportional to the square-root of the number of scans. The main drawback to this procedure is the time required. Typically, a scan requires 5 minutes, so 500 minutes are required to produce a ten-fold improvement in the signal-to-noise ratio. An n.m.r. spectrometer is an extremely expensive piece of equipment—by far the most expensive of the spectrometers described in this book—and it would not be economical to tie up the instrument for so long for a single spectrum. The same improvement in signal-to-noise ratio could be achieved, incidentally, by scanning so slowly that 500 minutes were required for a single scan, but this requires a high degree of long-term stability which is difficult to achieve.

An n.m.r. spectrum can be obtained more rapidly if the resonances at all

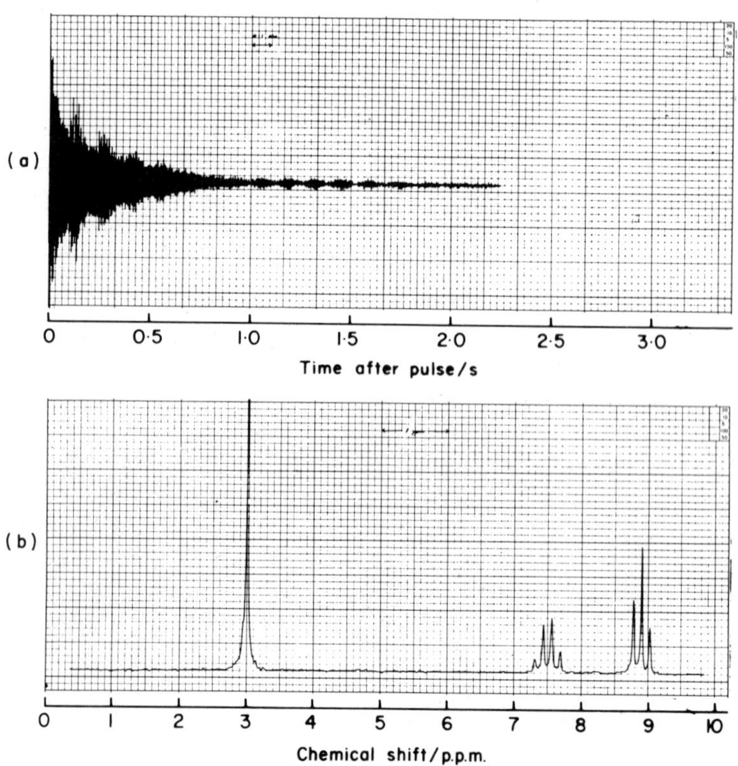

FIG. 13.2 Fourier transform nuclear magnetic resonance spectra. (a) Radiant intensity from a 2.5 % v/v sample of ethylbenzene after a single pulse. (b) Absorption spectrum obtained by Fourier transformation of the signal displayed in (a). (Reproduced by permission of the Perkin-Elmer Corporation.)

frequencies in the range of interest are observed simultaneously. The problem is exactly that encountered in far infrared spectroscopy, and the solution is similar to that described in Section 5.B. The sample is irradiated with a short pulse of high-intensity radiation, of duration of the order of 10 μs, containing frequencies covering the whole range of interest. The radiation emitted as the system relaxes to equilibrium is monitored for the next second or so. The total radiant intensity decays exponentially with a time constant of T_2. The radiation may be thought of as the sum of a set of damped simple harmonic oscillations, each at some frequency in the range of interest, and of amplitude proportional to the energy absorbed at that frequency from the initial pulse. As discussed in Section 5.B, the resulting waveform is complex, but can be decoded by a mathematical technique known as Fourier transformation, to obtain the original spectrum. A small computer is permanently linked to the n.m.r. spectrometer to perform this calculation. Figure 13.2 illustrates the application of this technique. Figure 13.2(a) shows the intensity of the radiation emitted by the sample as a function of time after irradiation by the high-intensity pulse. Figure 13.2(b) shows the spectrum obtained by Fourier transformation of the signal shown in Fig. 13.2(a), which is a conventional display of absorption as a function of frequency.

The spread in frequency of the excitation radiation is automatically provided by the briefness of the pulse. Radiation which has, say, exactly 90 000 000 cycles per second if observed for a duration of many seconds, only has 900 cycles if observed for 10 μs. Because one cannot count a fraction of a cycle, there is an uncertainty of ± 1 cycle in this figure of 900 cycles, so the frequencies in the nominally 90 MHz radiation are spread by ± 100 kHz when the radiation is in a 10 μs pulse.

13.C The Interpretation of n.m.r. Spectra

13.C.1 Chemical shifts

The extent to which the electrons around a nucleus shield it from the effect of a magnetic field, thereby giving rise to the phenomenon of chemical shift introduced in Section 13.A.1, is measured by the *shielding constant*, σ, where

$$B = B_{obs}(1-\sigma) \tag{13.4}$$

B_{obs} being the measured flux density, and B that experienced by the nucleus. Nuclei in different environments have different values of σ. Since σ for ^1H nuclei is of the order of 10^{-5}, it is more convenient to express the chemical shift in p.p.m. From the point of view of the theorist it would be most satisfactory to take σ as zero for a sample consisting entirely of bare protons, for which B_{obs} would equal B. However, the obvious practical difficulties result in the choice of another reference. Tetramethylsilane, TMS, is chosen for this purpose. The ^{1}H nuclei in TMS resonate at a given frequency at a higher applied flux density than those in nearly all other compounds; i.e. nearly all other ^{1}H nuclei are less well-shielded than those in TMS. Chemical shifts may be expressed in the δ scale, defined by:

$$\delta = [(B_{obs} - B_{obs,TMS})/B_{obs,TMS}] \times 10^6 \text{ p.p.m.} \qquad (13.5)$$

where $B_{obs,TMS}$ is the observed flux density for ^{1}H resonance in TMS. Because n.m.r. spectra are now often observed by scanning the frequency, it is now customary to express chemical shifts in terms of the fractional difference in frequency for resonance in a constant applied field. The modern δ scale is defined by

$$\delta = \frac{\nu - \nu_{TMS}}{\nu_{TMS}} \times 10^6 \text{ p.p.m.} \qquad (13.5a)$$

Values of δ defined by Eqn. 13.5a are equal to, but opposite in sign to, those defined by Eqn. 13.5. Chemical shifts for nuclei other than ^{1}H are now expressed in terms of δ values defined by Eqn. 13.5a. For ^{1}H chemical shifts the τ scale is commonly used. The τ value is defined by

$$\tau = 10.000 - \delta \qquad (13.6)$$

where δ is as defined by Eqn. 13.5a. Some typical τ values for ^{1}H nuclei in various environments are listed in Table 13.2. As for infrared characteristic group frequencies, extensive compilations are available for use in elucidation of the structure of organic molecules. Chemical shifts may also be expressed in terms of shift in frequency or applied flux density, but this is less satisfactory for tabulation purposes, as the numerical values depend on the operating frequency and flux density, by virtue of Eqn. 13.4. The greater the shielding of the nucleus, the greater the τ value. Because early n.m.r. spectrometers scanned the spectrum by varying the magnetic field rather than the frequency, resonances at high or low values of τ are often referred to respectively as high or low field signals.

The origin of chemical shift is the local magnetic field induced by circular motion of the electrons around the nucleus, this motion being itself induced by the applied magnetic field. The directions of the induced field and of the electron motion are described by Flemings right- and left-hand rules, and are illustrated in Fig. 13.3(a). The induced magnetic field opposes the applied magnetic field. The more electrons that are available to move around the nucleus, the greater the shielding effect and the larger the τ value. Electron-withdrawing groups adjacent to the nucleus thus cause the nucleus to have a

TABLE 13.2

Typical chemical shifts for ¹H

Molecule	τ
C^1H_3R	9.10
$RR'C^1HR$	8.50
$^1HC{\equiv}CR$	8.3
C^1H_3COR	7.90
C^1H_3I	7.84
C^1H_3Br	7.35
C^1H_3Cl	6.98
C^1H_3F	5.74
$^1H_2C{=}CR_2$	5.0
$C_6H_5NH_2$ (*para*-¹H)	3.28
C_6H_5	2.73
$C_6H_5NO_2$ (*para*-¹H)	2.18

small τ value, as can be seen in Table 13.2. N.m.r. chemical shifts differ from the analogous isomer shifts observed in Mössbauer spectra in that the former represent the total effect of all the electrons around the nucleus, whereas Mössbauer isomer shifts are due solely to s electrons.

A rather different phenomenon modifies the chemical shift for nuclei attached to atoms involved in unsaturated bonds. The applied magnetic field causes the π electrons to circulate around the internuclear axis, inducing a magnetic field which augments or diminishes the applied field at the nucleus under observation, depending on its position. As can be seen in Fig. 13.3(b), ^1H in $^1HC{\equiv}CR$ is shielded, but ^1H in $^1H_2C{=}CR_2$ is deshielded. Similarly the delocalised π electrons in an aromatic ring are caused to circulate around the ring, inducing a magnetic field which augments the applied field experienced by hydrogen atoms attached to the ring. These currents are termed *ring currents*. The effects are illustrated in Fig. 13.3(b) and (c).

A fairly simple proton magnetic resonance spectrum, that of ethanol, is shown in Fig. 13.4. It can be seen that the peaks for the ^1H nuclei in the CH_2 and CH_3 groups are multiplets, and have fine-structure. This is due to *spin–spin coupling* which is discussed in Section 13.C.2. The area under each multiplet is proportional to the total number of nuclei in that environment. As a further guide to the elucidation of molecular structure, most n.m.r. spectrometers have an integrating facility, and display the integrated signal as a cumulative total from the start of a scan. The spectrum in Fig. 13.4 shows that the numbers of ^1H nuclei in the CH_3, CH_2, and OH environments are, as expected, in the ratio 3:2:1. This proportionality between number of nuclei and peak area is occasionally used for quantitative analysis but standardisation is rather difficult.

Chemical shifts are solvent-dependent. The ring currents in the molecules of aromatic solvents, such as benzene, set up local magnetic fields which affect nuclei in adjacent molecules. Highly polar solvents also affect ^{1}H spectra, since the electron distribution around the ^{1}H nucleus is modified; this effect is especially significant with hydrogen-bonding solvents such as water or

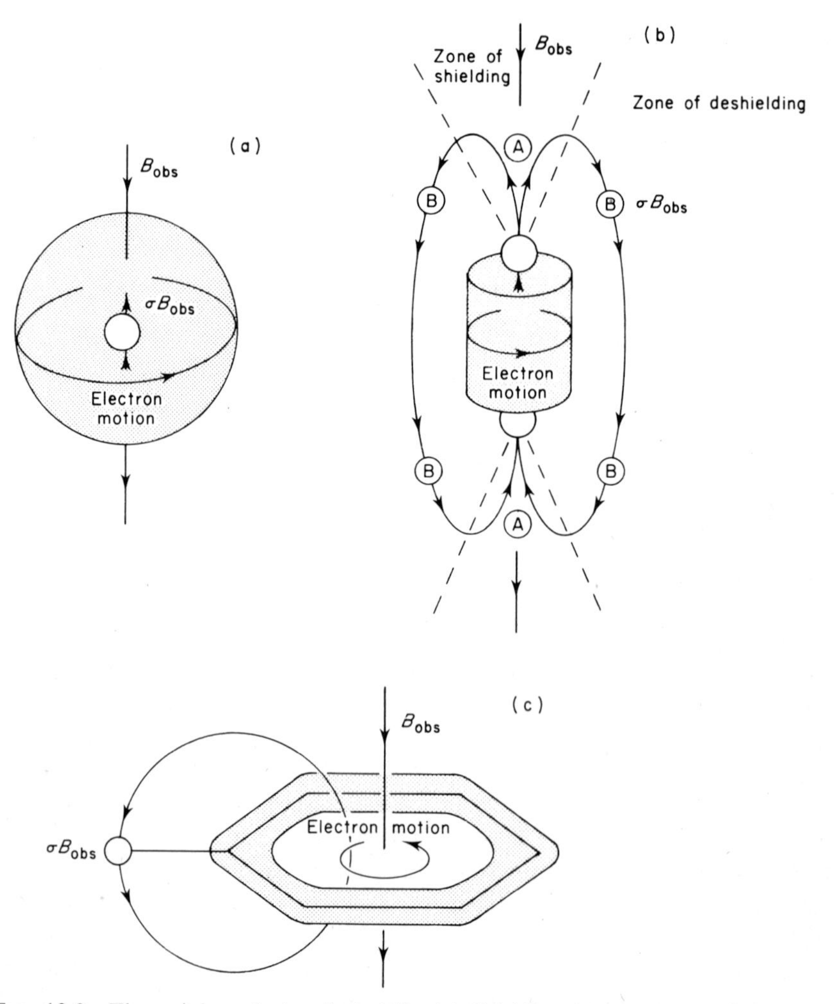

FIG. 13.3 The origins of chemical shift. (a) Shielding by electrons around the nucleus. (b) Shielding and deshielding zones around a multiple carbon–carbon bond; A indicates ^{1}H attached to C≡C, and B indicates ^{1}H attached to C=C. (c) Deshielding of ^{1}H attached to an aromatic ring.

methanol. In general, it is advisable to use carbon tetrachloride or cyclohexane as solvent so that the observed chemical shifts should correspond to the tabulated values. Solvent effects may be exploited to obtain a greater separation between the peaks for nuclei in different environments. Peaks that happen to coincide for samples in carbon tetrachloride solution may be well-separated if benzene is the solvent.

An important, recently developed, technique useful in the elucidation of complex spectra is the use of *lanthanide shift reagents*. Paramagnetic ions produce large changes in local magnetic fields, and hence large chemical shifts in the resonances for adjacent nuclei. Not all paramagnetic ions are suitable for this purpose. Paramagnetic species in general reduce the spin–spin lattice relaxation time, as discussed in Section 13.A.2, so, by virtue of Eqn. 13.3, resonances modified by these ions have unacceptably large line-widths. However, lanthanide ions have a very short electron spin–lattice relaxation time, of the order of 10^{-11} s, so coupling with the nuclear spins of adjacent nuclei is weak and does not facilitate spin–spin relaxation. The Eu^{3+} ion, which produces a downfield shift (to low τ) in adjacent ^{1}H nuclei, and the Pr^{3+} ion, which produces an upfield shift, are commonly employed.

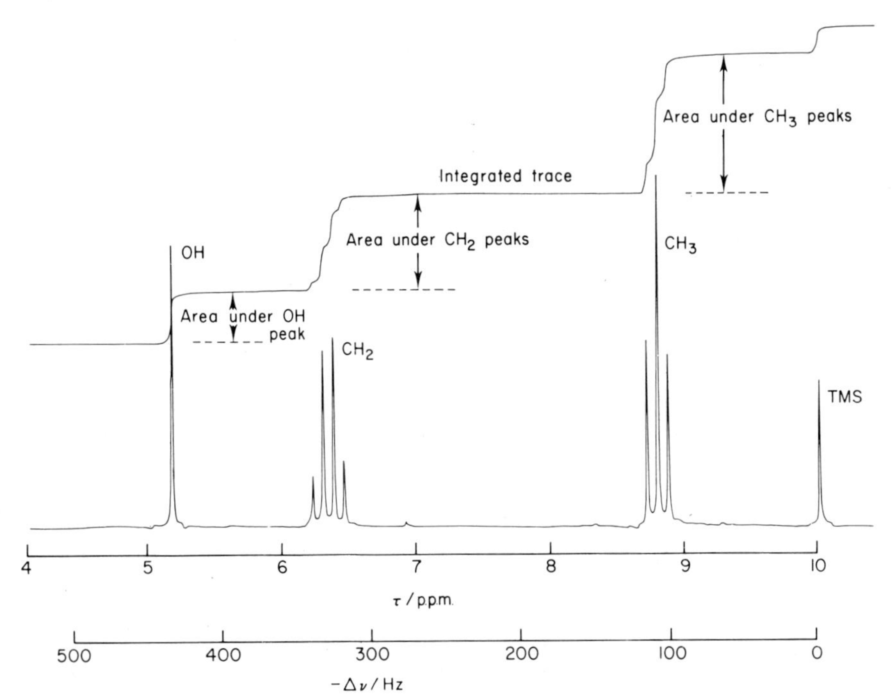

FIG. 13.4 The 90 MHz ^{1}H n.m.r. spectrum of C_2H_5OH (50% solution in C^2HCl_3 + trace of $^2HSO_4$–^2H_2O).

In order to get these ions into solution in organic solvents they are complexed with such ligands as dipivaloylmethane, usually known as DPM. Three DPM anions of the type:

$$\begin{array}{c} (CH_3)_3C-CH=C(CH_3)_3 \\ \quad\ \| \qquad\quad\ | \\ \quad\ O \qquad\quad O^- \end{array}$$

complex a Eu^{3+} ion to form the widely used shift reagent $Eu(DPM)_3$. This coordinates with groups containing lone pairs, e.g. OH, NH_2, C=O, in the sample molecule. The ^1H nuclei nearest the functional group are most influenced by the field of the Eu^{3+} ion, and so show the greatest chemical shift. Shifts of up to 20 τ can be produced in this way. The magnitude of the shift varies approximately with the inverse cube of the distance of the ^1H nucleus from the Eu^{3+} ion. Thus, whereas in the absence of a lanthanide shift reagent the C^1H_2 groups in, say, a steroid molecule give a tightly packed cluster of peaks, the addition of a shift reagent causes these peaks to be spread out over a range of 10 or so τ units. It is thus possible to determine the number of CH_2 groups at various distances from a functional group, which is of great value in the elucidation of complex molecular structures.

Chemical shifts for other nuclei are greater than for ^1H. The larger the number of electrons around the nucleus, the greater the possible variation in shielding effects. The most studied nucleus other than ^1H is ^{13}C. Since ^{13}C is present in only 1.1% abundance in naturally occurring carbon compounds, the main isotopic nucleus ^{12}C being of zero magnetic moment, and since the relative sensitivity of ^{13}C is low, ^{13}C spectra are weak and their observation is difficult. It was not until the advent of Fourier transform n.m.r. spectrometers that ^{13}C n.m.r. spectroscopy became a routine tool of the organic

TABLE 13.3

Typical chemical shifts (δ values) for ^{13}C (relative to TMS)

Molecule	δ
$^{13}C-CR_3$	0–40
$^{13}C\equiv CR$	70–90
$^{13}C\equiv N$	100–125
$^{13}C=CH_2$	110–120
$^{13}C=CR_2$	120–140
$R^{13}CO_2R$	150–180
$R_2{}^{13}C=O$	200–220

chemist. One advantage over ^{1}H spectroscopy for the elucidation of molecular structures is that the ^{13}C nucleus forms part of the skeleton of the molecule. Complicating effects discussed in Section 13.C.2 do not appear in ^{13}C spectra. Some values for ^{13}C chemical shifts are given in Table 13.3. The naturally occurring ^{13}C nuclei in TMS enable this substance to be used as a reference, other ^{13}C nuclei having resonances downfield from TMS.

13.C.2 Spin–spin coupling

In the discussion so far it has been assumed that a nucleus is not affected by other nuclei in the same molecule, but interaction between the spins does occur. This phenomenon is termed spin–spin coupling. A covalent bond joining two nuclei consists of two electrons, which tend to keep away from each other owing to charge correlation. At any instant one may picture one electron as being near one nucleus, and the other electron near the other nucleus. The energy of the states in which a nucleus and its adjacent electron are of opposing spin is lower than that for which the spins are parallel, and the two electrons are necessarily of opposite spins for the formation of the covalent bond. The net effect of this is that the state in which the two nuclei are of opposite spin is of lower energy than the state in which the spins are parallel. The energy gap between these two states is called the *coupling constant*, J, and is usually expressed in frequency units, Hz. In this instance, J is said to be negative. If the state in which the two nuclei are of opposite spin is of higher energy, J is positive. It is often, but not invariably, found that, as expected from the simple picture described above, negative coupling constants are observed for two ^{1}H nuclei joined to the same carbon atom and positive coupling constants are observed for ^{1}H nuclei attached to a pair of adjacent carbon atoms.

The effect of spin–spin coupling on the observed n.m.r. transitions is illustrated in Fig. 13.5. Two ^{1}H nuclei in different chemical environments in the same molecule are denoted by A and X. Each nucleus has a spin whose quantum number may be labelled α or β. There are thus four possible energy levels, corresponding to $\alpha\alpha$, $\alpha\beta$, $\beta\alpha$, and $\beta\beta$ (the quantum label for A being given first), and four possible transitions allowed by the selection rules, namely $\beta\beta \rightarrow \beta\alpha$ (denoted by X_1), $\beta\beta \rightarrow \alpha\beta$ (denoted by A_1), $\alpha\beta \rightarrow \alpha\alpha$ (denoted by X_2), and $\beta\alpha \rightarrow \alpha\alpha$ (denoted by A_2). If A and X are sufficiently far apart in the molecule for coupling to be negligible, the energy levels are as depicted on the left-hand side of Fig. 13.5. The magnitude of A_1 equals that of A_2, and the magnitude of X_1 equals that of X_2. The difference between A_1 and X_1 is the chemical shift, δ_{AX}, expressed in Hz. The student should note that Fig. 13.5 is not to scale. It would be difficult to draw it to scale, since at an applied magnetic field of flux density 1.4 T, as commonly employed, the spacing between $\beta\beta$ and $\alpha\beta$, and between $\alpha\beta$ and $\alpha\alpha$, is 60 MHz, whereas the spacing between $\beta\alpha$ and $\alpha\beta$ is at most of the order of a few hundred Hz, depending on A and X. The observed spectrum consists of two lines separated by the chemical shift δ_{AX}. If the nuclei are close enough for spin–spin coupling to occur, the energies of the states $\alpha\alpha$ and $\beta\beta$ are raised by an amount $J/4$ and the energies of the states $\alpha\beta$ and $\beta\alpha$ are lowered by the same amount. This situation is depicted in the central portion of Fig. 13.5. Transitions A_1

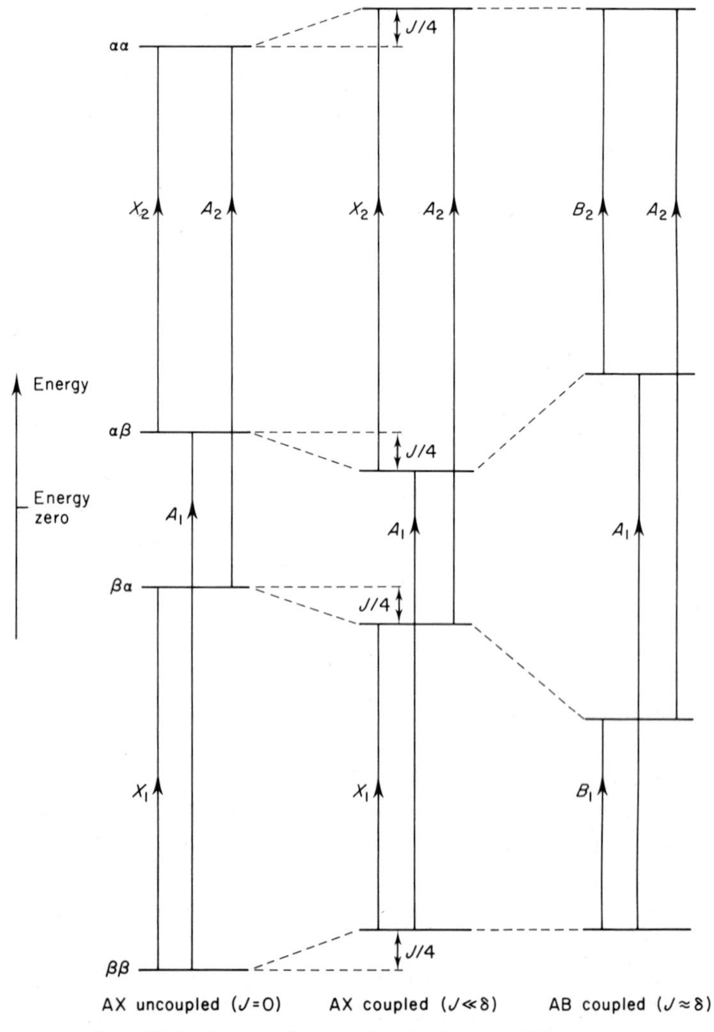

FIG. 13.5 Spin–spin coupling in the two-¹H system.

and A_2 then differ in magnitude by an amount J, as do transitions X_1 and X_2. The observed spectrum then consists of two pairs of lines; A_1,A_2 and X_1,X_2. The splitting within each pair, or doublet, is J, and the centres of the doublets are separated by δ_{AX}. Some values of J for ^1H nuclei in various positions are given in Table 13.4. The sign of J cannot be determined from the magnitude of the splitting, but can be found from observation of the effect of a strong electric field on the spectrum.

TABLE 13.4

Proton–proton spin coupling constants

Type	Typical J/Hz
H—C—H	−12 to −15
H—C—C—H (*trans*)	7
Benzene: *ortho*-H	8
meta-H	3
para-H	0
R′HC═CHR (*trans*)	18–12*
R′HC═CHR (*cis*)	14–4*
R′RC═CH_2	4–3*

* The greater the electronegativity of the substituent groups R,R′ the lower the value of J.

For some pairs of nuclei, denoted by AB, the magnitude of the chemical shift, δ_{AB}, is not much greater than J, and the $\alpha\beta$ and $\beta\alpha$ states tend to have similar energies; the two nuclei are said to be *strongly coupled.* Since these states are of the same symmetry, an interaction, explainable only in terms of quantum mechanics, takes place which causes a mutual repulsion of the energy levels of the $\alpha\beta$ and $\beta\alpha$ states. The effect of this interaction is shown on the right-hand side of Fig. 13.5. The difference between A_1 and A_2, and between B_1 and B_2 (replacing X_1 and X_2) is still J, but the difference between A_1 and B_2 is greater than it would be if this interaction did not occur. Spectra exhibiting this effect are termed *second order*, spectra of AX type being termed *first order.*

Energy levels and transition energies for the AX and AB systems are given in Table 13.5, and the theoretical spectra deduced from insertion of numerical values into the formulae are shown in Fig. 13.6. Whereas for AX spectra the line heights are all equal for all four transitions, this is not so for AB spectra. The ratio of the heights of the A_2 and A_1 lines (which corresponds to the areas under the peaks in the observed spectrum) is given by

$$(A_2 \text{ height})/(A_1 \text{ height}) = (1-a/b)^2/(1+a/b)^2 \tag{13.7}$$

where

$$a/b = [\delta_{AB} \pm (J^2+\delta_{AB}{}^2)^{\frac{1}{2}}]/J$$

An alternative formula is

$$(A_2 \text{ height})/(A_1 \text{ height}) = (A_1-B_2)/(A_2-B_1) \tag{13.8}$$

Since values of δ_{AB} are linearly proportional to the flux density of the applied field, whereas values of J are independent of the field, increasing the applied magnetic field has the effect of separating the A transitions from the B transitions, and making the spectrum more first-order in character. This is one reason for employing as powerful a magnet as possible. Even in the spectrum in Fig. 13.6, for which δ_{AB}/J is as great as 12.5, the lines in the doublets can be seen to be of different heights, a sure sign of second-order interaction. The value of J is readily found from the spectra, since it is always given by the spacing between A_1 and A_2, or B_1 and B_2, however great the second-order character of the spectrum. Values of δ_{AB} cannot be determined

TABLE 13.5

Spin-coupling energies

Level	AX uncoupled	AX coupled	AB coupled
	Spin-Coupled State Energies (referred to $B = 0$ as zero energy)		
$\alpha\alpha$	$\mu B(2-\sigma_A-\sigma_X)$	$J/4+\mu B(2-\sigma_A-\sigma_X)$	$J/4+\mu B(2-\sigma_A-\sigma_B)$
$\alpha\beta$	$\mu B(\sigma_X-\sigma_A)$	$-J/4+\mu B(\sigma_X-\sigma_A)$	$-J/4+\frac{1}{2}(J^2+\delta^2_{AB})^{\frac{1}{2}}$
$\beta\alpha$	$-\mu B(\sigma_X-\sigma_A)$	$-J/4-\mu B(\sigma_X-\sigma_A)$	$-J/4-\frac{1}{2}(J^2+\delta^2_{AB})^{\frac{1}{2}}$
$\beta\beta$	$-\mu B(2-\sigma_A-\sigma_X)$	$J/4-\mu B(2-\sigma_A-\sigma_X)$	$J/4-\mu B(2-\sigma_A-\sigma_B)$
	Energies of Allowed Transitions		
$(\alpha\beta \rightarrow \alpha\alpha)$ i.e. X_2	$2\mu B(1-\sigma_X)$	$J/2+2\mu B(1-\sigma_X)$	$J/2+\mu B(2-\sigma_A-\sigma_B)$ $-\frac{1}{2}(J^2+\delta^2_{AB})^{\frac{1}{2}}$
$(\beta\alpha \rightarrow \alpha\alpha)$ i.e. A_2	$2\mu B(1-\sigma_A)$	$J/2+2\mu B(1-\sigma_A)$	$J/2+\mu B(2-\sigma_A-\sigma_B)$ $+\frac{1}{2}(J^2+\delta^2_{AB})^{\frac{1}{2}}$
$(\beta\beta \rightarrow \alpha\beta)$ i.e. A_1	$2\mu B(1-\sigma_A)$	$-J/2+2\mu B(1-\sigma_A)$	$-J/2+\mu B(2-\sigma_A-\sigma_B)$ $+\frac{1}{2}(J^2+\delta^2_{AB})^{\frac{1}{2}}$
$(\beta\beta \rightarrow \beta\alpha)$ i.e. X_1	$2\mu B(1-\sigma_X)$	$-J/2+2\mu B(1-\sigma_X)$	$-J/2+\mu B(2-\sigma_A-\sigma_B)$ $-\frac{1}{2}(J^2+\delta^2_{AB})^{\frac{1}{2}}$

(1) $\delta_{AB} = 2\mu B(\sigma_B-\sigma_A)$.
(2) B in this table means B_{obs}, the applied flux density.

by inspection, but can be calculated from an equation derived from formulae in Table 13.5.

$$\delta_{AB} = [(A_2 - B_1)(A_1 - B_2)]^{\frac{1}{2}} \tag{13.9}$$

If second-order effects did not exist, A_1 and B_2 would coincide at $J = \delta_{AB}$, but, as can be seen in Fig. 13.6, second-order interaction causes a splitting still. Only if δ_{AB} is zero, i.e. if B is in the identical environment to A, do A_1 and B_2 coincide, and A_2 and B_1 have zero intensity. The spectrum of A_2 is thus a single line, of double the height for a single nucleus A. Nuclei for which the relative chemical shift is zero, and which couple to the same extent with all other nuclei, are said to be *equivalent*, and give singlet lines. For

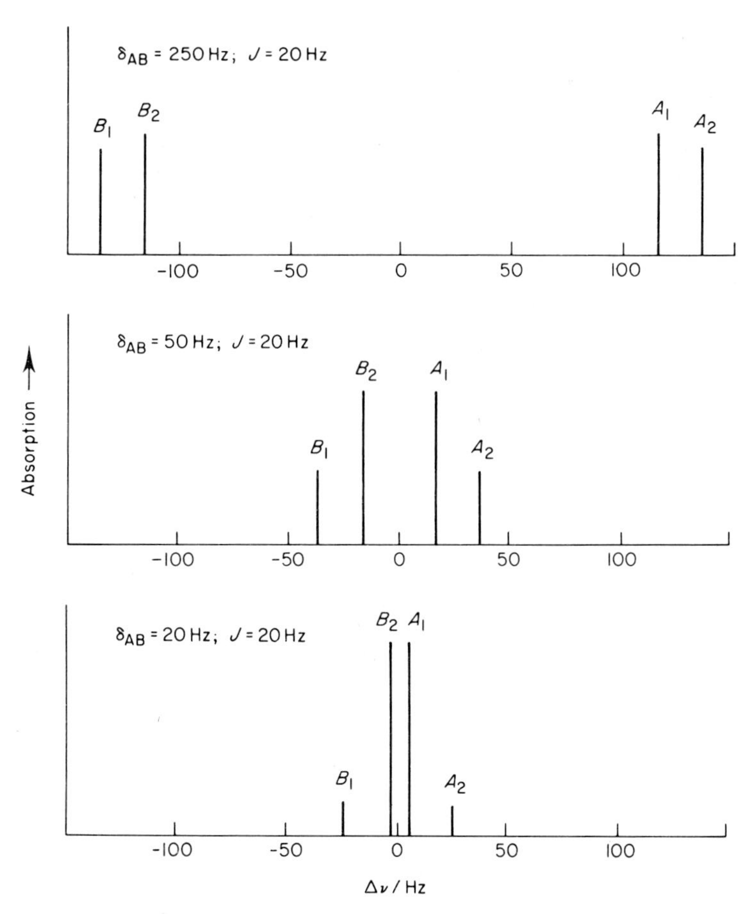

FIG. 13.6 Theoretical n.m.r. spectra for the AB system; the frequency scale zero is set at the centre, and the peak heights are normalised to constant total.

example, the twelve ^1H nuclei in TMS are equivalent, so the proton magnetic resonance spectrum of TMS consists of a single line only.

A system consisting of three nuclei, two of which are equivalent and have a large chemical shift relative to the third, is denoted by AX_2. The nucleus A is influenced by the combined effect of the two X nuclei. The spins of these two, which may be distinguished as X and X′, may be combined in four different ways as below:

Spin of X	Spin of X′	Combined spin
$+\frac{1}{2}$	$+\frac{1}{2}$	1
$+\frac{1}{2}$	$-\frac{1}{2}$	0
$-\frac{1}{2}$	$+\frac{1}{2}$	0
$-\frac{1}{2}$	$-\frac{1}{2}$	-1

There are thus three possible total spins for X_2 of 1, 0, and -1 respectively. The total spin of the X_2 pair combines with the spin of A, which may be either in the α or β state, to give six energy levels which may be represented as $\alpha(1)$, $\alpha(0)$, $\alpha(-1)$, $\beta(1)$, $\beta(0)$, $\beta(-1)$. The $\alpha(0)$ and $\beta(0)$ levels are doubly degenerate because the total spin state of 0 for the X_2 pair is doubly degenerate. These levels are shown in Fig. 13.7. Spin–spin coupling causes the energy of the $\alpha(1)$ and $\beta(1)$ states to be $J/2$ above that for the uncoupled spins, and similarly causes the energy of the $\alpha(-1)$ and $\beta(-1)$ states to be $J/2$ below that for the uncoupled spins. The permitted transitions for magnetic resonance of nucleus A are $\beta(1) \rightarrow \alpha(1)$, $\beta(0) \rightarrow \alpha(0)$, and $\beta(-1) \rightarrow \alpha(-1)$, so a triplet of lines is observed, spaced at intervals of J Hz. The central line is twice the height of the two lines flanking it, because of the double degeneracy of the $\alpha(0)$ and $\beta(0)$ levels, and is at the frequency corresponding to uncoupled spins. This argument can be extended for any number of nuclei coupled in first-order fashion. If a nucleus is coupled to n equivalent nuclei, all of spin quantum number $\frac{1}{2}$, to give an AX_n system, the spectrum due to resonance of nucleus A is a $(n+1)$ multiplet. The relative line heights (or peak areas) are given by *Pascal's triangle*, in which each new row of numbers is formed by adding adjacent numbers in the preceding row:

n	Relative heights								
0					1				
1				1		1			
2			1		2		1		
3		1		3		3		1	
4	1		4		6		4		1
etc.					etc.				

The permitted transitions for resonance of nucleus A for n up to 4, and the corresponding spectra, are shown in Fig. 13.7. The spacing between adjacent lines in each multiplet is J Hz. The student should note that it is not just the splitting of the energy levels that causes the multiplicity of the spectra, but rather the difference in the magnitude of splitting for the α and β spin states of nucleus A, due to spin–spin coupling. If spin–spin coupling were not operative, i.e. if J were zero, the energy levels would be those shown as broken lines in Fig. 13.7. The energy levels would still be $(n+1)$ multiplets,

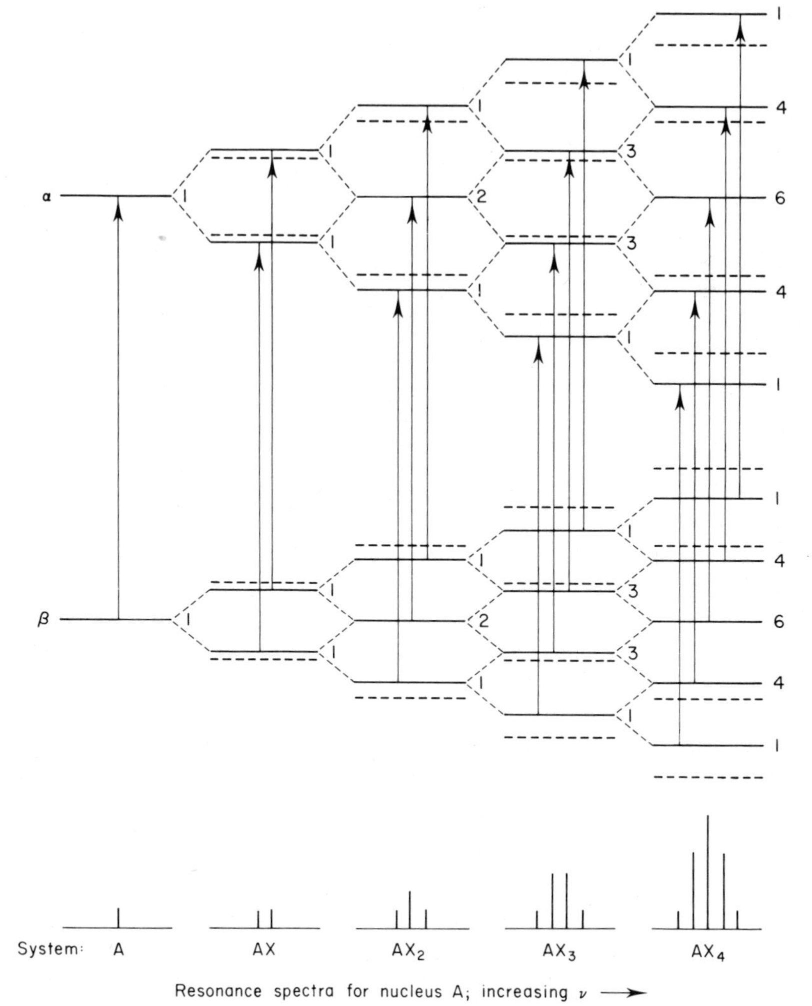

FIG. 13.7 Spin–spin coupling for AX_n systems; levels denoted by broken lines refer to zero spin–spin coupling, and numbers beside each level indicate the degeneracy.

but all the transitions would coincide and the observed spectrum would consist of a single line.

If nucleus A is coupled to two sets of nuclei represented by M and X respectively, the coupling to set M may be much greater than to set X. If nucleus A is coupled to i equivalent nuclei M and j equivalent nuclei X, the spectrum consists of an $(i+1)$ multiplet, the components of which are not single lines but are themselves $(j+1)$ multiplets. The spacing between the adjacent lines in each multiplet cluster is J for A–X coupling, and the spacing between the centres of adjacent multiplet clusters is J for A–M coupling.

The spectrum in Fig. 13.4 can now be interpreted further. The three equivalent ^1H nuclei in C^1H_3 are affected by the two equivalent ^1H nuclei in C^1H_2, and so give a triplet absorption, whereas the two equivalent ^1H nuclei in C^1H_2 are affected by the three equivalent ^1H nuclei in C^1H_3, and so give a quartet. The difference in intensity between the two flanking peaks in the C^1H_3 triplet is due to second-order interaction. The ^1H in O^1H gives a singlet in ethanol containing a trace of acid impurity because of chemical exchange. This nucleus is an acidic proton, and has a short lifetime on one particular molecule owing to the rapid, acid-catalysed, reaction:

$$C_2H_5O^1H + C_2H_5\overset{+}{O}H_2 \rightleftharpoons C_2H_5\overset{+}{O}{}^1HH + C_2H_5OH$$
$$\rightleftharpoons C_3H_5OH + C_2H_5\overset{+}{O}{}^1HH$$

The rate constant for this reaction is much greater than the spin–spin coupling constant, so the orientation of the nuclear spin of the ^1H is indeterminate over the time span required for spin–spin coupling to be observed.

The mathematical analysis of systems containing more than two nuclei, some or all of which are strongly coupled so that second-order spectra are produced, is extremely complex. The use of computers has enabled theoretical spectra to be constructed for all the spin systems generally encountered.

Spin–spin coupling is of course observed for nuclei other than ^1H. On the average, an organic molecule has one, but only one, ^{13}C nucleus in it, and it rarely happens that two ^{13}C nuclei are to be found close together. Thus ^{13}C–^{13}C coupling is not observed, but ^{13}C–^1H coupling is, giving rise to *satellite lines* in proton magnetic resonance spectra. Because of the low abundance of ^{13}C, satellite lines are of low intensity, but it is sometimes worthwhile taking pains to observe them as a guide to molecular structure. ^{13}C–^1H coupling constants are an order of magnitude greater than ^1H–^1H coupling constants, but the large difference in resonance frequency between ^{13}C and ^1H (44.9 MHz at 1.4 T) means that second-order effects are insignificant. Spin–spin coupling constants for other pairs of nuclei are greater than for ^1H–^1H coupling, but the proportional increase is not as great as that in the chemical shifts. N.m.r. spectra for other nuclei are thus simpler than for ^{1}H since they show only first-order, not second-order, coupling.

Nuclei with a spin quantum number greater than 1/2 possess an electric quadrupole moment, as discussed in Section 2.E.2. If the electric field due to the electrons around the nucleus is not symmetrical the quadrupole tends to orient itself in the electric field, so the orientation of the nucleus depends on the orientation of the molecule as a whole. Thermal agitation causes the orientation of molecules in liquids to be random, so quadrupolar nuclei will not orient themselves in an applied magnetic field. The coupling between ^{1}H and ^{35}Cl, for example, does not lead to a splitting of the ^{1}H lines in an alkyl chloride in solution because the ^{35}Cl nucleus has an indeterminate orientation. For the ^{14}N quadrupolar nucleus in certain environments, the quadrupole interaction is comparatively weak, and the orientation of the nucleus is not completely random in an applied magnetic field. The ^{14}N nucleus has a spin quantum number of 1, so the ^{1}H resonance due to $^{14}N^{1}H$ would be a triplet if quadrupole effects were absent, the ^{14}N having three possible orientations $(+1, 0, -1)$ with respect to the ^{1}H. If, on the other hand, quadrupole interaction rendered the ^{14}N orientation completely random, the ^{1}H resonance would be a single line. In practice a broad peak is usually observed, showing incipient splitting of the singlet into a triplet. The loss of spin energy as a consequence of the interaction of the nuclear quadrupole with the rest of the molecule is known as *quadrupole relaxation.*

13.D Double Resonance Techniques

13.D.1 Spin-decoupling

Although nuclear magnetic energy levels are so closely spaced that thermal energies at room temperature are amply sufficient to excite nuclei into upper levels, spin flipping is a comparatively slow process, as discussed in Section 13.A.2. Irradiation at the resonance frequency greatly facilitates spin-flipping by the mechanism described in Section 13.A.1; one may think of the radiation as acting as a catalyst for the process. Thus, if a sample is irradiated at a resonance frequency for one particular nucleus, or group of nuclei, in a molecule, the orientation of this nucleus, or group of nuclei, becomes indeterminate, and the resonance of other adjacent nuclei will not show splitting due to spin–spin coupling. This process is known as spin-decoupling. A double-coil spectrometer, as described in Section 13.B.3, is used to observe spin-decoupling. One coil supplies the radiation to scramble the orientation of one set of nuclei, and the other coil monitors the absorption due to other nuclei. Spin-decoupling is a useful technique for the simplification of spectra that are rendered complex by a large number of spin–spin couplings. The two nuclei may be of the same type, differing only in their chemical shift, the technique being called *homonuclear spin-decoupling*, or of different type, the technique then being called *heteronuclear spin-decoupling*.

Heteronuclear spin-decoupling finds an important application in ^{13}C n.m.r. spectroscopy. The large number of ^{13}C–^{1}H couplings in the average organic molecule would make the ^{13}C spectrum extremely complex if the spins were not decoupled. To do this, the sample is irradiated at a range of frequencies spanning the whole ^{1}H spectrum, a technique known as *broad-band decoupling*. A further advantage of spin-decoupling is that the intensity of the single line thus produced is the sum of the intensities of all the lines in the original multiplet, so the observed line height is much greater. This is especially beneficial for ^{13}C spectra, which tend to be extremely weak. Heteronuclear spin-decoupling is also used to reduce line-broadening due to ^{14}N quadrupole relaxation. ^{14}N–^{1}H coupling is ineffective, so the ^{1}H gives broad, low, peaks which may merge into background noise. Irradiation of ^{14}N to give complete decoupling causes the ^{1}H to give narrow, intense, lines which are easily observed.

It has so far been assumed that the irradiation has been sufficiently intense to produce complete decoupling, i.e. that the nuclei have been induced to flip so rapidly than their mean lifetime in a given orientation is much less than J^{-1} s, the time required for observation of spin–spin coupling. Useful information is also obtained if the applied radiation is weak. This causes further splitting in the observed spectrum of one nucleus which is only symmetrical if the irradiation frequency is exactly equal to the resonance frequency of the other nucleus. This technique is known as *spin-tickling*.

13.D.2 INDOR

In the spin-decoupling technique the decoupling frequency is kept constant and the spectrum is scanned by varying the detection frequency. An alternative procedure is to monitor the signal at a fixed frequency while varying the decoupling frequency. If, for example, the signal from a sample of AX type is monitored at the frequency of one of the components of the A doublet, the signal shows a constant absorption as the decoupling frequency is varied until the decoupling frequency coincides with one or other of the X lines. The A doublet then collapses into a singlet, and the monitored absorption disappears. The observed spectrum is thus that of the X nuclei, but inverted, the peaks indicating decreased absorption. This technique, called INter-nuclear DOuble Resonance, or INDOR, gives the spectrum of all the nuclei coupled to one particular nucleus, which is simpler and easier to interpret than that of the molecule as a whole. INDOR may be performed using either homonuclear or heteronuclear coupling. Heteronuclear INDOR makes it possible to observe resonance frequencies and measure chemical shifts for other nuclei using a proton magnetic resonance spectrometer. This is a convenient way of studying nuclei other than ^{1}H because the detector frequency is still that for ^{1}H whatever the other nucleus, and also because the signal intensity from ^{1}H is higher than for any other nucleus.

13.D.3 The nuclear Overhauser effect

If one nucleus is irradiated at its resonance frequency so intensely that the spin populations become equal, i.e. the spins are saturated, the intensity of n.m.r. absorption by other nuclei in the system may be altered. This is known as the nuclear Overhauser effect, or NOE. The mechanism of the NOE may be understood by consideration of the simple AX system, the energy levels of which are shown in Fig. 13.5. The populations of these levels are given in Table 13.6. The $\alpha\alpha$ state, having the highest energy, has the lowest population, by virtue of Eqn. 13.2. The energy gap between the $\alpha\beta$ and $\beta\alpha$ states is of the order of 10^{-5} times the gap between the $\alpha\alpha$ and $\alpha\beta$ states, as explained in Section 13.C.2, and so the population of the $\alpha\beta$ and $\beta\alpha$ states may be taken to be equal. The intensity of absorption is proportional to the difference in populations between the upper and lower states, as explained in Section 13.A.2, and so is as listed in Table 13.4. This argument only applies to weakly coupled AX systems, the line intensities being affected by strong coupling. If the AX system is intensely irradiated at frequencies X_1 and X_2 the spins of X become saturated, so the population of the $\alpha\alpha$ level tends to equal that of the $\alpha\beta$ level, and the population of the $\beta\alpha$ level tends to equal that of the $\beta\beta$ level. However, two other transitions counteract this tendency, namely $\alpha\alpha \rightarrow \beta\beta$ and $\alpha\beta \rightarrow \beta\alpha$, which tend to bring the distribution back to that for no applied radiation. The corresponding line intensities are given in Table 13.4. As a consequence of spin-decoupling the nucleus A gives a singlet, and the observed line intensity is the sum of that for A_1 and for A_2. In practice the effect of the $\alpha\alpha \rightarrow \beta\beta$ transition outweighs the effect of the $\alpha\beta \rightarrow \beta\alpha$ transition, and the absorption of A is enhanced. Since the enhancement is due to spin interaction of dipole–dipole type, the closer the nuclei the greater the enhancement. Plots of fractional enhancement, which may be as great as 30% for ^{1}H–^{1}H interaction, against the inverse sixth power of the internuclear distance are linear (Bell and Saunders, 1970). The magnitude of the NOE thus gives information about molecular geometry. The relevant distance is that through space, not following the line of the bonds, so molecular conformations can

TABLE 13.6

The nuclear Overhauser effect for the AX system

Level	Population in absence of radiation	Population if intense radiation at X_1 and X_2 frequencies applied	Population if transition $\alpha\alpha \rightarrow \beta\beta$ is significant	Population if transition $\beta\alpha \rightarrow \alpha\beta$ is significant
$\alpha\alpha$	$P-\delta$	$P-\delta/2$	$P-\delta/2-d$	$P-\delta/2$
$\alpha\beta$	P	$P-\delta/2$	$P-\delta/2$	$P-\delta/2+d$
$\beta\alpha$	P	$P+\delta/2$	$P+\delta/2$	$P+\delta/2-d$
$\beta\beta$	$P+\delta$	$P+\delta/2$	$P-\delta/2+d$	$P+\delta/2$

Relative Intensities of Transitions

Transition	Relative intensities in absence of radiation	Relative intensities if intense radiation at X_1 and X_2 frequencies applied	Relative intensities if transition $\alpha\alpha \rightarrow \beta\beta$ is significant	Relative intensities if transition $\beta\alpha \rightarrow \alpha\beta$ is significant
$\alpha\beta \rightarrow \alpha\alpha$ (i.e. X_2)	δ	0	d	d
$\beta\alpha \rightarrow \alpha\alpha$ (i.e. A_2)	δ	δ	$\delta+d$	$\delta-d$
$\beta\beta \rightarrow \alpha\beta$ (i.e. A_1)	δ	δ	$\delta+d$	$\delta-d$
$\beta\beta \rightarrow \beta\alpha$ (i.e. X_1)	δ	0	d	d

be determined. The enhancement also depends on the magnetic moments of the interacting nuclei; the greater the ratio between the magnetic moments, the greater the enhancement.

13.E Kinetic Effects

The chemical shift of a particular nucleus depends on its environment. If the nucleus can be in either of two environments, changing from one to the other at a rate whose first-order rate constant is much less than the chemical shift in Hz, the n.m.r. spectrum shows two peaks, separated by the chemical shift. If the rate of exchange is much greater than the chemical shift, the effects of the two environments are averaged out, and the spectrum shows a single peak, somewhere between the two positions for the peaks in the absence of exchange. A simple example is provided by the conformation change of cyclohexane. Any one hydrogen atom can be in either an equatorial or an axial position, as shown in Fig. 13.8, and interconversion between these two occurs as a consequence of internal rotation as the ring twists. There is a small chemical shift for ^1H between the axial and equatorial environments,

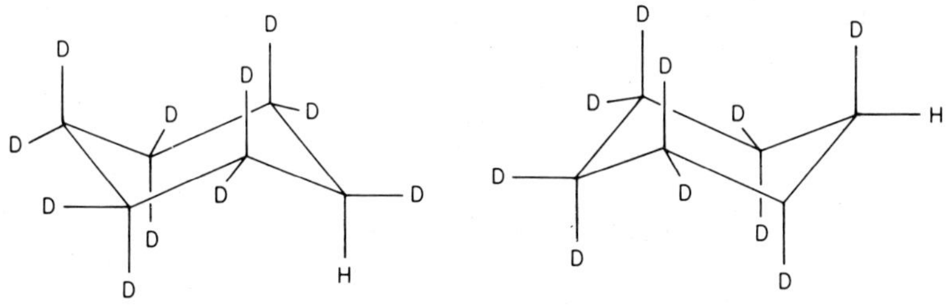

H axial

H equatorial

FIG. 13.8 Conformation changes of cyclohexane.

amounting to 29 Hz at 60 MHz, the equatorial protons having the lower τ value. An activation energy of 44 kJ mol^{-1} is required for the interconversion, so at sufficiently low temperatures thermal energy is insufficient for this process. The ^1H n.m.r. spectrum of $C_6{}^1H^2H_{11}$ (used instead of $C_6{}^1H_{12}$ to simplify the spectrum) at -89 °C thus shows two separate peaks for the proton in its two possible environments, as shown in Fig. 13.9. As the temperature is raised, the interconversion rate increases. At -68 °C the proton spends such a short time in each environment that each peak is broadened by virtue of the Heisenberg Uncertainty Principle. The interconversion is a first-order rate process of rate constant τ^{-1}. The value of τ* is related to the average lifetime of the proton in one particular environment. This quantity can be calculated from the magnitude of the line-broadening by

$$\tau^{-1} = 2\pi(\delta\nu - \delta\nu_0) \tag{13.10}$$

where $\delta\nu$ is the observed line half-width at half-height, and $\delta\nu_0$ is that observed in the absence of exchange. As can be seen from Eqn. 13.3, spin–spin relaxation is closely connected with the effect of dynamic processes on n.m.r. spectra. As the temperature is raised further, the value of τ decreases, and the two peaks move closer together. If the frequency interval between the peaks is $\Delta\nu$, compared with a value of $\Delta\nu_0$ in the absence of exchange, then an approximate value of τ^{-1} is given by

$$\tau^{-1} = 2^{\frac{1}{2}}\pi(\Delta\nu_0{}^2 - \Delta\nu^2)^{\frac{1}{2}} \tag{13.11}$$

* This τ must not be confused with the τ used for the scale of chemical shifts.

If interconversion is sufficiently rapid, the two peaks coalesce into one, as happens at $-60\ °C$ for the cyclohexane conformation change. At this temperature $\Delta\nu$ is zero, so

$$\tau^{-1} = 2^{\frac{1}{2}}\pi\Delta\nu_0 \tag{13.12}$$

At still higher temperatures the peak narrows as the averaging of the two environments becomes more effective. For spectra of this type, an approximate value of τ^{-1} is given by

$$\tau^{-1} = 4\pi\Delta\nu_0{}^2/(\delta\nu - \delta\nu_0) \tag{13.13}$$

$\delta\nu$ being the observed line half-width. More sophisticated calculations of τ^{-1} from the observed peak shapes may be achieved by fitting these to theoretical curves with the aid of a computer. The activation energy is obtained from the Arrhenius plot of log τ^{-1} against the reciprocal absolute temperature.

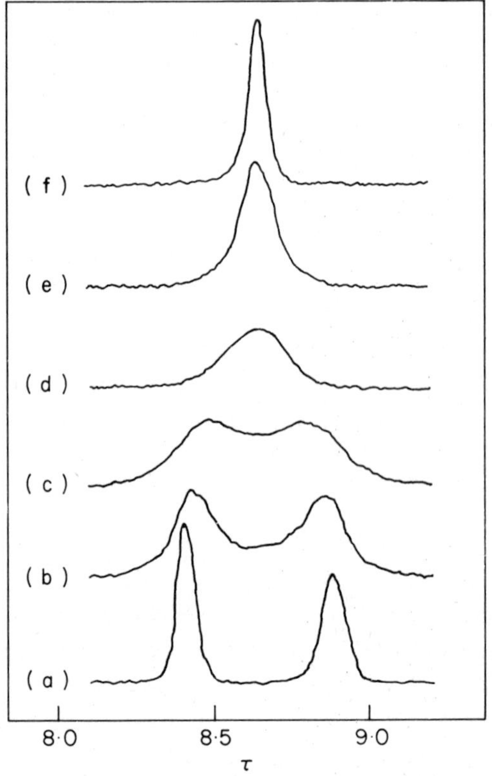

FIG. 13.9 N.m.r. spectra showing ring inversion in $C_6{}^1H^2H_{11}$ (Bovey *et al.*, 1964). Spectra run at (a) $-89\ °C$, (b) $-67.8\ °C$, (c) $-63.2\ °C$, (d) $-60.3\ °C$, (e) $-57.0\ °C$, (f) $-48.7\ °C$. All except (a) have quadrupole broadening effect of ^2H eliminated by spin decoupling.

Rates and energy barriers for internal rotation can be found in a similar fashion. The rotation of the methyl groups around the C–N bond in dimethylformamide is restricted because this bond has partial double-bond character due to partial polarisation

$$\begin{array}{ccc} \overset{\delta-}{O} & & CH_3 \\ & \diagdown \quad \overset{\delta+}{} \diagup & \\ & C\text{—}N & \\ & \diagup \quad\quad \diagdown & \\ H & & CH_3 \end{array}$$

The protons in one methyl group have a different chemical shift from those in the other methyl group because of the different distances from the oxygen atom. The n.m.r. spectrum at 35 °C shows two distinct peaks corresponding to the two sets of protons. As the temperature is raised the spectrum changes in the manner illustrated in Fig. 13.9. Coalescence occurs at 118 °C. The energy barrier to rotation is found to be 30 $kJ\ mol^{-1}$ (Gutowsky and Holm, 1956). Many other systems exhibiting internal rotation have been studied in this way.

A nucleus may have the choice of two different environments as the result of a dynamic chemical equilibrium. For example, in a mixture of acetic acid and acetylacetone any individual proton may be on either the acid or the diketone in its enolic tautomer:

$$CH_3CO_2{}^1H + CH_3COCH{=}C(OH)CH_3$$
$$\rightleftharpoons CH_3CO_2H + CH_3COCH{=}C(O^1H)CH_3$$

At −8 °C the ^1H n.m.r. spectrum of an equimolar mixture shows two peaks due to O^1H, separated by 153 Hz at 40 MHz. As the temperature is raised these two peaks broaden and approach each other, coalescing at 64.5 °C, in the same fashion as illustrated in Fig. 13.9 (Schneider and Reeves, 1958). Analysis of such spectra enables the rate and activation energy of the chemical reaction causing the exchange to be evaluated.

Rates of exchange reactions can also be measured from the line-broadening in a multiplet, which collapses into a singlet at high exchange rates. A classic example is the measurement of the rate of proton exchange, on protonated methylamine (Grunwald *et al.*, 1957). In strongly acid solution, the ^1H n.m.r. spectrum of $CH_3NH_3{}^+$ shows a quadruplet due to the C^1H_3 group split by spin–spin coupling with the ^1H nuclei in the $NH_3{}^+$ group, and a triplet due to the N^1H_3 group split by spin–spin coupling with the ^{14}N nucleus. The N^1H_3 peaks are broadened by ^{14}N quadrupole relaxation, and so do not show a multiplet fine structure due to coupling with the ^1H nuclei in the CH_3 group. Conversely, splitting of the C^1H_3 peaks by spin–spin coupling with the ^{14}N nucleus is not seen because of the ^{14}N quadrupole relaxation. As the pH is raised, the lines in the C^1H_3 quadruplet broaden, owing to exchange of the ^1H nuclei in the NH_3^+ group by the reactions

$$CH_3\overset{+}{N}{}^1HH_2 + CH_3NH_2 \overset{k_6}{\rightleftharpoons} CH_3NH_2 + CH_3\overset{+}{N}{}^1HH_2$$

$$CH_3\overset{+}{N}{}^1HH_2 + HOH + CH_3NH_2 \overset{k_7}{\rightleftharpoons} CH_3NH_2 + HOH + CH_3\overset{+}{N}{}^1HH_2$$

Coalescence occurs at pH 4.0, and the ^1H nuclei in the CH_3 group give a single narrow peak at pH 8.6. The mean lifetime τ of the protons on the NH_3^+ group at the pH at which coalescence occurs is given by

$$\tau^{-1} = 2\pi J \tag{13.14}$$

where J is the separation between the peaks in the quadruplet in the absence of exchange effects, which has the value 6.2 Hz. Values of τ^{-1} at other pH values were obtained by fitting theoretical curves to the observed peak shapes. The dependence of τ^{-1} on pH showed that $(k_6 + k_7)$ had the value 6.8×10^8 l mol^{-1} s^{-1} at 25 °C. The student should note that this technique differs from that based on chemical shift in that the splitting, and hence the condition for coalescence, is independent of the applied magnetic field. Another point of difference is that τ^{-1} is varied by variation of the concentration of one of the reacting species, here CH_3NH_2, rather than by variation of the temperature. For a first-order kinetic process, of course, such as a conformation change, the only way in which τ can be varied is by varying the temperature.

REFERENCES

Bell, R. A. and Saunders, J. K. (1970) *Canad. J. Chem.* **48**, 1114.

Bovey, F. A., Hood, F. P., III, Anderson, E. W. and Kornegay, R. L. (1964) *J. Chem. Phys.* **41**, 2041.

Grunwald, E., Loewenstein, A. and Meiboom, S. (1957) *J. Chem. Phys.* **27**, 630.

Gutowsky, H. S. and Holm, C. H. (1956) *J. Chem. Phys.* **25**, 1228.

Schneider, W. G. and Reeves, L. W. (1958) *Ann. New York Acad. Sci.* **70**, 858.

BIBLIOGRAPHY

Elvidge, J. A. (1970) in "An Introduction to Spectroscopic Methods for the Identification of Organic Compounds" (F. Scheinmann, ed.), Vol. I, pp. 41–58, Pergamon Press, Oxford.

Elvidge, J. A. (1974) in "An Introduction to Spectroscopic Methods for the Identification of Organic Compounds" (F. Scheinmann, ed.), Vol. II, pp. 211–311, Pergamon Press, Oxford.

Emsley, J. W., Feeney, J. and Sutcliffe, L. H. (1965) "High Resolution Nuclear Magnetic Resonance Spectroscopy", Pergamon Press, Oxford.

Feeney, J. and Walker, S. M. (1970) in "An Introduction to Spectroscopic Methods for the Identification of Organic Compounds" (F. Scheinmann, ed.), Vol. I, pp. 1–40. Pergamon Press, Oxford.

McFarlane, W. (1972) in "Techniques of Chemistry, Vol. IV, Elucidation of Organic Structures by Physical and Chemical Methods, Part I (K. W. Bentley and G. W. Kirby, eds.), pp. 225–322, Wiley-Interscience, New York.

Muller, N. (1972) in "Techniques of Chemistry, Vol. I, Physical Methods of Chemistry, Part IIIA, Optical, Spectroscopic and Radioactivity Methods" (A. Weissberger and B. W. Rossiter, eds.), pp. 599–714, Wiley-Interscience, New York and London.

Murray, M. and Schmutzler, R. (1972) in "Techniques of Chemistry, Vol. IV, Elucidation of Organic Structures by Physical and Chemical Methods, Part I (K. W. Bentley and G. W. Kirby, eds.), pp. 355–376, Wiley-Interscience, New York.

Nachod, F. C. and Zuckerman, J. J. (eds.) (1971) "Determination of Organic Structures by Physical Methods", Vol. IV. Academic Press, New York and London.

Noggle, J. H. and Schirmer, R. E. (1971) "The Nuclear Overhauser Effect", Academic Press, New York and London.

Phillips, L. (1972) in "Techniques of Chemistry, Vol. IV, Elucidation of Organic Structures by Physical and Chemical Methods, Part I (K. W. Bentley and G. W. Kirby, eds.), pp. 323–354, Wiley-Interscience, New York.

Pople, J. A., Schneider, W. G. and Bernstein, H. J. (1959) "High-Resolution Nuclear Magnetic Resonance", McGraw-Hill, New York.

Roberts, J. D. (1962) "An Introduction to the Analysis of Spin–Spin Splitting in High-Resolution Nuclear Magnetic Resonance Spectra", Benjamin, New York.

Rushworth, F. A. and Tunstall, D. P. (1973) "Nuclear Magnetic Resonance", Gordon and Breach, New York.

Sutherland, I. O. (1973) in "MTP International Review of Science, Organic Chemistry, Series One", Vol. I (W. D. Ollis, ed.), pp. 123–158.

14 Electron Spin Resonance Spectroscopy

14.A The Principles of Electron Spin Resonance

14.A.1 Electron spin magnetic energy levels

Many chemical species contain an odd number of electrons, with the consequence that the effects of electron spin are not cancelled out by the pairing of electrons. Such species are paramagnetic. Examples include organic free radicals, some transition-metal ions, and the molecules NO_2 and NO. Other species have two unpaired electrons, so they are also paramagnetic even though they have an even number of electrons. These are triplet states. Some transition-metal ions and the O_2 molecule have triplet states as their ground state, and many organic molecules may be excited into the triplet state by irradiation, as described in Chapter 7. Species containing unpaired electrons have a resultant electronic angular momentum of magnitude $J(J+1)\hbar$, where J is the quantum number introduced in Section 6.A.2. The unpaired electrons interact with an applied magnetic field to create a set of $(2J+1)$ energy levels. For organic free radicals there is only one unpaired electron, for which the angular momentum due to rotation about the nucleus interacts to an extremely small extent with the angular momentum due to its spin. There are thus only two electron spin energy levels for organic free radicals, given by

$$E = M_S g \beta B \qquad (14.1)$$

where M_S is the electron spin quantum number, which can be in this instance either $+\frac{1}{2}$ or $-\frac{1}{2}$, and the field is of flux density B. The constant β is the *Bohr magneton*, and g is a number characteristic of the particular radical, the *Landé g-factor*, or *spectroscopic splitting factor*. The Bohr magneton is the magnetic moment for a spinning sphere carrying a charge of one electron, as calculated by classical electromagnetic theory:

$$\beta = eh/4\pi m \qquad (14.2)$$

where e and m are the charge and mass of the electron. The value of β is 9.2732×10^{-24} J T^{-1}. It can be shown by the methods of quantum mechanics that the value of the Landé g-factor, sometimes called the *g-factor* for short, for any paramagnetic species is given by

$$g = \frac{3J(J+1) + S(S+1) - L(L+1)}{2J(J+1)} \tag{14.3}$$

where L and S are the total orbital and total spin quantum numbers respectively, as discussed in Section 6.A.2. For organic free radicals in the liquid and solid states, which contain a single unpaired electron whose intrinsic and orbital spin moments are not coupled, Eqn. 14.3 shows that the g-factor has the value 2. This is the value for a single, isolated, free electron. A small correction for relativistic effects increases this to 2.00232.

The energy gap between the two electron spin magnetic energy levels in an organic free radical is thus given by

$$\Delta E = g\beta B \tag{14.4}$$

Absorption of radiation of frequency $\Delta E/h$ causes excitation of the molecule from the lower to the upper state, and the spectroscopic technique based on this phenomenon is known as electron spin resonance, or e.s.r. for short. The magnitude of the absorption frequency depends on the applied field, and so can be chosen for experimental convenience. For reasons similar to those discussed in Section 13.B.1, the higher the value of B that can be used the better. It is standard practice to use a flux density of around 0.33 T corresponding to a frequency of around 9 GHz which is in the microwave region of the spectrum. There are in general several electron-spin magnetic energy levels for each species due to various quantised interactions between the spins of the electrons and the nuclei in the molecule, so a set of lines is observed, spanning a range of the order of 1% to 10% of the total.

14.A.2 Relaxation and linewidths

The gap between the electron-spin magnetic energy levels is much less than kT but much larger than the corresponding energy gaps for n.m.r. For e.s.r. in a 0.33 T field at 9 GHz, ΔE is 6.12×10^{-24} J. Substitution of this value into Eqn. 13.2 shows that the upper state is less populated than the lower by a factor of 1.5×10^{-3}. The signal intensity is proportional to this fraction, as discussed in Section 13.A.2, so e.s.r. is inherently a more sensitive technique than n.m.r., for which the corresponding factor is only 10^{-5}. It is possible to detect as little as 10^{-11} mole of a paramagnetic species, so e.s.r. is an extremely sensitive analytical technique. However, as with n.m.r., standardisation is often a problem. The difference between the spacing of energy levels for e.s.r. and n.m.r. can be eventually traced to the difference in mass between the electron and the proton, which causes the nuclear magneton to be less than the Bohr magneton by a factor of 1836.

Spin relaxation is a more efficient process for excited electron spin states than for the analogous nuclear states because the electrons are more exposed to external influences than the nuclei in a molecule. The spin–spin relaxation time is of the order of 10^{-6} s at room temperature, although as long as several seconds in solids at 77 K. As shown by Eqn. 13.3, typical linewidths for e.s.r. are thus much wider than for n.m.r., and are of the order of 500 kHz. This means that the problems of stabilising the magnetic field and monitoring frequency are less acute than for n.m.r., but on the other hand the resonance frequency cannot be determined to the same high degree of accuracy.

14.B Apparatus

14.B.1 E.s.r. spectrometers

Since e.s.r. spectra are observed in the microwave region of the spectrum, the components of an e.s.r. spectrometer are of the type discussed in Section 2.B. The microwave radiation is generated by a klystron and transmitted down a waveguide. One major problem which must be solved is that the sample probably also absorbs microwave energy by the processes described in Chapter 2, so it is necessary to distinguish the absorption due to spin resonance. To improve sensitivity, it is common to use a bridge technique. The microwaves are led into a waveguide component called a *magic-T*, which is illustrated in Fig. 14.1. The sample cell is located in a resonant cavity on one arm of the T, and an adjustable load, i.e. absorber of microwave energy, is located on the opposite arm. When the load is adjusted to absorb the same amount of energy as the sample, no microwave radiation passes down the leg of the T. This adjustment is made away from conditions for spin resonance. The sample cell is between the poles of a magnet of the appropriate strength, so as to give a flux density of 0.33 T for resonance at 9 GHz for example. The spectrum is scanned by altering the magnetic field by sweep coils, mounted on the pole-pieces of the magnet. When the field is such that the resonance condition is satisfied, the sample absorbs slightly more energy, and throws the magic-T out of balance. Some microwave radiation then passes down the leg of the T, and is monitored by the detector.

As discussed in Section 2.B, signal modulation is essential for the production of a high-quality spectrum. This is conveniently done by modulating the

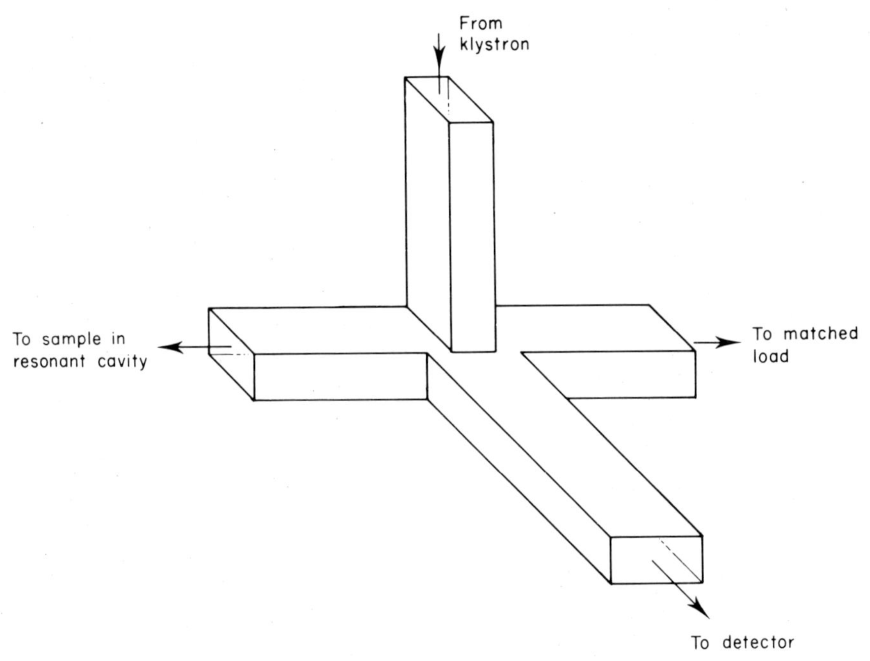

FIG. 14.1 The Magic-T waveguide component.

applied field at a frequency high enough to reduce flicker noise in the detector. A modulation frequency of 100 kHz is often chosen. If the amplitude of the field oscillation is sufficiently small, the observed signal is the first derivative of the absorption curve, i.e. the signal at any field is equal to the rate of change of absorption with field at that value of the field. The line shape is often Lorentzian and may be described by the equation:

$$G(\nu) = T_2/\pi(1+4\pi^2 T_2{}^2(\nu-\nu_0)^2) \tag{14.5}$$

where $G(\nu)$ is the absorption at frequency ν, and T_2 is the spin–spin relaxation time. A plot of this function, for a value for T of 1.6 μs, is shown in Fig. 14.2. In practice, as has been said, it is the magnetic field which is varied rather than the frequency, and the corresponding values of the flux density, assuming the free-electron value of g, are also shown along the horizontal axis. If the field is modulated with an amplitude of, say, 1 μT, there will be zero signal if the absorption is independent of field over that small interval. The greater the change in absorption over that interval, i.e. the steeper the slope of the absorption curve, the larger will be the signal. The observed signal is thus the first derivative, $\mathrm{d}G(\nu)/\mathrm{d}\nu$, which is also shown in Fig. 14.2. It can be seen that the first derivative curve crosses the baseline level when the absorption is at a maximum. In order to present a spectrum in which absorption peaks are seen as peaks, some e.s.r. spectrometers incorporate a further electronic differentiating circuit. The resulting second-derivative spectrum is also shown in Fig. 14.2. If this is inverted, a maximum is seen at the absorption maximum. First derivative spectra are, however, more commonly seen, and are very characteristic of e.s.r. studies.

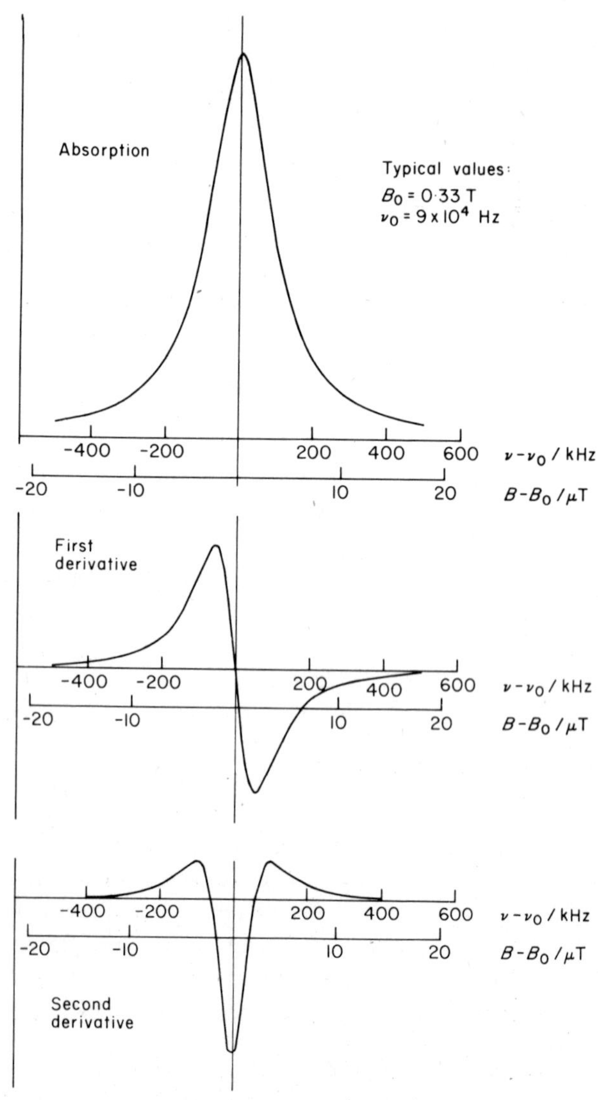

FIG. 14.2 Lorentzian line-shapes.

14.B.2 Sample preparation

Samples may be in the gaseous, liquid, or solid states. A gas sample occupies the whole of the resonant cavity, whereas a liquid sample is held in a small glass tube at the position of maximum change of the magnetic vector of the radiation. Single-crystal samples may be mounted on a goniometer head, so that the dependence of the spectrum on the orientation of the crystal axes can be observed.

Although stable paramagnetic inorganic species are quite numerous, very few such organic species are known. Diphenylpicrylhydrazyl, known as DPPH for short, is an example

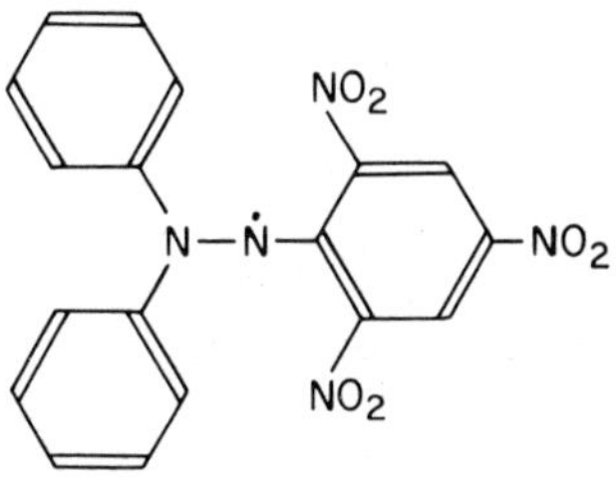

of a stable paramagnetic organic species, and is often used for testing and standardising e.s.r. spectrometers. Most organic free radicals are prepared shortly before their e.s.r. spectrum is run, by oxidation or reduction of a stable precursor. Both chemical and electrolytic processes have been widely used. For example, semiquinone radical anions, of type

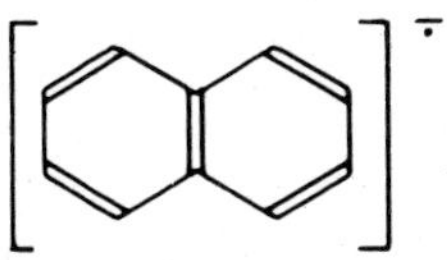

have been prepared by reduction of quinones in alkaline solution by glucose, and the naphthalene radical anion, and those of other polynuclear aromatic hydrocarbons, have been prepared in tetrahydrofuran by reduction with metallic potassium. Such radicals as these, although unstable over a period of time, are sufficiently long-lived for their spectra to be run. Aliphatic free radicals, e.g. $\cdot CH_2OH$, have lifetimes of the order of only milliseconds. In order to observe their spectra it is necessary to generate them continuously so that a steady-state concentration is maintained in the sample cell. This is achieved by rapid mixing of the solution of radical precursor, e.g. CH_3OH if $\cdot CH_2OH$ is required, with an oxidising or reducing solution, and passing the freshly mixed solution directly into the sample cell in a steady stream. Throughout the scanning of the spectrum the solution in the sample cell is only a few milliseconds old. The most commonly used oxidising solution is a mixture of Ti^{3+} with hydrogen peroxide in aqueous acid. This reagent generates $\cdot OH$ radicals, which abstract hydrogen atoms from the radical precursor.

Triplet states of organic molecules are produced by intense ultraviolet irradiation of the sample. It is customary for the organic species to be frozen into a glassy matrix so that the triplet lifetime is long enough to enable the spectrum to be run. The sample must be irradiated before insertion into the resonant cavity, since it is difficult in practice to mount an ultraviolet radiation source within the cavity.

Free radicals in a crystalline matrix may be generated by irradiation of the crystal with X-rays or γ-rays. The primary process occurring as the result of such irradiation is the ejection of electrons from the core orbitals, as discussed in Chapter 11. These high-energy electrons travel a considerable distance through the crystal, leaving a trail of radicals and ions. The radicals formed by this process are comparatively long-lived, since they are trapped in the solid matrix.

14.C The Interpretation of E.s.r. Spectra

14.C.1 The *g*-factor for paramagnetic metal ions

If the orbital momentum of the electrons makes no contribution to the total electron spin momentum of the molecule, the *g*-factor has the value 2.00232. Although this is a good approximation for organic free radicals, it is not generally valid for paramagnetic metal ions; values ranging from less than unity up to 18 have been measured for lanthanide ions. There is a tendency for the magnetic dipole due to the orbital momentum of the electrons to be aligned with the magnetic dipole due to the intrinsic spin. As the unpaired electron changes its spin state as a consequence of absorbing a quantum of electromagnetic radiation, the spin of the ion as a whole tends to follow.

Owing to this spin–orbit coupling the g-factor may be closer to the value for the ion as a whole, which can be calculated from Eqn. 14.3, than to the value for the free electron. The effect of spin–orbit coupling is fully operative only if the ion is completely removed from all external influences, or is in a perfectly symmetrical environment. Spin–orbit coupling tends to be destroyed, or quenched, if the ion is in a field that lifts the degeneracy of the d orbitals. An oversimplified explanation of this effect may be given as follows.

The angular momentum of an electron in a hydrogen-like atom is given by the quantum number l. Solution of the Schrödinger equation corresponding to $l = 2$ gives functions describing the five degenerate d orbitals, each having a different value of the magnetic quantum number m in the range $+l$ to $-l$, i.e. one of the integers $+2$, $+1$, 0, -1, -2. The sign of m represents the sense of the rotation of the orbital motion of the electron. These solutions are complex, i.e. they involve i, the square-root of minus unity. Real d orbitals are constructed as combinations of the complex d orbitals.

$$\mathrm{d}_{z^2} = d_0 \tag{14.6}$$

$$\mathrm{d}_{x^2-y^2} = 2^{-\frac{1}{2}}(d_{+2}+d_{-2}) \tag{14.7}$$

$$\mathrm{d}_{xy} = i2^{-\frac{1}{2}}(d_{-2}-d_{+2}) \tag{14.8}$$

$$\mathrm{d}_{yz} = i2^{-\frac{1}{2}}(d_{+1}+d_{-1}) \tag{14.9}$$

$$\mathrm{d}_{zx} = 2^{-\frac{1}{2}}(d_{-1}-d_{+1}) \tag{14.10}$$

In the presence of an external field which may be caused by adjacent molecules or ions perturbing the symmetry of the environment, a set of energy levels is produced corresponding to the real d orbitals. The field is conventionally considered to act along the z direction. An unpaired electron will then occupy the lowest available real d orbital. As can be seen from Eqns. 14.6–14.10, each real d orbital contains an equal contribution from the two senses of electron rotation so the orbital momentum of an electron in a real d orbital averages to zero in the direction of the applied field. The angular momentum of the electron is then due solely to its intrinsic spin. In practice the observed g-factor is the net result of two opposing tendencies: spin–orbital coupling favouring the free-ion value, and environmental quenching favouring the free-electron value. The greater the splitting between occupied and empty orbitals, i.e. the greater the lifting of the degeneracy of the d orbitals, the more effective is the quenching of spin–orbital coupling and the closer does the observed g-factor approach the free-electron value of 2.00232. The more protected the unpaired electrons are from the environment, the less is the lifting of the degeneracy and the more likely it is that the g-factor will approach the value for the free ion. Thus g values for lanthanide ions, in which the unpaired electrons occupy 4f orbitals well-shielded from the environment, deviate most widely from the free-electron value. Some illustrative values are given in Table 14.1.

The extent to which a magnetic field interacts with a paramagnetic molecule depends on the orientation of the molecule with respect to the field. The g-factor is thus anisotropic. In solution the random tumbling of the molecules averages out the anisotropy, but it is observed for samples in the solid phase, especially for crystals containing paramagnetic ions. Mathematically speaking, the g-factor is not scalar (a scalar being a quantity that is defined solely by its magnitude) but a *tensor*. A tensor is a mathematical operator, having the form of a determinant, which when applied to a vector changes both its magnitude and its direction. The g-tensor relates to the total spin of the unpaired electron to the magnetic moment of the molecule by the vector equation

$$\boldsymbol{\mu} = -\beta \boldsymbol{g} \cdot \boldsymbol{S} \tag{14.11}$$

where $\boldsymbol{\mu}$ and $\boldsymbol{S}$ are vectors describing the orientation of the magnetic moment of the molecule and the angular momentum of the electron with respect to the applied field. The g-factor is the scalar version of the g-tensor, and is calculated as the average of the diagonal elements in the determinant describing the g-tensor.

TABLE 14.1

g-Factors for transition-metal and lanthanide ions (Ayscough, 1967)

Ion	g	$g_{\parallel}$*	$g_{\perp}$*
V^{2+} in $(NH_4)_2V(SO_4)_2$	1.951	—	—
Cr^{3+} in $CsCr(SO_4)_2.12H_2O$	1.98	—	—
Mn^{2+} in $(NH_4)_2Mn(SO_4)_2.6H_2O$	2.00	—	—
Ni^{2+} in $K_2Ni(SO_4)_2.6H_2O$	2.25	—	—
Ce^{3+} in $Ce(NO_3)_3$ at 4.2 K	—	0.25	1.84
Dy^{3+} in $Dy(NO_3)_3$ at 4.2 K	—	4.28	8.92
Tb^{3+} in $LaCl_3$ matrix at 4.2 K	—	17.78	<0.1

* $g_{\parallel}$ and $g_{\perp}$ show the effect of the orientation of a sample, in the form of a single crystal, with respect to the magnetic field.

14.C.2 The g-factor for organic free radicals

The g-factors found for organic free radicals are found to deviate from the free-electron value of 2.00232 by less than 0.5%. This may be considered as being due to an almost complete quenching of spin–orbit coupling by a mechanism similar to that discussed in Section 14.C.1. The unpaired electron in an organic free radical is usually in a molecular orbital derived from a p orbital. The degeneracy of the three p orbitals on the carbon atom is very strongly lifted by covalent bond formation, which creates a highly asymmetric environment. Small deviations from the free-electron value arise as a consequence of participation by molecular orbitals other than that containing the unpaired electron. A hypothetical excitation of the unpaired electron into an empty upper molecular orbital gives an excited state which makes a small contribution to the complete quantum-mechanical description of the molecule. This causes the radical to have a net orbital angular momentum in such a sense as slightly to reduce the g-factor below the free-electron value. The less the energy required for this hypothetical excitation, the greater the contribution of the excited state, and the greater the reduction in the g-factor. If, on the other hand, excitation of an electron from a full molecular orbital into the orbital occupied by the unpaired electron requires less energy than the excitation of the unpaired electron, this hypothetical excitation causes the radical to have a net orbital momentum in the opposite sense, and the g-factor is greater than the free-electron value. As before, the less the excitation energy, the greater the increase in the g-factor. Some g-factors for substituted 1,4-benzoquinone radical anions are listed in Table 14.2. Electron-withdrawing substituents on the ring reduce the energy gap between the full and half-empty π orbitals and cause the g-factor to increase slightly.

Although the g-factor is analogous to the chemical shift observed in n.m.r. spectroscopy, it is by no means as useful, and values of the g-factor are rarely used to identify unknown radicals. Whereas nuclei in different environments in the same molecule have different chemical shifts, the g-factor relates to the electron in the molecule as a whole. However complex the radical, it has only one g-factor (if it is in solution), so little information about its structure is obtained. Much more information is obtained by the study of hyperfine splitting, described in Section 14.C.3.

TABLE 14.2

g-Factors for free radicals

Radical anion	g	Ref.
2,6-Dimethyl-1,4-benzosemiquinone	2.00445	*a*
1,4-Benzosemiquinone	2.00470	*b*
2-Bromo-1,4-benzosemiquinone	2.00510	*b*
2,3-Dichloro-1,4-benzosemiquinone	2.0054	*c*
Tetrabromo-1,4-benzosemiquinone	2.008	*b*

a, Adams *et al.* (1958). *b*, Ayscough (1967). *c*, Wertz and Vivo (1955).

14.C.3 Hyperfine splitting

In Section 13.C.2 it has been shown how coupling between nuclear spins causes a splitting of the nuclear magnetic energy levels and hence a splitting of the lines in n.m.r. spectra. Coupling between the intrinsic spin of the unpaired electron in a free radical and the nuclear spins similarly gives rise to a splitting of the e.s.r. lines, termed *hyperfine splitting*. The type of splitting most generally observed for organic free radicals is due to spin–spin coupling with the protons in the radical. Since both the electron and the proton have spin quantum number $\frac{1}{2}$, the splitting patterns resemble those described for proton–proton coupling in Section 13.C.2. Figure 13.7 may also be used to illustrate electron–proton spin–spin coupling if the electron is denoted by A, and levels α and β refer to M_S having the values $+\frac{1}{2}$ and $-\frac{1}{2}$ respectively. The permitted transitions are those for which $\Delta M_S = 1$, and so are those illustrated in Fig. 13.7. There are, however, two significant differences between electron–proton and proton–proton coupling. Because the electron can approach the proton much more closely than can another proton, the coupling is much stronger and the splitting is much greater. Whereas values of the spin–spin coupling constant for protons, J^{HH}, are of the order of 10 Hz, electron–proton coupling constants, a^{H}, are of the order of 10 MHz. Coupling between protons is strongly dependent on their separation, and usually only the coupling between protons on adjacent carbon atoms need be considered. By contrast, the effect of the unpaired electron often extends over a large part of the free radical, and secondary splitting is often observed; each line in the multiplet caused by interaction with one set of protons is itself a multiplet due to interaction with a more weakly coupled set of protons. The spectrum of the naphthalene radical anion, shown in Fig. 14.3, provides a good example. At any instant the electron is located in one of the rings, and couples with the four protons attached to that ring to give a 1, 4, 6, 4, 1 line pattern. Coupling with the four more distant protons in the other ring causes each line to be itself split into a 1, 4, 6, 4, 1 pattern of lines. Figure 14.3 shows the construction of the theoretical spectrum. It can be seen that the agreement with the observed spectrum is excellent, if it is remembered that a maximum in the absorption spectrum corresponds to the first derivative crossing the baseline.

Because e.s.r. spectra are obtained by scanning the magnetic field rather than the frequency, hyperfine splitting constants are usually quoted in units of magnetic flux density rather than frequency. The old-fashioned centimetre-gram-second system of electromagnetic units, as discussed in the Appendix, is still in common use among e.s.r. spectroscopists, so hyperfine splitting constants are usually quoted in units of gauss (10^4 gauss = 1 tesla). The two hyperfine splitting constants for the naphthalene radical anion, for example,

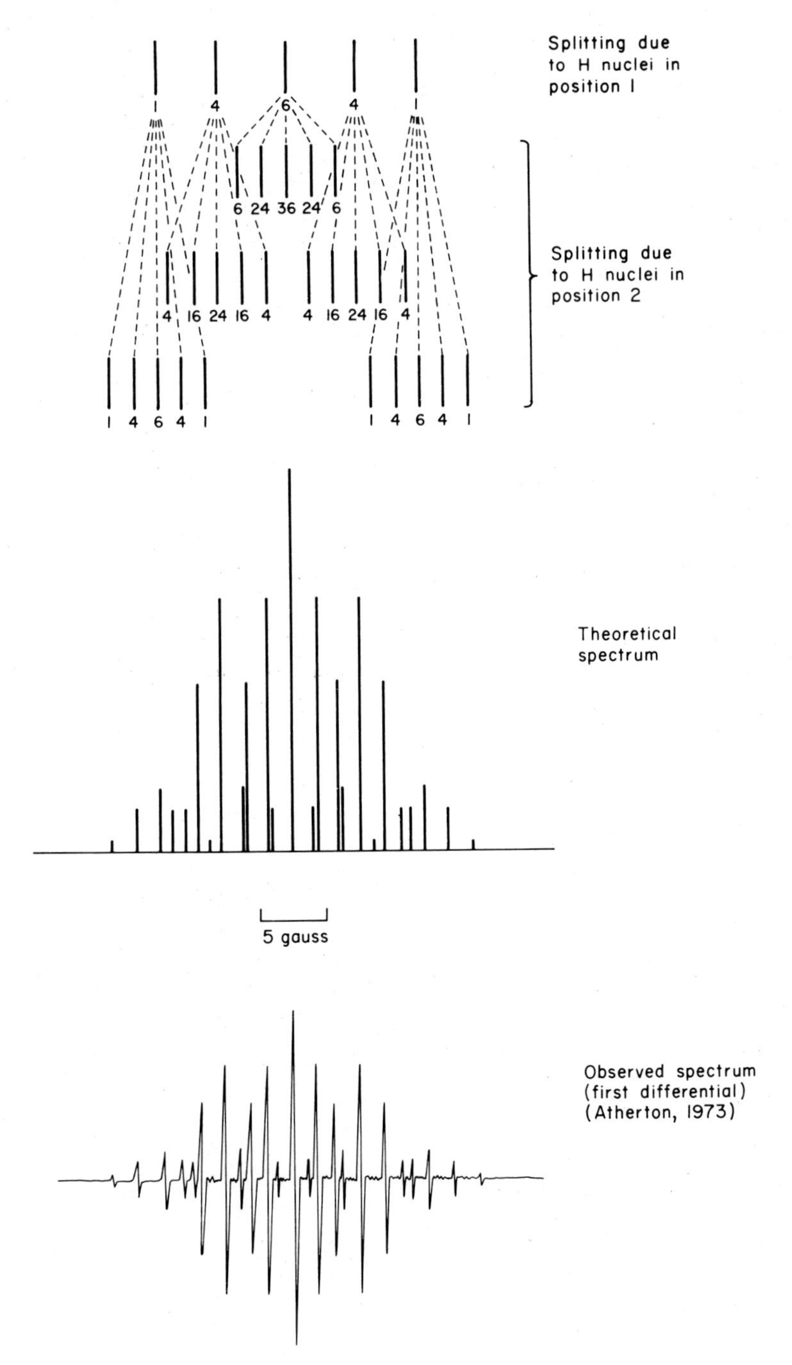

FIG. 14.3 E.s.r. spectrum of naphthalene radical anion.

are 4.90 and 1.83 gauss. Some typical values are listed in Table 14.3. If g is taken to have the free-electron value, substitution of numerical values into Eqn. 14.4 shows that a splitting of 1 gauss is equivalent to a splitting of 2.8024 MHz. Occasionally the c.g.s. unit of magnetic field strength, the oersted, is used instead of the gauss, since in the c.g.s. system of units the gauss and the oersted are almost exactly interchangeable except for ferromagnetic substances. The wavenumber, cm^{-1}, is also used as an energy unit by e.s.r. spectroscopists, especially to describe energy levels in triplet states. A splitting of 1 gauss is equivalent to a splitting of 9.3478×10^{-5} cm^{-1} if g is given its free-electron value.

TABLE 14.3

Typical hyperfine splitting constants (values in gauss) (Ayscough, 1967)

Radical	a_α^H	a_β^H	a_γ^H
$\dot{C}H_3$	23.07	—	—
$CH_3\dot{C}H_2$	22.38	26.87	—
$CH_3CH_2\dot{C}H_2$	22.08	33.2	0.38
$CH_2{=}\dot{C}H$	13.4	102.4	—
$\dot{C}H_2OH$	17.2	—	—
$HCO_2\dot{C}H_2$	20.9	—	2.51

Radical anion		Ring protons	
	a_1^H	a_2^H	a_3^H
1,4-Benzosemiquinone	2.368	—	—
2-Methyl-1,4-benzosemiquinone	2.7	2.44	1.95
Naphthalene	4.90	1.83	—
Toluene	5.12	5.45	0.59

Coupling with nuclei of spin greater than $\frac{1}{2}$ produces more complex spectra. The most commonly occurring nucleus of this type is ^{14}N, of spin 1. The ^{14}N nucleus thus has three possible orientations with respect to the electron spin corresponding to values of M_I of 1, 0, and -1. The e.s.r. spectrum of $K_2NO(SO_3)_2$ is shown in Fig. 14.4. The $NO(SO_3)_2^{2-}$ radical anion contains only one nucleus with spin, ^{14}N, so the spectrum consists of three lines of equal intensity. Although the multiplicity is that found for coupling to two equivalent nuclei of spin $\frac{1}{2}$, the intensity pattern is different since the $M_I = 0$ state for ^{14}N is not doubly degenerate. In general, coupling to a nucleus of spin I produces a $(2I+1)$ multiplet, all lines being equally intense. Coupling to a group of n such equivalent nuclei produces a $(n+1)(2I+1)$ multiplet. It can be seen that very complex spectra are possible.

There are two possible mechanisms for spin–spin coupling, one isotropic and the other anisotropic. Isotropic coupling occurs as a consequence of the electron having a finite density at the interacting nucleus. All orbitals except those of s type have a node at the nucleus, so isotropic coupling is possible only if the unpaired electron has some s character. The magnitude of the *isotropic hyperfine splitting constant*, a_{iso}, is then a measure of this s character.

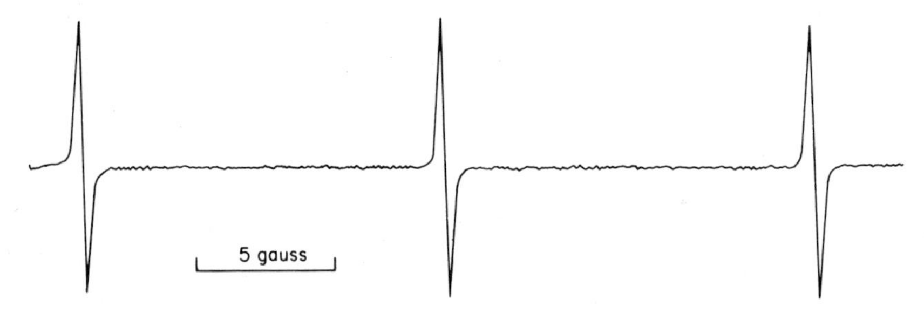

FIG. 14.4 E.s.r. spectrum of $K_2(NO)(SO_3)_2$ (Atherton, 1973).

Values of a_0, the isotropic hyperfine splitting constant that would be observed if the unpaired electron were in an orbital of pure s character, can be calculated by the methods of quantum mechanics, and compared with the observed values of a_{iso}. It is found that organic radicals may be classified according to the magnitude of the a_{iso}/a_0 ratio. Radicals in which the unpaired electron is in a σ orbital, termed *σ-radicals*, have values of a_{iso}/a_0 greater than 0.1; radicals in which the unpaired electron is in a π orbital, *π-radicals*, have values of a_{iso}/a_0 less than 0.05.

Isotropic hyperfine interaction of α protons has been much studied theoretically. The system considered is $>\dot{C}$–H, which may be part of an aliphatic or aromatic free radical. It is found that the isotropic hyperfine splitting constant for an electron in an orbital containing the carbon atom is proportional to the unpaired spin density at the proton. It is important to distinguish clearly between the *spin population* in an orbital, denoted by ρ, and the *spin density* at a nucleus; it is the latter that is inevitably directly related to the isotropic hyperfine splitting constant. The spin population, sometimes called the orbital spin density, is a measure of the extent to which an orbital is occupied by an unpaired electron. The theory predicts that

$$a^{H}_{iso} = Q^{H}_{CH}\rho_{C} \qquad (14.12)$$

where a^{H}_{iso} is the isotropic hyperfine splitting constant for interaction with the proton, and ρ_C is the spin population in the sp^2 orbital on the carbon atom. Q_{CH}^{H} is a constant; experimental values for Q_{CH}^{H} for π radicals are around -23 gauss. The negative sign of Q_{CH}^{H} indicates that the unpaired spin induced in the 1s orbital on the hydrogen atom is opposite in sign to that induced in the sp^2 orbital on the carbon atom. This is because the energy of two adjacent electrons is lower if they have opposing spins than if their spins are parallel; the analogous situation for nuclear spins has been discussed in Section 13.C.2. As for nucleus–nucleus spin–spin coupling, the sign of a_{iso} is not directly observable from the e.s.r. spectrum, and both positive and negative values of ρ_C are possible. A negative spin population may be interpreted as due to the effect of the adjacent nuclear spin forcing the electrons to pair up more than they would otherwise do.

Anisotropic hyperfine splitting is due to the interaction between the electronic and nuclear magnetic dipole moments, which can be treated by classical electromagnetic theory. The magnitude of the interaction is inversely proportional to the cube of the distance between the electron and the nucleus, and is also dependent on the relative orientation of the magnetic dipoles. If the molecule is in solution, its orientation changes so rapidly that the anisotropic interaction averages to zero. The relative contribution of isotropic and anisotropic interactions to the hyperfine splitting observed for a radical in a solid matrix, e.g. a glass of frozen solvent at liquid-nitrogen temperature, can be found by comparison with the hyperfine splitting observed in liquid solution. The magnitude of the *anisotropic hyperfine splitting constant* gives a measure of the non-s character of the orbital containing the unpaired electron. For organic radicals, contributions from d and f orbitals may be ignored, so the anisotropic hyperfine splitting constant effectively measures the p character of the orbital.

14.C.4 Fine splitting in triplet spectra

Many species are known that contain two unpaired electrons, including biradicals and also atoms and molecules in triplet states, as discussed in Chapter 7. The electron spins in such species are oriented so as to give a set of three energy levels, corresponding to spins $+\frac{1}{2}, +\frac{1}{2}$; $+\frac{1}{2}, -\frac{1}{2}$; $-\frac{1}{2}, +\frac{1}{2}$; and $-\frac{1}{2}, -\frac{1}{2}$, so M_S has one of the values $+1$, 0, and -1 respectively. The level for which M_S is zero is doubly degenerate. Since the selection rule for absorption of electromagnetic radiation is $\Delta M_S = +1$, the permitted transitions are $M_S = -1 \rightarrow 0$ and $M_S = 0 \rightarrow 1$. Biradicals are organic species with unpaired electrons on two well-separated sites. If, as for such species, the two electrons are not spin-coupled, the two permitted transitions are of equal energy, and a single absorption line is observed. If, on the other hand, the electrons are spin-coupled, as in triplet state species, the transitions are of different energies, and two lines are observed. This is termed *fine splitting*, since electron–electron spin-coupling was originally observed as the origin of fine structure in atomic spectra in the visible/ultraviolet region. Hyperfine splitting is much less than fine splitting, and gives a multiplet structure to each of the two lines in the e.s.r. spectrum. Orders of magnitude of the energies of the various types of spin–spin coupling are: electron–electron, 2×10^{-25} J; electron–nucleus, 10^{-26} J; nucleus–nucleus, 10^{-32} J.

Organic molecules in the triplet state are produced by irradiation of frozen solutions, often at liquid-nitrogen temperature, by ultraviolet radiation as discussed in Chapter 7. Electron–electron spin–spin coupling in triplet species is of dipole–dipole type, and is thus anisotropic. The magnitude of the quantum of electromagnetic radiation required to cause a transition depends on the orientation of the molecule with respect to the magnetic vector of the radiation. Microwave radiation is plane polarised, so the orientation of the magnetic vector is clearly defined. Fine splitting is thus characterised by two parameters, namely D, which measures the magnitude of the spin-coupling interaction, and E, which measures the anisotropy. It is usual to quote values of D and E in units of cm^{-1}. The energy levels of a triplet molecule depend on whether it is oriented with its major symmetry axis along the x, y, or z axes of the incident radiation, i.e. has X, Y, or Z orientation. The energy levels for the X orientation are

$$M_S = +1: E_+ = (Y+Z)/2+\{[(Y-Z)/2]^2+(g\beta B/hc)^2\}^{\frac{1}{2}} \quad (14.13)$$

$$M_S = 0: E_0 = X \quad (14.14)$$

$$M_S = -1: E_- = (Y+Z)/2-\{[(Y-Z)/2]^2+(g\beta B/hc)^2\}^{\frac{1}{2}} \quad (14.15)$$

where

$$X = D-E \quad (14.16)$$

$$Y = D+E \quad (14.17)$$

$$Z = -D \quad (14.18)$$

Equations 14.13–14.15 give the energies in units of cm^{-1}, c being the velocity of light in cm s^{-1}, if D and E are in the conventional units of cm^{-1}. The corresponding energy levels for the Y and Z orientations are obtained by interchanging X, Y, and Z in Eqns. 14.13–14.15 in a systematic cyclic fashion. Figure 14.5 shows the energy levels for triplet naphthalene in solid solution in a single crystal of durene. All the naphthalene molecules have the same orientation in the crystal, so it is possible to distinguish experimentally between the X, Y, and Z orientations with respect to the microwave radiation. The values found for D and E are $+0.1006$ and -0.0138 cm^{-1} respectively (Hutchison and Mangum, 1961). These are fairly typical values for organic molecules in the triplet state.

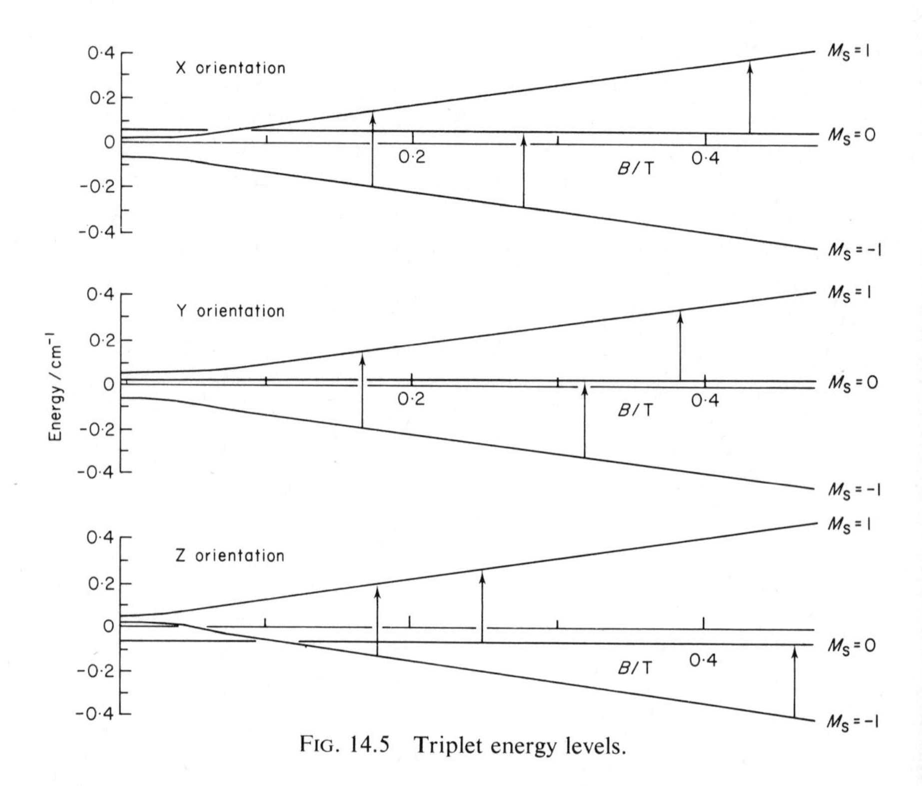

FIG. 14.5 Triplet energy levels.

Inspection of Eqns. 14.13–14.15 and also of Fig. 14.5 shows that there is a difference in the energy levels for the different values of M_S even at zero applied magnetic field. It is sometimes stated that this zero-field splitting is found only for triplet species. This is not strictly correct; spin–spin coupling causes zero-field splitting for all systems containing more than one particle with spin. However, only if both spin-coupled particles are electrons is the splitting sufficiently large to be observable on an energy diagram drawn to scale, such as Fig. 14.5. Zero-field splitting is also shown in Fig. 13.5, for proton–proton coupling, but, as explained in Section 13.C.2, Fig. 13.5 is not drawn to scale.

Resonance at some given frequency ν for the molecule in the X orientation is obtained for magnetic flux density B tesla for the permitted transitions:

$$M_S;\ 0 \rightarrow +1:\ B = (hc/g\beta)\{[(\nu/c)+(3X/2)]^2-[(Y-Z)/2]^2\}^{\frac{1}{2}} \qquad (14.19)$$

$$M_S;\ -1 \rightarrow 0:\ B = (hc/g\beta)\{[(\nu/c)-(3X/2)]^2-[(Y-Z)/2]^2\}^{\frac{1}{2}} \qquad (14.20)$$

Values of B for other orientations are obtained by interchanging X, Y, and Z as for Eqns. 14.13–14.15. Transitions for a resonance frequency of 10 GHz for triplet naphthalene in the three orientations are shown in Fig. 14.5. It can be seen that the magnitudes both of the resonance field and the splitting are very dependent on the orientation of the molecule. Unless great care is taken to ensure that all the molecules are in the same orientation, only a small proportion will be correctly oriented for any one resonance condition; the signal will be weak and the spectrum difficult to observe. This was the cause of much difficulty in the early days of e.s.r. spectroscopy of organic triplet states, since the molecules are usually frozen into a glassy matrix in random orientations. However the transition $M_S; -1 \rightarrow +1$ is weakly permitted if the microwave magnetic vector is parallel to the applied magnetic field. The resonant field for molecules in the X orientation is then given by

$$M_S; -1 \rightarrow +1: \ B = (hc/g\beta)\{(\nu/2c)^2 - [(Y-Z)/2]^2\}^{\frac{1}{2}} \qquad (14.21)$$

As before, values of B for the other orientations are obtained by systematic cyclic interchange of X, Y, and Z. The $\Delta M_S = 2$ transitions for a resonance frequency of 10 GHz for triplet naphthalene are shown in Fig. 14.5. It can be seen that the magnitude of the resonance field is much less dependent on molecule orientation for the $\Delta M_S = 2$ transitions than for the $\Delta M_S = 1$ transitions. It is thus possible to obtain better spectra for samples containing randomly oriented molecules for the $\Delta M_S = 2$ transition, even though this is only weakly permitted. Because the $\Delta M_S = 2$ transition occurs at approximately half the magnetic flux density as the $\Delta M_S = 1$ transitions, it is often referred to as the *half-field transition*.

14.D Kinetic Effects

As discussed in Section 13.D, the exchange of nuclei between two different environments causes a broadening of the n.m.r. line, and the rate of the exchange process can be calculated from the magnitude of the line-broadening. In a similar fashion, electron exchange between molecules leads to broadening of e.s.r. lines. The classic study of this type was of the electron exchange between naphthalene molecules and naphthalene radical anions (Ward and Weissman, 1957):

$$C_{10}H_8^- + C_{10}H_8 \rightleftharpoons C_{10}H_8 + C_{10}H_8^-$$

Rate constants of the order of 10^8 l mol^{-1} s^{-1} were observed, the exact value depending on the solvent and the alkali-metal counterion. As for n.m.r., significant line-broadening is observed when the residence time of the electron on a particular molecule is of the order of T_2. Because T_2 is much smaller for e.s.r. than for n.m.r., it is in principle easier to observe very fast reactions by e.s.r., but the technique is by no means as widely employed for kinetic studies as is n.m.r.

REFERENCES

Adams, M., Blois, M. S., Jr. and Sands, R. H. (1958) *J. Chem. Phys.* **28**, 774.
Hutchison, C. A., Jr. and Mangum, B. W. (1961) *J. Chem. Phys.* **34**, 908.
Ward, R. L. and Weissman, S. I. (1957) *J. Amer. Chem. Soc.* **79**, 2086.
Wertz, J. E. and Vivo, J. L. (1955) *J. Chem. Phys.* **23**, 2441.

BIBLIOGRAPHY

Alger, R. S. (1968) "Electron Paramagnetic Resonance", Wiley-Interscience, New York and London.
Atherton, N. M. (1973) "Electron Spin Resonance", Ellis Horwood (Wiley), Chichester.
Ayscough, P. B. (1967) "Electron Spin Resonance in Chemistry", Methuen, London.
McMillan, J. A. (1968) "Electron Paramagnetism", Reinhold, New York and Amsterdam.
Reiger, P. H. (1972) in "Techniques of Chemistry, Vol. I, Physical Methods of Chemistry, Part IIIA, Optical, Spectroscopic and Radioactivity Methods" (A. Weissberger and B. W. Rossiter, eds.), pp. 499–598, Wiley-Interscience, New York and London.
Russell, G. A. (1972) in "Techniques in Chemistry, Vol. IV, Elucidation of Organic Structures by Physical and Chemical Methods, Part I" (K. W. Bentley and G. W. Kirby, eds.), pp. 441–480, Wiley-Interscience, New York.
Wardale, H. W. (1974) in "An Introduction to Spectroscopic Methods for the Identification of Organic Compounds" (F. Scheinmann, ed.), Vol. II, pp. 153–194, Pergamon Press, Oxford.

Appendix: SI Units

The International System of units, known by the initials of its French name as the SI, was adopted by a general conference on weights and measures in 1960. The standard units are: the metre, m; the kilogram, kg; the second, s; the ampere, A; the kelvin, K; and the candela, cd.

The *metre* is the unit of length, equal to 1 650 762.73 wavelengths in vacuum of the radiation corresponding to the transition between the 2p and 5d lines of the ^{86}Kr atom.
The *kilogram* is the unit of mass, defined as the mass of a certain cylinder of platinum–iridium alloy kept in Paris.
The *second* is the unit of time, equal to the duration of 9 192 631 770 periods of the radiation corresponding to the transition between two hyperfine lines of the ^{137}Cs atom.
The *ampere* is the steady current which, if maintained in two straight parallel conductors of infinite length and negligible cross-section, 1 metre apart in vacuum, produces a force of 2×10^{-7} newtons per metre between the conductors.
The *kelvin* is the unit of temperature, and is 1/273.16 of the thermodynamic temperature of the triple point of water.
The *candela* is the unit of luminous intensity, and is that of 1/0.6 mm^2 of a black body at the temperature of freezing platinum.

Multiples of these units are denoted by the prefixes:

$\times 10^{-3}$	milli	m	$\times 10^{3}$	kilo	k
$\times 10^{-6}$	micro	μ	$\times 10^{6}$	mega	M
$\times 10^{-9}$	nano	n	$\times 10^{9}$	giga	G
$\times 10^{-12}$	pico	p	$\times 10^{12}$	tera	T

Some derived units have special names, e.g.

Quantity	SI unit name	SI unit symbol	Definition
Frequency	hertz	Hz	s^{-1}
Energy	joule	J	$kg\ m^2\ s^{-2}$
Force	newton	N	$J\ m^{-1}$
Power	watt	W	$J\ s^{-1}$
Pressure	pascal	Pa	$N\ m^{-2}$
Electric charge	coulomb	C	A s
Electric potential difference	volt	V	$J\ A^{-1}\ s^{-1}$
Capacitance	farad	F	$C\ V^{-1}$
Magnetic flux	weber	Wb	V s
Magnetic flux density	tesla	T	$V\ s\ m^{-2}$
Inductance	henry	H	$V\ A^{-1}\ s$

Readers who have recently studied physics in British schools will probably be accustomed to the use of these units, and their use is now encouraged in the scientific literature. Unfortunately one finds a range of other units in the older scientific literature, mostly based on the c.g.s. (centimetre-gram-second) system, and one cannot use most of the published data without an appreciation of the older units.

In some cases, conversion is trivial, e.g.

1 Ångstrom	= 100 pm
1 international calorie	= 4.1868 J
1 mm Hg ≈ 1 Torr	= 133.3 Pa
1 atmosphere	= 101 325 Pa
1 dyne	= 10^{-5} N
1 erg	= 10^{-7} J
1 litre	= 1 dm^3
1 electron-volt	= 1.60207×10^{-19} J

However, the conversions for electrical and magnetic units are much more difficult, because the SI faces up to problems which the older systems had "swept under the carpet" and ignored for reasons of convenience. There are two systems of units in which the c.g.s. system has been extended to include electric and magnetic units: the electrostatic system, or e.s.u., and the electromagnetic system, or e.m.u. The description of the electrostatic system of units starts by defining a unit of charge as that which attracts a charge of equal magnitude and opposite sign, at a distance r cm, with a force of $1/r^2\varepsilon$ dynes, where ε is the permittivity of the medium separating the charges. Permittivity is taken to be dimensionless. The permittivity of a vacuum, ε_0, is set at unity. The e.s.u. of current is defined by the rate of flow of charge. The description of the electromagnetic system of units starts by defining a unit magnetic pole which attracts an opposite unit pole, at a distance of r cm, with a force of $1/r^2\mu$ dynes, where μ is the permeability of the medium, taken to be dimensionless. The permeability of a vacuum, μ_0, is set at unity. The e.m.u. of current is defined by comparing the magnetic field produced around a wire carrying the current with the field produced by a magnetic pole of unit strength. It is found by experiment that:

$$1 \text{ practical or SI ampere} = 10 \text{ e.m.u. of current} = 1/(3 \times 10^9) \text{ e.s.u. of current}$$

It can be seen that the e.s.u. of current is larger than the e.m.u. of current by a factor of 3×10^{10}, and it is no mere coincidence that this is the numerical value of the velocity of light in cm s^{-1}. In what is generally regarded as the outstanding piece of work in theoretical physics of the nineteenth century, Maxwell showed that $(\varepsilon_0\mu_0)^{-\frac{1}{2}}$ equals c, the velocity of light in a vacuum. Einstein later showed that magnetism is a consequence of the effect of relativity on the motion of electrons. (The velocity of light plays an important part in the Special Theory of Relativity.) Electrical effects are thus in a sense more fundamental than magnetic effects, and magnetic units should be derived from electric units, rather than the other way round. A clue that this is the right way round is provided by the experimental observation that bodies carrying a single electric charge exist, but single magnetic poles do not exist; the e.m.u. system is thus based on the concept of a non-existent entity. The electrical units of the SI are based on the ampere, the rate of flow of electrons. By taking the ampere as the fundamental unit, the SI requires that both ε and μ have dimensions, and furthermore that ε_0 and μ_0 do not have value unity. Their values are:

$$\varepsilon_0 = 10^7/4\pi c^2 = 8.85416 \times 10^{-12} \text{ F m}^{-1} \text{ (i.e. A s V}^{-1}\text{ m}^{-1}\text{)}$$

$$\mu_0 = 4\pi \times 10^{-7} \text{ H m}^{-1} \text{ (i.e. N A}^{-2}\text{)}$$

Units of magnetic quantities present the most difficulty, and so will be discussed here in some detail. When the magnetic field threading a circuit is made to vary, e.g. by moving a bar magnet to or from a coil of wire, an electromotive force is induced in the circuit. The product of the induced voltage at time t and the time taken to change the field, i.e. $\int V\,dt$, is found to be the same for all rates of change, and so can be used to define a parameter of the field, the magnetic flux, ϕ. Increasing the flux produces an induced e.m.f. in such a direction as to oppose the increase, so that

$$\phi_B - \phi_A = -\int_A^B V \, dt$$

for a change from state A to state B. Magnetic flux thus has units V s. The unit V s is given a special name, the weber, or Wb for short. The magnetic flux per unit area is called the magnetic flux density, or magnetic induction. Since the area is defined in a plane perpendicular to some axis, the magnetic flux density is a vector quantity, and is denoted by $\boldsymbol{B}$, whereas the magnetic flux is a scalar quantity. $\boldsymbol{B}$ has units Wb m^{-2}, which unit is given the name the tesla, or T for short. Magnetic properties may also be related to electrical quantities by considering the magnetic field produced by a current flowing in a solenoid. The magnetic force, or magnetic field intensity, denoted by $\boldsymbol{H}$, is defined as a vector, along the axis of a long solenoid, of magnitude equal to the number of ampere-turns per unit length of the solenoid. $\boldsymbol{H}$ thus has units A m^{-1}. There is no special name for this unit. A magnetic field is thus defined by two vectors, $\boldsymbol{B}$ and $\boldsymbol{H}$. These are related by the equation

$$\boldsymbol{H} = \boldsymbol{B}/\mu$$

where μ is the permeability of the medium. μ thus has units V s $(\text{A m}^{-1})^{-1}$, i.e. N A^{-2} or H m^{-1}. The corresponding equation for electric field vectors is

$$\boldsymbol{D} = \varepsilon \boldsymbol{E}$$

where $\boldsymbol{E}$ is the electric field strength in units V m^{-1} and $\boldsymbol{D}$ is the electric displacement in units C m^{-2}. ε thus has units C m^{-2} (V m^{-1}), i.e. C V^{-1} m^{-1} or F m^{-1}. Since $\boldsymbol{H}$ is analogous to $\boldsymbol{D}$, and $\boldsymbol{E}$ is analogous to $\boldsymbol{B}$, it would be more satisfactory to call $\boldsymbol{B}$ the magnetic field strength, but these quantities were named before their interrelationships were fully appreciated.

The e.s.u. system does not have any magnetic units. The e.m.u. of magnetic flux density, $\boldsymbol{B}$, is the gauss, 1 gauss being equal to 10^{-4} T. The e.m.u. of magnetic field intensity, $\boldsymbol{H}$, is the oersted, 1 oersted being $10^3/4\pi$ A m^{-1}. Since the permittivity of a vacuum in the e.m.u. system is both dimensionless and unity, $\boldsymbol{B}$ in gauss is exactly equal, both numerically and in dimensions, to $\boldsymbol{H}$ in oersted for a vacuum. For air, and even paramagnetic and diamagnetic liquids and solids, the numerical value of $\boldsymbol{B}$ in gauss is still very close to that of $\boldsymbol{H}$ in oersted. As a consequence there has arisen a widely spread confusion in the literature of chemistry, particularly in that pertaining to n.m.r., between $\boldsymbol{B}$ and $\boldsymbol{H}$, and hence between the units gauss and oersted. In many textbooks the student will find that Eqn. 13.1 contains $\boldsymbol{H}$ rather than $\boldsymbol{B}$, $\boldsymbol{H}$ being termed the magnetic field strength, and measured in gauss. To add further to the confusion, the student will occasionally find ***H*** given its correct units of oersteds, but used in place of ***B***.

Further insight into the conceptual difficulties inherent in the e.m.u. and e.s.u. system can be achieved by consideration of an unusual system of units, in which the velocity of light in a vacuum is defined to be dimensionless and of value unity. In this set of units, velocities are dimensionless numbers, and there is no distinction between the units of length and time. The statement:

> Since the distance between A and B is 50 km, a car travelling at 100 km h^{-1} would travel from A to B in 30 minutes

would be replaced by either of the two equivalent statements:

> Since the distance between A and B is 50 km, a car travelling at 9.25×10^{-9} would travel from A to B in 5.4×10^{11} km.
>
> Since the distance between A and B is 1.66×10^{-7} s, a car travelling at 9.25×10^{-9} would travel from A to B in 1800 s.

The confusion that would arise from the general adoption of such a system is obvious, and is exactly analogous to the confusion that arises from a similarly arbitrary assignment of ε_0 or μ_0 as a dimensionless quantity of value unity. This system has in fact been adopted by spectroscopists for one application: the use of the cm^{-1} unit to describe frequencies is founded on this system. Many authors, especially in the older literature, skip lightly from e.m.u. to e.s.u., from gauss to oersted, with an agility born of long experience and deep understanding. Students who attempt to follow their reasoning must do so with care, or they will stumble into confusion.

Revision Problems

The numerical values below may be assumed:

$$8\pi^2 = 78.957 \qquad k = 1.3806 \times 10^{-23}\ \text{J K}^{-1}$$
$$h = 6.6262 \times 10^{-34}\ \text{J s} \qquad N_0 = 6.0222 \times 10^{23}\ \text{mol}^{-1}$$
$$c = 2.9979 \times 10^{8}\ \text{m s}^{-1} \qquad R = 8.31432\ \text{J mol}^{-1}\ \text{K}^{-1}$$

Relative mass of:

$$^{1}\text{H} = 1.0078 \qquad ^{35}\text{Cl} = 34.969 \qquad ^{11}\text{B} = 11.009$$
$$^{2}\text{H} = 2.0141 \qquad ^{37}\text{Cl} = 36.966 \qquad ^{12}\text{C} = 12.000$$

(1) Rearrange Eqn. 2.23 so as to show that a graphical plot of $(B_0 - B_v)/v$ against v may be used to evaluate α_e and γ_e. Use the data in Table 2.2 to calculate α_e and γ_e for $^{39}\text{K}^{127}\text{I}$ in the rotational level $J = 5$.

Equation 2.24 shows that, if allowance is made for the effect of centrifugal stretching, the rotational energy of a diatomic molecule is given by

$$E_{\text{rot}} = B_v J(J+1) - D_j J^2(J+1)^2 \qquad (15.1)$$

Hence show that the frequency, ν, for a transition from rotational level J to rotational level $(J+1)$ for a molecule in vibrational state of quantum number v is given by

$$\nu = 2B_v(J+1) - 4D_j(J+1)^3 \qquad (15.2)$$

Use a graphical plot based on this equation to derive B_v and D_j for $^{39}\text{K}^{127}\text{I}$ in the vibrational state $v = 5$. Assuming that α_e and γ_e are independent of J, and that D_j is independent of v, calculate B_e for $^{39}\text{K}^{127}\text{I}$ and hence calculate the equilibrium internuclear distance.

(2) Rosenblum *et al.* (1958) find that the frequencies for the $J = 0 \rightarrow 1$ transition in carbon monoxide are 115 271.204, 112 359.276, and 109 782.182 MHz for $^{12}\text{C}^{16}\text{O}$, $^{12}\text{C}^{17}\text{O}$, and $^{12}\text{C}^{18}\text{O}$ respectively. If the mass of ^{12}C is exactly 12 atomic units, and that of ^{16}O is 15.99491 atomic units, evaluate

the masses of ^{17}O and ^{18}O to six significant figures. Ignore the effects of bond elasticity.

(3) Murty and Curl (1969) studied the Stark effect on the $J = 2 \rightarrow 3$ transition on a sample of $^{39}K^{35}Cl$ vapour. The shift, $\Delta\nu$, from the original frequency ν_0, was found to vary with the applied voltage, V, as below:

$\Delta\nu$/MHz :	3.3	6.9	10.5	13.9
V/volt :	1410	2000	2450	2830

The frequency for the $J = 2 \rightarrow 3$ transition was found to be 23 067 MHz in the absence of an applied electric field. For the observed transition, $M = 1$, $\Delta M = 0$. It can be shown, starting from Eqn. 2.37, that the frequency shift is related to μ, the dipole moment of the molecule, by

$$\Delta\nu = \frac{\mu^2 E^2}{\nu_0 h^2} \times \frac{6M^2(8J^2+16J+5)-8J(J+1)^2(J+2)}{J(J+2)(2J-1)(2J+1)(2J+3)(2J+5)} \qquad (15.3)$$

where J is the rotational quantum number for the lower state, and E is the electric field strength in units V m^{-1}. Verify the square-law dependence of $\Delta\nu$ on the applied voltage. Assuming that the effective electrode spacing is 3.9 mm, calculate the dipole moment of $^{39}K^{35}Cl$ in debye units.

(4) The absorption spectrum of hydrogen chloride gas at atmospheric pressure shows two sets of lines in the region of wavelength around 3.4 μm. One set is about three times the intensity of the other. The wavenumbers of lines in the more intense set were found by Mills *et al.* (1953) to be as listed below.

2544.24	2727.75	2906.25	3030.12
2571.86	2752.01	2925.92	3045.15
2599.02	2775.77	2944.99	3059.38
2625.75	2799.00	2963.35	3072.90
2651.94	2821.59	2981.05	3085.67
2677.77	2843.63	2998.05	3097.76
2702.95	2865.14	3014.50	3109.11

The wavenumbers of lines in the less intense set were found to be

2542.75	2725.90	2904.07	3027.84
2570.28	2750.13	2923.74	3042.80
2597.36	2773.82	2942.79	3057.06
2624.04	2797.01	2961.13	3070.55
2650.17	2819.56	2978.80	3083.33
2675.98	2841.56	2995.78	3095.39
2701.15	2863.06	3012.23	3106.76

It is suggested that these two sets of lines are the vibration–rotation spectra of $^{1}H^{35}Cl$ and $^{1}H^{37}Cl$ respectively. Assign each line to the correct branch and J number.

Starting from Eqn. 15.1, show that if the wavenumbers of lines due to transitions from rotational level J in the R and P branches are $R(J)$ and $P(J)$ respectively, then:

$$\Delta_2 F'(J) = R(J) - P(J) = (4\tilde{B}_1 - 6\tilde{D}_j)(J+\tfrac{1}{2}) - 8\tilde{D}_j(J+\tfrac{1}{2})^3 \quad (15.4)$$

$$\Delta_2 F''(J) = R(J-1) - P(J+1) = (4\tilde{B}_0 - 6\tilde{D}_j)(J+\tfrac{1}{2}) - 8\tilde{D}_j(J+\tfrac{1}{2})^3 \quad (15.5)$$

Consideration of Fig. 4.7 shows that, whatever the equation describing the rotational energy, $\Delta_2 F''(J)$ is always equal to the spacing between the rotational levels $(J-1)$ and $(J+1)$ of the lower vibrational state. For example, the difference between wavenumbers for P(4) and R(2) is equal to the difference between levels $J = 2$ and $J = 4$ of the lower vibrational state. Similarly, $\Delta_2 F'(J)$ gives the spacing between rotational levels in the upper vibrational state. The values of $\tilde{D}_j$ obtained from Eqns. 15.4 and 15.5 thus refer the molecule in lower and upper vibrational states respectively, although $\tilde{D}_j$ is generally found to be equal for vibrational levels v and $(v+1)$ within the limits of experimental error.

Use graphical plots based on these equations to derive values of $\tilde{B}_0$, $\tilde{B}_1$, and $\tilde{D}_j$ for $^{1}H^{35}Cl$ and $^{1}H^{37}Cl$. Hence evaluate $\tilde{\alpha}_e$, $\tilde{B}_e$, and the equilibrium internuclear distance r_e for the two molecules assuming the molecules are $^{1}H^{35}Cl$ and $^{1}H^{37}Cl$.

Calculate the vibration frequency ω (uncorrected for anharmonicity) for the two molecules, and use Eqn. 2.18 to discover whether the ratio of the two frequencies is in accordance with the assumption that the molecules are $^{1}H^{35}Cl$ and $^{1}H^{37}Cl$.

(5) The infrared spectrum of gaseous carbon monoxide at 298 K was run on a spectrophotometer of resolution insufficient to show the rotational fine structure. The separation between the maxima of the humps representing the P and R branches was 56 cm^{-1}. Apply Eqn. 2.14 to obtain an estimate of $\tilde{B}$. The $-\tfrac{1}{2}$ term in Eqn. 2.14 may be ignored, as may the difference between $\tilde{B}_0$ and $\tilde{B}_1$. What resolution would be required to show the fine structure of the P and R branches?

(6) The infrared spectra of the following compounds, whose molecular formulae are given, show absorption bands as listed. Identify the compounds as far as possible.

(i) $C_6H_{10}O$: 2900, 1720, 1400 cm^{-1}. Nothing above 3000 cm^{-1}.

(ii) $C_8H_6O_4$: 3300–2500 (broad band), 1700, 1600–1450 (several weak bands), 740, 700 cm^{-1}.

(iii) $C_{12}H_{10}O$: 3100, 1600–1450 (several bands), 750 cm^{-1}. Nothing around 1700 or above 3200 cm^{-1}.

(7) Mecke (1923) observed absorption maxima at the following wavenumbers in the visible spectrum of iodine vapour. These peaks are due to excitation from the ground vibrational level of the ground electronic state ($v'' = 0$) to upper vibrational levels, of quantum number v', of an upper electronically excited state.

v'	$\tilde{\nu}/cm^{-1}$	v'	$\tilde{\nu}/cm^{-1}$	v'	$\tilde{\nu}/cm^{-1}$	v'	$\tilde{\nu}/cm^{-1}$
56	19821.2	46	19515.9	36	19024.1	26	18320.9
55	19797.2	45	19473.3	35	18961.6	25	18239.1
51	19691.3	41	19292.7	31	18697.8	21	17892.8
50	19657.8	40	19245.2	30	18626.6	20	17801.6

Starting from Eqn. 2.22, derive an expression for $\Delta\tilde{\nu}$, the difference between the wavenumbers for the transitions to vibrational levels v' and $(v'+1)$ for the electronically excited state. With the aid of a graphical plot, evaluate $\tilde{\omega}_e'$ and x_e', the equilibrium vibration frequency and anharmonicity factor for the iodine molecule in the electronically excited state.

The wavenumber for the transition $(v'' = 1) \rightarrow (v' = 26)$ was found to be 18107.5 cm^{-1}. Hence evaluate $\tilde{\omega}_e''$, the equilibrium vibration frequency for the iodine molecule in its ground electronic state.

Derive an expression for the wavenumber of a peak in the $v'' = 0$ series in terms of v', $\tilde{\omega}'$, $\tilde{\omega}_e''$, x_e', x_e'', and E_{el}, where E_{el} is the electronic excitation energy. Evaluate E_{el} by substituting in known values for $v' = 26$. It may be assumed that $x''\omega_e''/4$ is negligibly small, and that $\tilde{\omega}_e''$ is equal to $\tilde{\omega}_0''$.

The dissociation limit for transitions from $v'' = 0$ is reached at 20 020 cm^{-1}. The excited iodine molecule dissociates into two atoms, one in the ground state, $^2P_{2/3}$ and one in an excited state, $^2P_{1/2}$. The spectrum of atomic iodine shows that the energy of the $(^2P_{1/2} - {}^2P_{3/2})$ transition is 7600 cm^{-1}. With the aid of Fig. 6.2, evaluate D_0 and D_0', the dissociation energies of the iodine molecule in ground and electronically excited states.

(8) Bauer *et al.* (1964) generated the species BH by flash photolysis of borine carbonyl at atmospheric pressure. The ultraviolet absorption spectrum of BH was observed by use of a vacuum grating spectrograph. The radiation source was a xenon flash lamp giving an intense pulse 5 μs in duration, 30 μs after the photolysis flash. Sets of lines around 190 nm were observed, and assigned to the transitions listed (values in cm^{-1}).

Electronic transition: $B'\Sigma^+ - X'\Sigma^+$

$(v'' = 0) \rightarrow (v' = 0)$

J	R(J)	P(J)	J	R(J)	P(J)
0	52371.22	—	12	52689.16	52105.32
1	395.48	52322.99	13	716.66	090.09
2	420.89	299.86	14	743.98	074.71
3	446.38	277.29	15	771.07	060.50
4	472.31	255.70	16	798.16	046.92
5	498.61	234.51	17	824.63	033.64
6	525.30	214.04	18	850,73	021.69
7	552.26	194.14	19	876.24	009.94
8	579.40	174.92	20	901.25	51998.60
9	606.73	156.51	21	925.30	987.53
10	634.19	138.71			
11	661.65	121.74			

	$(v'' = 0) \rightarrow (v' = 1)$			$(v'' = 1) \rightarrow (v' = 1)$	
J	R(J)	P(J)	J	R(J)	P(J)
0	54617.78	—	0	52348.46	—
1	640.20	54571.29	1	371.22	52302.64
2	661.83	547.02	2	395.48	279.35
3	682.89	522.16	3	418.57	257.85
4	702.93	496.78	4	442.16	235.91
5	722.48	471.34	5	465.66	214.04
6	740.94	—	6	489.20	192.82
			7	512.47	171.72
			8	535.50	—

	$(v'' = 2) \rightarrow (v' = 2)$			$(v'' = 3) \rightarrow (v' = 3)$	
J	R(J)	P(J)	J	R(J)	P(J)
0	52247.28	—	0	—	—
1	268.61	—	1	52060.41	—
2	289.71	52181.25	2	078.36	51977.50
3	310.46	158.80	3	095.58	954.46
4	330.80	136.35	4	—	930.68
5	350.50	113.42			
6	—	090.56			

Equations 15.4 and 15.5 may be used to interpret these data if the values of $\tilde{B}$ and $\tilde{D}_j$ are taken to be those of the molecule in its upper and ground electronic states respectively. Evaluate $\tilde{B}_v'$ and $\tilde{B}_v''$ the rotational constants for BH in the zeroth rotational level of the upper and ground electronic states in

vibrational levels 0, 1, 2, and 3. Evaluate $\tilde{D}_j'$ and $\tilde{D}_j''$, which may be assumed to be independent of v. Use the graphical plot introduced in Problem 1 to evaluate $\tilde{B}_e'$ and $\tilde{B}_e''$, $\tilde{\alpha}_e'$ and $\tilde{\alpha}_e''$, $\tilde{\gamma}_e'$ and $\tilde{\gamma}_e''$.

Calculate the internuclear distance in the BH molecule, assuming it to be $^{11}B^1H$, in the following states:

(i) Ground electronic state, $X'\Sigma^+$; at equilibrium distance r_e''.
(ii) Upper electronic state, $B'\Sigma^+$; at equilibrium distance r_e'.
(iii) Ground electronic state, $X'\Sigma^+$; $v = 0, J = 0$.
(iv) Ground electronic state, $X'\Sigma^+$; $v = 0, J = 20$.
(v) Ground electronic state, $X'\Sigma^+$; $v = 3, J = 0$.

Evaluate the wavenumber, $\tilde{\nu}_{00}$, for the hypothetical transition ($J = 0$, $v'' = 0$) $\rightarrow$ ($J = 0$, $v' = 0$). Show that this quantity is related to E_{el}, the energy for electronic excitation, the vibration frequencies $\tilde{\omega}_e''$ and $\tilde{\omega}_e'$ for the molecule in ground and upper electronic states, and x_e'' and x_e', the corresponding anharmonicity factor, by

$$\tilde{\nu}_{00} = E_{el} + \tfrac{1}{2}\omega_e' - \tfrac{1}{4}\omega_e' x_e' - \tfrac{1}{2}\omega_e'' + \tfrac{1}{4}\omega_e'' x_e'' \quad (15.6)$$

Derive similar expressions for the transitions ($v'' = 0$) $\rightarrow$ ($v' = 1$) and ($v'' = 1$) $\rightarrow$ ($v' = 1$). The data are insufficient to calculate both $\tilde{\omega}_e$ and x_e values, but the approximate, empirical, relationship

$$x_e = 0.6\, \tilde{\alpha}_e / \tilde{B}_e \quad (15.7)$$

may be used to obtain estimates for x_e' and x_e''. Hence evaluate $\tilde{\omega}_e'$ and $\tilde{\omega}_e''$.

Calculate E_{el} in wavenumber and in electron-volt units.

(9) The addition of nicotinamide to solutions of iodine in chloroform changes the characteristic violet colour to orange-red. This is attributed to the formation of a complex due to charge-transfer interaction, whereby the iodine molecule accepts an electron from the nitrogen atom in the pyridine ring of nicotinamide. Srivastava and Prasad (1966) found that the visible spectrum of such solutions showed a peak at 410 nm attributable to the complex. The absorbances at 410 nm of a set of such solutions in which the stoichiometric iodine concentration was kept fixed at 1.214×10^{-3} mol l^{-1} were found to be as below for solutions in a 1-cm cell.

10^2[Nicotinamide]/mol l^{-1}	Absorbance
0	0.095
1.48	0.387
2.20	0.505
2.97	0.619
4.19	0.714
5.59	0.845

The formation of the complex is described by the equation

$$I_2 + D = C$$

where D and C represent donor nicotinamide and complex respectively. The equilibrium constant for complex formation is given by:

$$K = [C]/[I_2][D]$$

In the experiment performed by Srivastava and Prasad, the stoichiometric iodine concentration, $[I]_0$, is kept constant and the stoichiometric donor concentration, $[D]_0$, may be assumed equal to the free donor concentration, [D], since the donor is present in large excess. The absorbance at 410 nm is due to both iodine and complex, the donor having no absorption in this region. Show that

$$\frac{[I_2]_0}{A - A'} = \frac{1}{[D]K(\varepsilon - \varepsilon')} + \frac{1}{(\varepsilon - \varepsilon')} \qquad (15.8)$$

where ε and ε' are the extinction coefficients for complex and iodine respectively, A is the observed absorbance, and A' is the absorbance for the solution containing no nicotinamide.

Evaluate ε and K by use of a graphical plot based on Eqn. 15.8.

(10) Brinen *et al.* (1969) investigated the triplet state of the aromatic polynuclear hydrocarbon, chrysene, in frozen deoxygenated 2-methyltetrahydrofuran solution at 77 K by luminescence and e.s.r. techniques. The quantum yield for fluorescence, ϕ_f, was found to be 0.19, and the ratio of the quantum yields for fluorescence and phosphorescence, ϕ_f/ϕ_p, was found to be 0.29. The lifetime of the upper singlet, τ_S, measured by monitoring the fluorescence intensity after a brief flash of light, was found to be 50 ns. The lifetime of the triplet state, τ_T, was measured by monitoring the phosphorescence intensity, and also the concentration of triplet state, by e.s.r. after a brief flash of light. These two methods gave the same value of 2.7 s.

The energy levels of chrysene are as illustrated in Fig. 7.1. The rate constants for fluorescence and phosphorescence emission are k_f and k_p s^{-1} respectively, and the rate constant for the radiationless transfer from singlet to triplet state is k_i s^{-1}. The rate of phosphorescence quenching, i.e. radiationless conversion of triplet into ground-state singlet, is k_n. It may be assumed that the rate of radiationless conversion of excited singlet into ground-state singlet is slow by comparison with the other processes. Show that:

$$k_p = \phi_p \tau_T^{-1}/(1 - \phi_f) \qquad (15.9)$$

$$k_n = \tau_T^{-1} - k_p \qquad (15.10)$$

$$k_f = \tau_s^{-1}/(1+\phi_p/\phi_f\tau_T k_p) \qquad (15.11)$$

$$k_i = (1-\phi_f)\tau_S^{-1} \qquad (15.12)$$

Hence evaluate these rate constants.

(11) Hann *et al.* (1974) found that the fluorescence intensity from a solution of 1-methylanthracene in methanol (2.5×10^{-5} mol l^{-1}) was decreased by the addition of 1-methylpyridinium toluene-*p*-sulphonate. The intensity of fluorescence, at 25 °C, in arbitrary units is tabulated below as a function of 1-methylpyridinium toluene-*p*-sulphonate concentration.

Intensity (arbitrary units)	10^2[1-methylpyridinium toluene-*p*-sulphonate]/mol l^{-1}
7.41	0
3.28	2.0
2.67	2.9
2.13	3.8
1.92	4.7
1.72	5.6

The lifetime of the upper singlet state of 1-methylanthracene in methanol at 25 °C is 4.9 ns. Calculate the rate of quenching of fluorescence by 1-methylpyridinium toluene-*p*-sulphonate, by use of the Stern–Volmer equation, Eqn. 7.3.

Compare the rate at which fluorescent and quencher molecules meet, as calculated from Eqn. 7.4. The viscosity of methanol at 25 °C is 5.47×10^{-4} kg m^{-1} s^{-1}.

(12) The high-resolution Raman spectrum of benzene vapour at a pressure of about half an atmosphere, as studied by Stoicheff (1954), shows a series of emission lines attributable to pure rotational transitions. Values of $\Delta\tilde{\nu}$, the wavenumber difference between excitation and emission, are tabulated below. The J value refers to the rotational level before excitation. The values quoted are averages for the Stokes and anti-Stokes lines.

J	$\Delta\tilde{\nu}$ for $^{12}C_6{}^1H_6$/cm^{-1}	$\Delta\tilde{\nu}$ for $^{12}C_6{}^2H_6$/cm^{-1}
25	20.093	—
30	23.888	19.756
35	27.674	22.891
40	31.461	26.019
45	35.255	29.159
50	39.034	32.283
55	42.816	35.422
60	46.600	38.560
65	50.370	41.692
70	54.166	44.824

The benzene molecule may be pictured as a flat hexagon, with a centre of symmetry at the centre of the hexagon. It is thus an oblate symmetric top. The rotational levels for an oblate symmetric top, including allowance for centrifugal stretching are given by:

$$E = BJ(J+1)+(C-B)K^2 - D_J J^2(J+1)^2 - D_{JK}J(J+1)K^2 - D_K K^4 \qquad (15.13)$$

where $B = (h/8\pi^2)I_B$ and $C = (h/8\pi^2)I_C$. D_J, D_{JK}, and D_K are centrifugal stretching constants. The selection rules for pure rotational transitions of a symmetric top are:

$$\Delta J = 0, \pm 1, \pm 2; \Delta K = 0 \qquad (15.14)$$

In addition to the lines tabulated above, another set is also observed, at exactly half the spacing. The tabulated values thus refer to the S and O branches for which $\Delta J = \pm 2$, the other set of lines, being the P and R branches, for which $\Delta J = \pm 1$. Show that for the S branch, the Stokes emission for $\Delta J = +2$, the wavenumbers are given by:

$$\Delta\tilde{\nu} = (4\tilde{B}-6\tilde{D}_J)(J+3/2)-4\tilde{D}_{JK}K(J+3/2)-8\tilde{D}_J(J+3/2)^3 \qquad (15.15)$$

where, as usual, the ~ superscript indicates quantities in wavenumber units.

Use graphical plots based on Eqn. 15.15 to evaluate $\tilde{D}_J$ and $\tilde{B}$ for $^{12}C_6{}^1H_6$ and $^{12}C_6{}^2H_6$. It may be assumed that $\tilde{D}_{JK}$ is similar in magnitude to $\tilde{D}_J$.

The B axis of the benzene molecule lies in the plane of the ring and passes through opposite carbon and hydrogen atoms. Show that the moment of inertia about this axis, I_B, is given by

$$I_B/3 = m_C r_{CC}{}^2 + m_H(r_{CC}+r_{CH})^2 \text{ kg m}^2 \qquad (15.16)$$

where m_C and m_H are the masses of the carbon and hydrogen atoms, and r_{CC} and r_{CH} are the carbon–carbon and carbon–hydrogen bond lengths respectively. Substitute the values of I_B for $^{12}C_6{}^1H_6$ and $^{12}C_6{}^2H_6$ into Eqn. 15.16 to evaluate r_{CC} and r_{CH}. These bond lengths refer to the molecule in the zeroth vibrational level, not at equilibrium, so the assumption that r_{CH} is equal for ^{12}C–^1H and ^{12}C–^2H is not exactly valid, although necessary for the solution of the problem.

(13) Iodine trichloride reacts with antimony pentachloride to form a deep-red crystalline compound. This substance may be a molecular complex, $ICl_3.SbCl_5$, or an ionic compound of structure either $ICl_4{}^+.SbCl_4{}^-$ or $ICl_2{}^+.SbCl_6{}^-$. Shamir and Rafaeloff (1973) found that the Raman spectrum excited by a He–Ne laser showed emission at $\Delta\tilde{\nu}$ of 149, 176, 271, 324, 366, and 372 cm^{-1}. The emission at $\Delta\tilde{\nu}$ of 366 cm was strongly polarised, whereas the emission at 372 cm^{-1} was depolarised. This distinction aided resolution

of the two peaks. The Raman spectrum of the ionic compound $NO^+ . SbCl_6^-$ shows emission at $\Delta\tilde{\nu}$ of 176, 283, and 331 cm^{-1}.

Identify the nature and shape of the polyhalide cation and assign the vibrational frequencies. Figure 4.6 will be of assistance.

(14) The photoelectron spectra of methyl isocyanide and deuteromethyl isocyanide in the gas phase were observed by Lake and Thompson (1971). Peaks were observed at the ionisation potentials listed below.

C^1H_3NC		C^2H_3NC	
I.P./eV	Relative peak height	I.P./eV	Relative peak height
11.240	210	11.250	225
11.410	25	11.378	30
11.518	45	11.528	55
12.228	170	12.274	180
12.364	185	12.383	165
12.460	235	12.500	205
12.596	170	12.609	180
12.692	210	12.726	185
12.828	130	12.835	130
12.924	120	12.952	110
15.570	40	15.720	90
15.665	80	15.810	115
15.760	110	15.900	155
15.855	140	15.990	185
15.950	165	16.080	200
16.045	180	16.170	220
16.140	170	16.260	210

Deduce the adiabatic and vertical ionisation potentials, and state whether these refer to ionisation from non-bonding or other molecular orbitals.

Use the vibrational fine structure to deduce wavenumbers of fundamental vibration modes of the various molecule-ions formed. A selection rule for a vibration of a molecule-ion to be observed in a photoelectron spectrum is that the vibration is totally symmetric. An observed line may be due to simultaneous excitation of two vibration modes, as illustrated for infrared spectra in Table 4.1. The infrared spectrum of gaseous C^1H_3NC shows bands centred on the wavenumbers listed below.

CH_3, sym. stretch: 2951 cm^{-1}
N≡C, sym. stretch: 2166 cm^{-1}
CH_3, sym. bend: 1455 cm^{-1}
N–C, stretch: 945 cm^{-1}

With the aid of these data, and by comparison of the wavenumbers for C^1H_3NC and C^2H_3NC, assign the observed vibrations to the fundamental modes of the molecule-ions. Hence establish whether the photoelectrons have been ionised from bonding or antibonding orbitals.

(15) The X-ray photoelectron spectrum of solid cystine, excited by Al K_α radiation of wavelength 833.9 pm, was observed by Axelson *et al.* (1967) to contain the following peaks:

Kinetic energy of photoelectrons observed/eV	Peak height in counts per 40 s
950	1870
1080	840
1192 }	880
1196 }	1460
1253	730

(The peaks linked by a bracket are not properly resolved.)

Is this spectrum consistent with the structure of cystine being as below?

$$HO_2C\cdot CH(NH_2)\cdot CH_2\cdot S\cdot S\cdot CH_2\cdot CH(NH_2)\cdot CO_2H$$

Assign the peaks to the atoms in the molecule from which they originate. In addition to the data in Table 10.2 it may be taken that the ionisation potential for an electron from the 2s orbital in the sulphur atom is around 225 eV, and that an energy of 5 eV is required to remove an electron from a crystal of cystine as a whole, independent of the ionisation energy from an individual atom.

Two possible structures of cystine dioxide have been suggested. In one, the thiosulphonate structure, both oxygen atoms are attached to the same sulphur atom, and in the other, the di-sulphoxide structure, each sulphur atom carries one oxygen. The X-ray photoelectron spectrum of cystine dioxide shows two peaks of equal heights at kinetic energies 1317 and 1313 eV. Deduce the structure of cystine dioxide, and assign these peaks.

(16) In an experiment performed by Stone *et al.* (1969), ozone was bubbled through a suspension of $Na_2Np_2O_7$ in aqueous alkali. The solid material was filtered off after 2 days, dried, and spread as a thin layer bonded with adhesive on to a brass disc. The Mössbauer spectrum was run, using as source ^{241}Am (3%) in thorium metal, which provides 59.6 keV γ-rays. Both source and absorber were maintained at 4.2 K by liquid helium. The isomer shift, δ, was

found to be -60 mm s^{-1} relative to NpO_2. Previously established isomer shifts relative to NpO_2 are:

Compound	Oxidation state of ^{237}Np	δ/mm s^{-1}
NpF_3	+3	41
$(NpO_2)C_2O_4H.2H_2O$	+5	−17
$KNpO_2CO_3$	+5	−12
$K_3NpO_2F_5$	+6	−46
$(NpO_2)(NO_3)_2.xH_2O$	+6	−36

Deduce the oxidation state of ^{237}Np in the ozonised sample. Is the nuclear radius of ^{237}Np greater or less in the excited state than in the ground state?

The half-life of the excited state of ^{237}Np is 63 ns. Calculate the theoretical line-width and compare with the observed value of 1.1 mm s^{-1} for the source used in this experiment.

(17) (i) The 60 MHz proton magnetic resonance spectra of two substances A and B show peaks as below:

A		B	
δ	Relative peak height	δ	Relative peak height
1.15	31	1.20	55
1.27	72	1.30	66
1.39	38	2.54	4*
3.37	48	2.66	19*
4.04	11	2.78	33*
4.16	32	2.90	40*
4.28	30	3.02	27*
4.40	10	3.14	12*
		3.26	3*
		7.25	62

Peaks marked * were recorded at a ten-fold greater sensitivity than the others. The molecular formulae of A and B were found to be $C_7H_{12}O_4$ and C_9H_{12} respectively. Deduce the structure of the compounds, and give values for the chemical shifts and spin–spin coupling constants for the hydrogen nuclei involved.

(ii) The ^{13}C proton-decoupled n.m.r. spectrum of a substance of formula $C_4H_9NO_2$ shows peaks at δ values of 14.7, 27.4, 60.7, and 157.8. Deduce the molecular structure of this substance.

(18) The proton magnetic resonance spectrum of *N*,*N*-dimethylformamide in dimethyl sulphoxide solution shows a peak at a δ value of 8 p.p.m., and two peaks, each of area about three times that of the first peak, around 3 p.p.m. The splitting between these two latter peaks depends on the temperature; at 102 °C it is 13.3 Hz, falling to 12.8 Hz at 112 °C, 10.0 Hz at 122 °C, and 6.4 Hz at 126 °C. The peaks broaden as they approach each other, until at 129 °C they coalesce into a single broad peak. As the temperature is raised still further this peak narrows, having a half-width at half-height of 2.2 Hz at 143 °C, 1.4 Hz at 153 °C, and 0.9 Hz at 163 °C. Explain the phenomenon giving rise to these observations.

Assuming that the limiting low-temperature and high-temperature behaviour is observed at 102 °C and 163 °C respectively, calculate values of τ, the mean lifetime of the molecule in a particular conformation, at intermediate temperatures. Hence evaluate the activation energy for rotation about the C–N bond. Equations 13.11 and 13.13 may be used, but the student should note the warning given by Allerhand *et al.* (1966) concerning the validity of these very approximate equations.

(19) The oxidation of phloroglucinol in acidic aqueous solution by Ce^{4+} gives a short-lived radical species of type (I). The e.s.r. spectrum of this has been studied by Dixon and Murphy (1976) using a flow-type sample cell. At pH 2 the spectrum shows a series of lines in the following pattern:

Relative peak height:	1		2		2		2		1
Space between peaks/mT:		0.600		0.590		0.590		0.600	

Show that this spectrum is to be expected for the radical shown, and evaluate a_o and a_p, the coupling constants for the electron with the nuclei of the hydrogen atoms in the *ortho* and *para* positions respectively.

In solutions of higher acidity this spectrum collapses and a new pattern emerges. In 2 mol l^{-1} sulphuric acid the spectrum shows a 1:3:3:1 pattern. This is due to acid-catalysed exchange of protons on the hydroxyl groups, by which the distinction between hydrogen atoms substituted *ortho* and *para* to the site of the odd electron disappears.

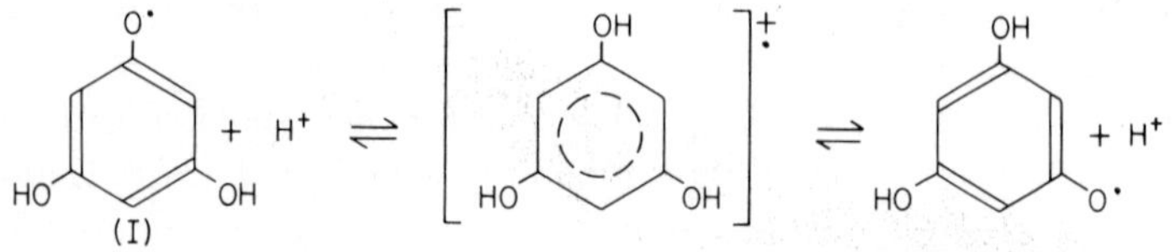

(I)

The proton exchange rate is $k[H^+]\,s^{-1}$, where k is the second-order rate constant for the reaction between radical and hydroxonium ion. The heights of the flanking peaks are constant with changing pH, but the height of the central peaks varies with pH, rising to a limiting value of three times that of the flanking peaks in highly acidic solution. This is because the width of the peak in an absorption spectrum is inversely proportional to the rate of proton exchange. In a differential display, as commonly employed in e.s.r. spectroscopy, the peak height is proportional to, among other things, the absorption peak width. Dixon and Murphy show that

$$(3h_0/h)^{\frac{1}{2}} = \frac{4\pi(\Delta\nu)^2}{27k\Delta\nu_0[H^+]} + 1 \qquad (15.17)$$

In this equation, h is the height of either of the two lines which have a relative height of 3 in the limit of fast exchange, and h_0 is the height of either of the flanking lines, which is independent of pH; $\Delta\nu$ is the difference in coupling constants $(a_p - a_o)$ in units of Hz, and $\Delta\nu_0$ is the natural line-width, also in units of Hz.

Values found for h/h_0 at various hydrogen ion concentrations are tabulated below.

$10^2[H^+]/mol\,l^{-1}$	h/h_0
5.00	0.25
6.67	0.38
10.0	0.58
14.3	0.83
25.0	1.26
50.0	2.04
100.0	2.44

By means of a graphical plot, show that these data are in accordance with Eqn. 15.17. The value of $\Delta\nu_0$ may be taken as 0.03 mT for this apparatus, and the g-factor may be assumed to have the free-electron value. Evaluate k.

Answers to Problems

(1) $\alpha_e = 8.037$ MHz; $\gamma_e = 13.5$ kHz.
Plot $\nu/(J+1)$ against $(J+1)^2$. Hence $B_v = 1821.04$ MHz; $D_j = 1.6$ kHz.
$B_e = 1821.04$ MHz; $r_e = 304.7$ pm.

(2) $^{17}O = 16.9965$; $^{18}O = 17.9938$.

(3) $\mu = 10.63$ debye.

(4) For $^1H^{35}Cl$: $\tilde{B}_0 = 10.441\ cm^{-1}$, $\tilde{B}_1 = 10.137\ cm^{-1}$, $\tilde{D}_j = 5.1 \times 10^{-4}\ cm^{-1}$, $\tilde{\alpha}_e = 0.304\ cm^{-1}$, $\tilde{B}_e = 10.593\ cm^{-1}$, $r_e = 1.279 \times 10^{-10}$ m.
For $^1H^{37}Cl$: $\tilde{B}_0 = 10.426\ cm^{-1}$, $\tilde{B}_1 = 10.126\ cm^{-1}$, $\tilde{D}_j = 5.2 \times 10^{-4}\ cm^{-1}$, $\tilde{\alpha}_e = 0.300\ cm^{-1}$, $\tilde{B}_e = 10.576\ cm^{-1}$, $r_e = 1.279 \times 10^{-10}$ m.
For $^1H^{35}Cl$: $\tilde{\omega} = 2886.02\ cm^{-1}$; for $^1H^{37}Cl$: $\tilde{\omega} = 2883.91\ cm^{-1}$, $\tilde{\omega}(^1H^{37}Cl)/\tilde{\omega}(^1H^{35}Cl) = 0.99927$.
(Reduced mass $^1H^{35}Cl$)/(Reduced mass $^1H^{37}Cl$) = 0.99924.

(5) $\tilde{B} = 1.89\ cm^{-1}$. Spacing between peaks is thus $\sim 4\ cm^{-1}$, so a resolution of about $2\ cm^{-1}$ would be required.

(6) (i) Cyclohexanone
(ii) Benzene-1,3-dicarboxylic acid
(iii) Diphenyl ether

(7) $\Delta\tilde{\nu} = \tilde{\omega}_e' - 2\tilde{\omega}_e' x_e' + 2v'\tilde{\omega}_e' x_e'$
$\tilde{\omega}_e' = 130\ cm^{-1}$; $x_e' = 7.3 \times 10^{-3}$.
$\tilde{\omega}_e'' = 213.4\ cm^{-1}$.
$\tilde{\nu} = E_{el} + (v' + \frac{1}{2})\tilde{\omega}_e' - (v' + \frac{1}{2})^2 x_e'\tilde{\omega}_e' - \frac{1}{2}\tilde{\omega}_e'' + \frac{1}{4}x_e''\tilde{\omega}_e''$
$E_{el} = 15\,643\ cm^{-1}$.
$D_0 = 12\,420\ cm^{-1}$, $D_0' = 4377\ cm^{-1}$.

(8) $\tilde{B}_0' = 12.09, \tilde{B}_1' = 11.53, \tilde{B}_2' = 10.87, \tilde{B}_3' = 10.10\ cm^{-1}; \tilde{B}_0'' = 11.82$, $\tilde{B}_1'' = 11.43$, $\tilde{B}_2'' = 11.01$, $\tilde{B}_3'' = 10.65\ cm^{-1}$; $\tilde{D}_j' = 1.29 \times 10^{-3}$, $\tilde{D}_j'' = 1.17 \times 10^{-3}\ cm^{-1}$;
$\tilde{B}_e' = 12.33$, $\tilde{B}_e'' = 12.02\ cm^{-1}$;
$\tilde{\alpha}_e' = +0.46$, $\tilde{\alpha}_e'' = +0.39\ cm^{-1}$;
$\tilde{\gamma}_e' = -0.05$, $\tilde{\gamma}_e'' = 0.00\ cm^{-1}$;
$r = 1.242, 1.224, 1.253, 1.279, 1.319$ pm, respectively.
$\tilde{\nu}_{00} = 52\,347.10$, $\tilde{\nu}_{01} = 54\,595.54$, $\tilde{\nu}_{11} = 52\,325.56\ cm^{-1}$;
$\tilde{\omega}_e' = 2353$, $\tilde{\omega}_e'' = 2361\ cm^{-1}$;
$E_{el} = 52\,352\ cm^{-1} = 6.4902$ eV.

(9) $\varepsilon = 1270\ l\ mol^{-1}\ cm^{-1}$; $K = 15.0\ l\ mol^{-1}$.

(10) $k_p = 2.52 \times 10^{-2}\ s^{-1}$; $k_n = 0.345\ s^{-1}$;
$k_f = 3.8 \times 10^6\ s^{-1}$; $k_i = 1.62 \times 10^7\ s^{-1}$.

(11) $k_c = 1.26 \times 10^{10}\ l\ mol^{-1}\ s^{-1}$.
Diffusion-controlled rate $= 1.21 \times 10^{10}\ l\ mol^{-1}\ s^{-1}$.

(12) $^{12}C^1H_6$: $\tilde{D}_j = 2.2 \times 10^{-8}\ cm^{-1}$; $\tilde{B} = 0.18961\ cm^{-1}$.
$^{12}C^2H_6$: $\tilde{D}_J = 8.2 \times 10^{-9}\ cm^{-1}$; $\tilde{B} = 0.15618\ cm^{-1}$.
$r_{CC} = 137.9$ pm; $r_{CH} = 108.5$ pm.

(13) Cl–I–Cl; bent.
$149\ cm^{-1}$, bending; $366\ cm^{-1}$, sym. stretch, $372\ cm^{-1}$, antisym. stretch.

(14) C^1H_3NC: Adiabatic I.P.s: 11.240, 12.228, 15.570.
Vertical I.P.s: 11.240, 12.460, 16.045.
C^2H_3NC: Adiabatic I.P.s: 11.250, 12.274, 15.720.
Vertical I.P.s: 11.250, 12.500, 16.170.
C^2H_3NC: First ionisation is from non-bonding orbital.
Spacings = 0.170 eV≡1370 cm^{-1},
and 0.278 eV≡2240 cm^{-1}.
C^2H_3NC: Spacings = 0.128 eV≡1030 cm^{-1},
and 0.278 eV≡2240 cm^{-1}.
The C^1H_3NC vibration at 1370 cm^{-1} is thus CH_3 bend and that at 2240 cm^{-1} is N≡C stretch. Ionisation from bonding orbital.
Second ionisation.
Ionisation from bonding or antibonding orbital.
C^1H_3CN: Spacings = 0.136 eV≡1100 cm^{-1},
and 0.232 eV≡1870 cm^{-1}.
C^2H_3CN: Spacings = 0.226 eV≡1820 cm^{-1},
and 0.109 eV≡880 cm^{-1}.
The C^1H_3NC vibration at 110 cm^{-1} is CH_3 bend and that at 1820 cm^{-1} is N≡C stretch. Ionisation from bonding orbital.
Third ionisation.
C^1H_3CN: Spacings = 0.095 eV≡770 cm^{-1}.
C^2H_3CN: Spacings = 0.090 eV≡730 cm^{-1}.
The vibration is C–N stretch. Ionisation from bonding orbital.

(15) Peaks are (top to bottom): O 1s, N 1s, C (carboxylic), C (alkyl), S 2s.
Thiosulphonate structure. The $-SO_2-$ peak is at 1313 eV.

(16) Oxidation number, +7.
Excited-state nuclear radius is greater.
Theoretical $\Gamma = 2.52 \times 10^6\ s^{-1} = 5.2 \times 10^{-2}$ mm s^{-1}.

(17) (i) A is $(\overset{a}{C}H_3\overset{c}{C}H_2CO_2)_2\overset{b}{C}H_2$.
Chemical shifts: a, 1.27; b, 3.37; c, 4.22.
HCCH spin–spin coupling constant = 60 × 0.12 = 7.2 Hz.
B is $\overset{a}{C}H_3\text{–}\overset{b}{C}H(C_6\overset{c}{H}_5)\text{–}\overset{a}{C}H_3$.
Chemical shifts: a, 1.25; b, 2.90; c, 7.25.
HCCH spin–spin coupling constant = 60 × 0.12 = 7.2 Hz.
(ii) $\overset{b}{C}H_3N\overset{c}{H}CO_2\overset{c}{C}H_2\overset{a}{C}H_3$.
Chemical shifts: a, 14.7; b, 27.4; c, 60.7; d, 157.8.

(18) Activation energy: 116 kJ mol^{-1}.

(19) $a_o = 0.600$ mT, $a_p = 1.180$ mT.
(Note that the two central lines, each of unit relative intensity, overlap.)
$k = 1.2 \times 10^9$ l mol^{-1} s^{-1}.

REFERENCES

Allerhand, A., Gutowsky, H. S., Jones, J. and Meinzer, R. A. (1966), *J. Amer. Chem. Soc.*, **88**, 3185.

Axelson, G., Hamrin, K., Fahlman, A. and Nordling, C. (1967), *Spectrochim. Acta*, **23A**, 2015.

Bauer, S. H., Herzberg, G. and Johns, J. W. C. (1964), *J. Mol. Spectroscopy*, **13**, 256.

Brinen, J. S., Orloff, M. K., Gallivan, J. B., Stamm, R. F. and Roberts, B. G. (1969), *J. Mol. Spectroscopy*, **32**, 368.

Dixon, W. T. and Murphy, D. (1976), *J. Chem. Soc. Faraday Trans. II*, 135.

Hann, R. A., Rosseinsky, D. R. and White, T. P. (1974), *J. Chem. Soc. Faraday Trans. II*, 70, 1522.

Lake, R. F. and Thompson, H. W. (1971), *Spectrochim. Acta.*, **27A**, 783.

Mecke, R. (1923), *Ann. Physik*, **71**, 104.

Mills, I. M., Thompson, H. W. and Williams, R. L. (1953), *Proc. Roy. Soc. A.*, **218**, 29.

Murty, A. N. and Curl, R. F., Jr. (1969), *J. Mol. Spectroscopy*, **30**, 102.

Rosenblum, B., Nethercot, A. H., Jr. and Townes, C. H. (1958), *Phys. Rev.*, **109**, 400.

Shamir, J. and Rafaeloff, R. (1973), *Spectrochim. Acta*, **29A**, 873.

Srivastava, R. D. and Prasad, G. (1966), *Spectrochim. Acta*, **22**, 825.

Stoicheff, B. P. (1954), *Canad. J. Phys.*, **32**, 339.

Stone, J. A., Pillinger, W. L. and Karraker, D. G. (1969), *Inorg. Chem.*, **8**, 2519.

Index

Many terms, which do not appear in the headings, are fully explained in the text, as it is expected that they will be unfamiliar to the student. Such terms appear in the index in bold type, and the numbers of the pages on which the explanations are to be found are similarly in bold type.

D

H

I

J

K

L

M

T